MODELING AND PRICING OF SWAPS FOR FINANCIAL AND ENERGY MARKETS WITH STOCHASTIC VOLATILITIES

MODELING AND PRICING OF SWAPS FOR FINANCIAL AND ENERGY MARKETS WITH STOCHASTIC VOLATILITIES

Anatoliy Swishchuk

University of Calgary, Canada

World Scientific

NEW JERSEY · LONDON · SINGAPORE · BEIJING · SHANGHAI · HONG KONG · TAIPEI · CHENNAI

Published by

World Scientific Publishing Co. Pte. Ltd.

5 Toh Tuck Link, Singapore 596224

USA office: 27 Warren Street, Suite 401-402, Hackensack, NJ 07601

UK office: 57 Shelton Street, Covent Garden, London WC2H 9HE

Library of Congress Cataloging-in-Publication Data
Svishchuk, A. V. (Anatolii Vital'evich)
  Modeling and pricing of swaps for financial and energy markets with stochastic volatilities / Anatoliy Swishchuk.
    pages cm
    Includes index.
    ISBN 978-9814440127 (hardcover : alk. paper) -- ISBN 978-9814440134 (electronic book)
    1. Swaps (Finance)--Mathematical models. 2. Finance--Mathematical models. 3. Stochastic processes.
I. Title.
  HG6024.A3S876 2013
  332.64'5--dc23

                                                                     2012047233

British Library Cataloguing-in-Publication Data
A catalogue record for this book is available from the British Library.

In-house Editor: Chye Shu Wen

Printed in Singapore by World Scientific Printers.

To my lovely and dedicated family: wife Mariya, son Victor and daughter Julia

Preface

> *One may think that stochastic volatilities like storms, hurricanes, tornadoes, etc. But the formers are chaotic. In my mind, stochastic volatility is a beautiful serenity (like this book's cover picture) that hides inside that chaos, ready to explode any time. To find an order in this chaos and to tame it is our goal. I've tried to find in this book an order in many chaotic structures inside the realm of chaos hidden in the serene beauty of stochastic volatility.*

The book is devoted to the modeling and pricing of various kinds of swaps, such as variance, volatility, covariance and correlation, for financial and energy markets with variety of stochastic volatilities.

In Chapter 1, we provide an overview of the different types of non-stochastic volatilities and the different types of stochastic volatilities. With respect to stochastic volatility, we consider two approaches to introduce stochastic volatility: (1) changing the clock time t to a random time $T(t)$ (subordinator) and (2) changing constant volatility into a positive stochastic process.

Chapter 2 is devoted to the description of different types of stochastic volatilities that we use in this book. They include, in particular: Heston stochastic volatility model; stochastic volatilities with delay; multi-factor stochastic volatilities; stochastic volatilities with delay and jumps; Lévy-based stochastic volatility with delay; delayed stochastic volatility in Heston model (we call it 'delayed Heston model'); semi-Markov modulated stochastic volatilities; COGARCH(1,1) stochastic volatility; stochastic volatilities driven by fractional Brownian motion; and continuous-time GARCH stochastic volatility model.

Chapter 3 deals with the description of different types of swaps and pseudo-swaps: variance, volatility, covariance, correlation, pseudo-variance, pseudo-volatility, pseudo-covariance and pseudo-correlations swaps.

In Chapter 4 we provide an overview on change of time methods (CTM), and show how to solve many stochastic differential equations (SDEs) in finance (geometric Brownian motion (GBM), Ornstein-Uhlenbeck (OU), Vasiček, continuous-time GARCH, etc.) using change of time methods. As applications of CTM, we present two different models: geometric Brownian motion (GBM) and mean-reverting models. The solutions of these two models are different. But the nice thing is that they can be solved by CTM as many other models mentioned in this chapter. Moreover, we can use these solutions to easily find the option pricing formulas: one is classic-Black-Scholes and another one is new — for a mean-reverting asset. These formulas can be used in practice (for example, in energy markets) because they all are explicit.

Chapter 5 considers applications of the change of time method to yet one more derive the well-known Black-Scholes formula for European call options. We mention that there are many proofs of this result, including PDE and martingale approaches, for example.

In Chapter 6, we study variance and volatility swaps for financial markets with underlying asset and variance following the Heston (1993) model. We also study covariance and correlation swaps for the financial markets. As an application, we provide a numerical example using $S\&P60$ Canada Index to price swap on the volatility.

Variance swaps for financial markets with underlying asset and stochastic volatilities with delay are modelled and priced in Chapter 7. We find some analytical close forms for expectation and variance of the realized continuously sampled variance for stochastic volatility with delay both in stationary regime and in general case. The key features of the stochastic volatility model with delay are the following: i) continuous-time analogue of discrete-time GARCH model; ii) mean-reversion; iii) contains the same source of randomness as stock price; iv) market is complete; v) incorporates the expectation of log-return. We also present an upper bound for delay as a measure of risk. As applications, we provide two numerical examples using $S\&P60$ Canada Index (1998–2002) and $S\&P500$ Index (1990–1993) to price variance swaps with delay.

Variance swaps for financial markets with underlying asset and multi-factor, i.e., two- and three-factors, stochastic volatilities with delay are modelled and priced in Chapter 8. We found some analytical close forms for expectation and variance of the realized continuously sampled variances for multi-factor stochastic volatilities with delay. As applications, we provide a numerical examples using $S\&P60$ Canada Index (1998–2002) to price variance swaps with delay for all these models.

In Chapter 9, we incorporate a jump part in the stochastic volatility model with delay proposed by Swishchuk (2005) to price variance swaps. We find some analytical closed forms for the expectation of the realized continuously sampled variance for stochastic volatility with delay and jumps. The jump part in our model is finally represented by a general version of compound Poisson processes and the expectation and the covariance of the jump sizes are assumed to be deterministic functions. We note that after adding jumps, the model still keeps those good features of the previous model such as continuous-time analogue of GARCH model, mean-reversion and so on. But it is more realistic and still quick to implement. Besides, we also present a lower bound for delay as a measure of risk. As applications of our analytical solutions, a numerical example using $S\&P60$ Canada Index (1998–2002) is also provided to price variance swaps with delay and jumps.

The valuation of the variance swaps for local Lévy–based stochastic volatility with delay (LLBSVD) is discussed in Chapter 10. We provide some analytical closed forms for the expectation of the realized variance for the LLBSVD. As applications of our analytical solutions, we fit our model to 10 years of $S\&P500$ data (2000-01-01–2009-12-31) with variance gamma model and apply the obtained analytical solutions to price the variance swap.

In Chapter 11, we present a variance drift adjusted version of the Heston model which leads to significant improvement of the market volatility surface fitting (compared to Heston). The numerical example we performed with recent market data shows a significant (44%) reduction of the average absolute calibration error[1] (calibration on 30th September 2011 for underlying EURUSD). Our model has two additional parameters compared to the Heston model and can be implemented very easily. The main idea behind our model is to take into account some past history of the variance process in its (risk-neutral) diffusion.

Following Chapter 11, we consider in Chapter 12 the variance and volatility swap pricing and dynamic hedging for delayed Heston model. We derived a closed formula for the variance swap fair strike, as well as for the Brockhaus and Long approximation of the volatility swap fair strike. Based on these results, we considered hedging of a position on a volatility swap with variance swaps. A closed formula — based on the Brockhaus and Long approximation — was derived for the number of variance swaps one should hold at each time t in order to hedge the position (hedge ratio).

In Chapter 13, we consider a semi-Markov modulated market consisting of a riskless asset or bond, B, and a risky asset or stock, S, whose dynamics depend on a semi-Markov process x. Using the martingale characterization of semi-Markov processes, we note the incompleteness of semi-Markov modulated markets and find the minimal martingale measure. We price variance and volatility swaps for stochastic volatilities driven by the semi-Markov processes. We also discuss some extensions of the obtained results such as local semi-Markov volatility, Dupire formula for the local semi-Markov volatility and residual risk associated with the swap pricing.

In Chapter 14, we price covariance and correlation swaps for financial markets with Markov-modulated volatilities. As an example, we consider stochastic volatility driven by two-state continuous Markov chain. In this case, numerical example is presented for VIX and VXN volatility indeces ($S\&P500$ and NASDAQ-100, respectively, since January 2004 to June 2012). We also use VIX (January 2004 to June 2012) to price variance and volatility swaps for the two-state Markov-modulated volatility and to present a numerical result in this case.

Chapter 15 presents volatility and variance swaps' valuations for the COGARCH (1,1) model. We consider two numerical examples: for compound Poisson COG-

---

[1]Average of the absolute differences between market and model implied BS volatilities.

ARCH(1,1) and for variance gamma COGARCH(1,1) processes. Also, we demonstrate two different situations for the volatility swaps: with and without convexity adjustment to show the difference in values.

In Chapter 16, we study financial markets with stochastic volatilities driven by fractional Brownian motion with Hurst index $H > 1/2$. Our models for stochastic volatility include new fractional versions of Ornstein-Uhlenbeck, Vasiček, geometric Brownian motion and continuous-time GARCH models. We price variance and volatility swaps for the above-mentioned models. Since pricing volatility swaps needs approximation formula, we analyze when this approximation is satisfactory. Also, we present asymptotic results for pricing variance swaps when time horizon increases.

Chapter 17 is devoted to the pricing of variance and volatility swaps in energy markets. We found explicit variance swap formula and closed form volatility swap formula (using change of time) for energy asset with stochastic volatility that follows continuous-time mean-reverting GARCH (1,1) model. Numerical example is presented for AECO Natural Gas Index (1 May 1998–30 April 1999).

In Chapter 18 we consider a risky asset S_t following the mean-reverting stochastic process. We obtain an explicit expression for a European option price based on S_t, using a change of time method from Chapter 4. A numerical example for the AECO Natural Gas Index (1 May 1998–30 April 1999) is presented.

In Chapter 19 we introduce new one-factor and multi-factor α-stable Lévy-based models to price energy derivatives, such as forwards and futures. For example, we introduce new multi-factor models such as Lévy-based Schwartz-Smith and Schwartz models. Using change of time method for SDEs driven by α-stable Lévy processes we present the solutions of these equations in simple and compact forms.

Chapter 20 deals with the Markov-modulated volatility and its application to generalize Black-76 formula. Black formulas for Markov-modulated markets with and without jumps are derived. Application is given using Nordpool weekly electricity forward prices.

The book will be useful for academics and graduate students doing research in mathematical and energy finance, for practitioners working in the financial and energy industries and banking sectors. It may also be used as a textbook for graduate courses in mathematical finance.

Anatoliy V. Swishchuk
University of Calgary
Calgary, Alberta, Canada

Acknowledgments

I would like to thank my many colleagues and students very much for fruitful and enjoyable cooperation: Robert Elliott, Gordon Sick, Tony Ware, Yulia Mishura, Nelson Vadori, Ke Zhao, Kevin Malenfant, Xu Li, Matt Couch and Giovanni Salvi.

My first experience with swaps was in Vancouver in 2002 at a 5-day Industrial Problems Solving Workshop organized by PIMS. The problem was brought up by RBC Financial Group and it concerned the pricing of swaps involving the so-called pseudo-statistics, namely pseudo-variance, -covarinace, -volatility, and -correlation. The team consisted of 9 graduate students, Andrei Badescu, Hammouda Ben Mekki, Asrat Fikre Gashaw, Yuanyuan Hua, Marat Molyboga, Tereza Neocleous, Yuri Petratchenko, Raymond K. Cheng, and Stephan Lawi, with whom we solved the problem and prepared our report. I'd like to thank them all for a very productive collaboration during this time. The idea of using the change of time method for solving this problem had actually occurred to me on this workshop.

My thanks also to Paul Wilmott who gave me many useful suggestions to improve my first paper on variance, volatility, covariance and correlation swaps for Heston model published by Wilmott Magazine in 2004.

I am very grateful to Yubing Zhai (WSP) who encouraged me to write this book and always helped when I needed it. I would also like to thank Agnes Ng (WSP) for reading the manuscript and for adding some valuable corrections and suggestions with respect to the style of the book.

Many thanks to Chye Shu Wen and Rajesh Babu (WSP) who helped me a lot in preparing the manuscript.

Last, but not least, thanks and great appreciation are due to my family, wife Mariya, son Victor and daughter Julia, who were patient enough to give me continuous support during the book preparation.

Contents

Chapter 1

Stochastic Volatility

1.1 Introduction

Volatility, as measured by the standard deviation, is an important concept in financial modeling because it measures the change in value of a financial instrument over a specific horizon. The higher the volatility, the greater the price risk of a financial instrument. There are different types of volatility: historical, implied volatility, level-dependent volatility, local volatility and stochastic volatility (e.g., jump-diffusion volatility). Stochastic volatility models are used in the field of quantitative finance. Stochastic volatility means that the volatility is not a constant, but a stochastic process and can explain: volatility smile and skew.

Volatility, typically denoted by the Greek letter σ, is the standard deviation of the change in value of a financial instrument over a specific horizon such as a day, week, month or year. It is often used to quantify the price risk of a financial instrument over that time period. The price risk of a financial instrument is higher the greater its volatility.

Volatility is an important input in option pricing models. The Black-Scholes model for option pricing assumes that the volatility term is a constant. This assumption is not always satisfied in real-world options markets because: probability distribution of common stock returns has been observed to have a fatter left tail and thinner right tail than the lognormal distribution (see Hull, 2000). Moreover, the assumption of constant volatility in financial model, such as the original Black-Scholes option pricing model, is incompatible with option prices observed in the market.

As the name suggests, stochastic volatility means that volatility is not a constant, but a stochastic process. Stochastic volatility models are used in the field of quantitative finance and financial engineering to evaluate derivative securities, such as options and swaps. By assuming that volatility of the underlying price is a stochastic process rather than a constant, it becomes possible to more accurately model derivatives. In fact, stochastic volatility models can explain what is known as the volatility smile and volatility skew in observed option prices.

In this chapter, we provide an overview of the different types non-stochastic volatilities and the different types of stochastic volatilities. There are two approaches to introduce stochastic volatility: (1) changing the clock time t to a random time $T(t)$ (subordinator) and (2) changing constant volatility into a positive stochastic process.

1.2   Non-Stochastic Volatilities

We begin by providing an overview of the different types of non-stochastic volatilities measures. These include: historical volatility; implied volatility; level-dependent volatility; local volatility.

1.2.1   *Historical Volatility*

Historical volatility is the volatility of a financial instrument or a market index based on historical returns. It is a standard deviation calculated using historical (daily, weekly, monthly, quarterly, yearly) price data. The annualized volatility σ is the standard deviation of the instrument's logarithmic returns over a one-year period:

$$\sigma = \sqrt{\frac{1}{n-1}\sum_{i=1}^{n}(R_i - \bar{R})^2},$$

where $R_i = \ln\frac{S_{t_i}}{S_{t_{i-1}}}$, $\bar{R} = \frac{1}{n}\sum_{i=1}^{n}\ln\frac{S_{t_i}}{S_{t_{i-1}}}$, S_{t_i} is a asset price at time t_i, $i = 1, 2, ..., n$.

1.2.2   *Implied Volatility*

Implied volatility is related to historical volatility. However, there are two important differences. Historical volatility is a direct measure of the movement of the price (realized volatility) over recent history. Implied volatility, in contrast, is set by the market price of the derivative contract itself, and not the undelier. Therefore, different derivative contracts on the same underlier have different implied volatilities. Most derivative markets exhibit persistent patterns of volatilities varying by strike. The pattern displays different characteristics for different markets. In some markets, those patterns form a smile curve. In others, such as equity index options markets, they form more of a skewed curve. This has motivated the name "volatility skew". For markets where the graph is downward sloping, such as for equity options, the term "volatility skew" is often used. For other markets, such as FX options or equity index options, where the typical graph turns up at either end, the more familiar term "volatility smile" is used. In practice, either the term "volatility smile" or "volatility skew" may be used to refer to the general phenomenon of volatilites varying by strike.

The models by Black and Scholes (1973) (continuous-time (B,S)-security market) and Cox, Ross and Rubinstein (1976) (discrete-time (B,S)-security market (binomial tree)) are unable to explain the negative skewness and leptokurticity (fat tail) commonly observed in the stock markets. The famous implied-volatility smile would not exist under their assumptions. Most derivatives markets exhibit persistent patterns of volatilities varying by strike. In some markets, those patterns form a smile. In others, such as equity index options markets, it is more of a skewed curve. This has motivated the name volatility skew. In practice, either the term 'volatility smile' or 'volatility skew' (or simply skew) may be used to refer to the general phenomena of volatilities varying by strike. Another dimension to the problem of volatility skew is that of volatilities varying by expiration, known *volatility surface*.

Given the prices of call or put options across all strikes and maturities, we may deduce the volatility which produces those prices via the full Black-Scholes equation.[1] This function has come to be known as *local volatility*. Local volatility-function of the spot price S_t and time t : $\sigma \equiv \sigma(S_t, t)$ (see Dupire (1994) formulae for local volatility).

1.2.3 *Level-Dependent Volatility and Local Volatility*

Level-dependent volatility (e.g., constant elasticity of variance (CEV) or Firm Model (see Beckers (1980), Cox (1975)) — a function of the spot price alone. To have a smile across strike price, we need σ to depend on $S : \sigma \equiv \sigma(S_t)$. In this case, the volatility and stock price changes are now perfectly negatively correlated (so-called "leverage effect").

Local volatility is a volatility function of the spot price and time. Volatility smile can be retrieved in this case from the option prices. Dupire (1994) derived the local volatility formula in continuous time and Derman and Kani (1994) used the binomial (or trinomial tree) framework instead of the continuous one to find the local volatility formula. The LV models are very elegant and theoretically sound. However, they present in practice many stability issues. They are ill-posed inversion problems and are extremely sensitive to the input data. This might introduce arbitrage opportunities and, in some cases, negative probabilities or variances.

1.3 Stochastic Volatility

Stochastic volatility means that volatility is not a constant, but a stochastic process. Black and Scholes (1973) made a major breakthrough by deriving pricing formulas for vanilla options written on the stock. The Black-Scholes model assumes that the volatility term is a constant. Stochastic volatility models are used in the field of quantitative finance to evaluate derivative securities, such as options, swaps. By

---

[1] Black and Scholes (1973), Dupire (1994), Derman and Kani (1994).

assuming that the volatility of the underlying price is a stochastic process rather than a constant, it becomes possible to more accurately model derivatives.

The above issues have been addressed and studied in several ways, such as:
(1) Volatility is assumed to be a deterministic function of the time[2]: $\sigma \equiv \sigma(t)$, with the implied volatility for an option of maturity T given by $\hat{\sigma}_T^2 = \frac{1}{T} \int_0^T \sigma_u^2 du$;
(2) Volatility is assumed to be a function of the time and the current level of the stock price $S(t)$: $\sigma \equiv \sigma(t, S(t))$[3]; the dynamics of the stock price satisfies the following stochastic differential equation:

$$dS(t) = \mu S(t)dt + \sigma(t, S(t))S(t)dW_1(t),$$

where $W_1(t)$ is a standard Wiener process;
(3) The time variation of volatility involves an additional source of randomness, besides $W_1(t)$, represented by $W_2(t)$, and is given by

$$d\sigma(t) = a(t, \sigma(t))dt + b(t, \sigma(t))dW_2(t),$$

where $W_2(t)$ and $W_1(t)$ (the initial Wiener process that governs the price process) may be correlated[4];
(4) Volatility depends on a random parameter x such as $\sigma(t) \equiv \sigma(x(t))$, where $x(t)$ is some random process.[5]
(5) Stochastic volatility, namely, uncertain volatility scenario. This approach is based on the uncertain volatility model developed in Avellaneda *et al.* (1995), where a concrete volatility surface is selected among a candidate set of volatility surfaces. This approach addresses the sensitivity question by computing an upper bound for the value of the portfolio under arbitrary candidate volatility, and this is achieved by choosing the local volatility $\sigma(t, S(t))$ among two extreme values $\sigma_{\min}$ and $\sigma_{\max}$ such that the value of the portfolio is maximized locally;
(6) The volatility $\sigma(t, S_t)$ depends on $S_t = S(t+\theta)$ for $\theta \in [-\tau, 0]$, namely, stochastic volatility with delay.[6]

In approach (**1**), the volatility coefficient is independent of the current level of the underlying stochastic process $S(t)$. This is a deterministic volatility model, and the special case where σ is a constant reduces to the well-known Black-Scholes model that suggests changes in stock prices are lognormal. Empirical test by Bollerslev (1986) seem to indicate otherwise. One explanation for this problem of a lognormal model is the possibility that the variance of $\log(S(t)/S(t-1))$ changes randomly.

In the approach (**2**), several ways have been developed to derive the corresponding Black-Scholes formula: one can obtain the formula by using stochastic calculus and, in particular, the Ito's formula (see Shiryaev (2008), for example).

---

[2]Wilmott *et al.* (1995), Merton (1973).
[3]Dupire (1994), Hull (2000).
[4]Hull and White (1987), Heston (1993).
[5]Elliott and Swishchuk (2007), Swishchuk (2000, 2009), Swishchuk *et al.* (2010).
[6]Kazmerchuk, Swishchuk and Wu (2005), Swishchuk (2005, 2006, 2007, 2009a, 2010).

A generalized volatility coefficient of the form $\sigma(t, S(t))$ is said to be *level-dependent*. Because volatility and asset price are perfectly correlated, we have only one source of randomness given by $W_1(t)$. A time and level-dependent volatility coefficient makes the arithmetic more challenging and usually precludes the existence of a closed-form solution. However, the arbitrage argument based on portfolio replication and a completeness of the market remain unchanged.

1.3.1 *Approaches to Introduce Stochastic Volatility*

The idea to introduce stochastic volatility is to make volatility itself a stochastic process. The aim with a stochastic volatility model is that volatility appears not to be constant and indeed varies randomly. For example, the situation becomes different if volatility is influenced by a second "non-tradable" source of randomness, and we usually obtain a *stochastic volatility model*, introduced by Hull and White (1987). This model of volatility is general enough to include the deterministic model as a special case. Stochastic volatility models are useful because they explain in a self-consistent way why it is that options with different strikes and expirations have different Black-Scholes implied volatilities (the volatility smile). These cases were addressed in the approaches **(iii)**, **(iv)** and **(v)**. Stochastic volatility is the main concept used in the fields of financial economics and mathematical finance to deal with the endemic time-varying volatility and co-dependence found in financial markets. Such dependence has been known for a long time, early comments include Mandelbrot (1963) and Officer (1973).

There are two approaches to introduce stochastic volatility: one approach is to change the clock time t to a random time $T(t)$ (change of time). Another approach-change constant volatility into a positive stochastic process. Continuous-time stochastic volatility models include: Ornstein-Uhlenbeck (OU) process (Ornstein-Uhlenbeck (1930)); geometric Brownian motion with zero correlation with respect to a stock price (Hull and White (1987)); geometric Brownian motion with noon-zero correlation with respect to a stock pric (Wiggins (1987)); OU process, mean-reverting, positive with noon-zero correlation with respect to a stock price (Scott (1987)); OU process, mean-reverting, negative, with zero correlation with respect to a stock price (Stein and Stein (1991)); Cox-Ingersoll-Ross process, mean-reverting, non-negative with noon-zero correlation with respect to a stock price (Heston (1993)).

Heston and Nandi (1998) showed that OU process corresponds to a special case of the GARCH model for stochastic volatility. Hobson and Rogers (1998) suggested a new class of non-constant volatility models, which can be extended to include the aforementioned level-dependent model and share many characteristics with the stochastic volatility model. The volatility is non-constant and can be regarded as an endogenous factor in the sense that it is defined in terms of the past behavior of the stock price. This is done in such a way that the price and volatility form a multi-dimensional Markov process.

1.3.2   *Discrete Models for Stochastic Volatility*

Another popular process is the continuous-time GARCH(1,1) process, developed by Engle (1982) and Bollerslev (1986) in discrete framework. The Generalized AutoRegressive Conditional Heteroskedacity (GARCH) model (see Bollerslev (1986)) is popular model for estimating stochastic volatility. It assumes that the randomness of the variance process varies with the variance, as opposed to the square root of the variance as in the Heston model. The standard GARCH(1,1) model has the following form for the variance differential:

$$d\sigma_t = \kappa(\theta - \sigma_t)dt + \gamma\sigma_t dB_t.$$

The GARCH model has been extended via numerous variants, including the NGARCH, LGARCH, EGARCH, GJR-GARCH, etc.

Continuous-time models provide the natural framework for an analysis of option pricing, discrete-time models are ideal for the statistical and descriptive analysis of the patterns of daily price changes. Volatility clustering, periods of high and low variance (large changes tend to be followed by small changes (Mandelbrot (1963)), led to using of discrete models, GARCH models. There are two main classes of discrete-time stochastic volatility models. First class-autoregressive random variance (ARV) or stochastic variance models-is a discrete time approximation to the continuous time diffusion models that we outlined above. Second class is the autoregressive conditional heteroskedastic (ARCH) models introduced by Engle (1982), and its descendents (GARCH (Bollerslev (1986)), NARCH, NGARCH (Duan, 1996), LGARCH, EGARCH, GJR-GARCH). General class of stochastic volatility models, that include many of the above-mentioned models, have been introduced by Ewald, Poulsen, and Schenk-Hoppe (2006). Gatheral (2006) introduced the Heston-like model for stochastic volatility that is more general than Heston model.

1.3.3   *Jump-Diffusion Volatility*

Jump-diffusion volatility is essential as there is evidence that assumption of a pure diffusion for the stock return is not accurate. Fat tails have been observed away from the mean of the stock return. This phenomenon is called *leptokurticity* and could be explained in different ways. One way to explain smile and leptocurticity is to introduce a jump-diffusion process for stochastic volatility (see Bates (1996)). Jump-diffusion is not a level-dependent volatility process, but can explain leverage effect.

1.3.4   *Multi-Factor Models for Stochastic Volatility*

One-factor SV models (all above-mentioned): 1) incorporate the leverage between returns and volatility and 2) reproduce the skew of implied volatility. However, it fails to match either the high conditional kurtosis of returns (Chernov *et al.* (2003)) or the full term structure of implied volatility surface (Cont *et al.* (2004)). Two primary generalizations of one-factor SV models are: 1) adding jump components in returns and/or volatility process, and 2) considering *multi-factor SV models.* Among multi-factor SV models we mention here the following ones: Fouque *et al.* (2005) SV model, Chernov *et al.* (2003) model (used efficient method of moments to obtain comparable empirical-of-fit from affine jump-diffusion mousiondels and two-factor SV family models); Molina *et al.* (2003) model (used a Markov Chain Monte Carlo method to find strong evidence of two-factor SV models with well-separated time scales in foreign exchange data); Cont *et al.* (2004) (found that jump-diffusion models have a fairly good fit to the implied volatility surface); Fouque *et al.* (2000) model (found that two-factor SV models provide a better fit to the term structure of implied volatility than one-factor SV models by capturing the behavior at short and long maturities); Swishchuk (2006) introduced two-factor and three-factor SV models with delay (incorporating mean-reverting level as a random process (geometric Brownian model, OU, continuous-time GARCH(1,1) model)).

We also mention SABR model (see Hagan *et al.* (2002)), describing a single forward under stochastic volatility, and Chen (1996) three-factor model for the dynamics of the instantaneous interest rate.

Multi-factor SV models have advantages and disadvantages. One of disadvantages–multi-factor SV models do not admit in general explicit solutions for option prices. One of the advantages–they have direct implications on hedges. As a comparison, a class of jump-diffusion models (Bates (1996)) enjoys closed-form solutions for option prices. But the complexity of hedging strategies increases due to jumps. In this way, there is no strong empirical evidence to judge the overwhelming position of jump-diffusion models over multi-factor SV models or vice versa.

The probability literature demonstrates that stochastic volatility models are fundamental notions[7] in financial markets analysis.

1.4   Summary

- Because it measures the change in value of a financial instrument over a specific horizon, volatility, as measured by the standard deviation, is an important concept in financial modeling.
- The different types of volatility are historical, implied, jump-diffusion, level-dependent, local, and stochastic volatilities.
- Stochastic volatility means that the volatility is not a constant, but a stochastic process. Stochastic volatility can explain the well documented volatility smile and skew observed in option markets.
- Stochastic volatility is the main concept used in finance to deal with the endemic time-varying volatility and co-dependence found in financial markets and

---

[7]Barndorff-Nielsen, Nicolato and Shephard (2002), Shephard (2005).

stochastic volatility models are used to evaluate derivative securities such as options, swaps.
– Two approaches to introduce stochastic volatility are: (1) changing the clock time to a random time and (2) changing constant volatility into a positive stochastic process.

Bibliography

Avellaneda, M., Levy, A. and Paras, A. (1995). Pricing and hedging derivative securities in markets with uncertain volatility. *Applied Mathematical Finance*, 2, 73-88.

Ahn, H. and Wilmott, P. (2006). *Stochastic Volatility and Mean-Variance Analysis*. New York: Wiley/Finance.

Barndorff-Nielsen, O.E., Nicolato, E. and Shephard, N. (2002). Some recent development in stochastic volatility modeling. *Quantitative Finance*, 2, 11-23.

Bates, D. (1996). Jumps and stochastic volatility: The exchange rate processes implicit in Deutschemark options. *Review Finance Studies*, 9, 69-107.

Beckers, S. (1980) The constant elasticity of variance model and its implications for option pricing. *Journal of Finance*, 35, 661-673.

Black, F. and Scholes, M. (1973). The pricing of options and corporate liabilities. *Journal of Political Economy*, 81, 637-54.

Bollerslev, T. (1986). Generalized autoregressive conditional heteroscedasticity. *Journal of Economics*, 31, 307-27.

Chen, L. (1996). Stochastic mean and stochastic volatility-A three-factor model of the term structure of interest rates and its application to the pricing of interest rate derivatives. *Financial Markets, Institutions and Instruments*, 5, 1-88.

Chernov, R., Gallant, E., Ghysels, E. and Tauchen, G. (2003). Alternative models for stock price dynamics. *Journal of Econometrics*, 116, 225-257.

Cont, R. and Tankov, P. (2004). *Financial Modeling with Jump Processes*. New York: Chapman & Hall/CRC Fin. Math. Series.

Cox, J. (1975). Notes on option pricing I: Constant elasticity of variance diffusions. Stanford, CA: Stanford University, Class notes.

Cox, J., Ingersoll, J. and Ross, S. (1985). A theory of the term structure of interest rate. *Econometrics*, 53, 385-407.

Cox, J. C., Ross, R. A. and Rubinstein, M. (1976). Option pricing: A simplified approach. *Journal of Financial Economics*, 7, 229-263.

Derman, E. and Kani, I. (1994). Riding on a smile. *Risk*, 7, 2, 32-39.

Duan, J. (1996). The GARCH option pricing model. *Mathematical Finance*, 5(1): 13-32.

Dupire, B. (1994). Pricing with a smile. *Risk*, 7, 1, 18-20.

Elliott, R. and Swishchuk, A. (2007). Pricing options and variance swaps in Markov-modulated Brownian markets. *Hidden Markov Models in Finance*. New York: Springer. 45-68.

Engle, R. (1982). Autoregressive conditional heteroscedasticity with estimates of the variance of United Kingdom inflation. *Econometrica*, 50, 4, 987-1007.

Fouque, J.-P., Papanicolaou, G. and Sircar, K. R. (2000). *Derivatives in Financial Markets with Stochastic Volatilities*. New York: Springer.

Fouque, J.-P. and Han, C.-H. (2003). A control variate method to evaluate option prices under multi-factor stochastic volatility models. Working Paper, Santa Barbara, CA: University of California.

Gatheral, J. (2006). *The Volatility Surface. A Practitioner's Guide*. New York: Wiley.

Hagan, P., Kumar, D., Lesniewski, S. and Woodward, D. (2002). Managing smile risk. *Wilmott Magazine*, 7/26/02, 84-108.

Heston, S. (1993). A closed-form solution for options with stochastic volatility with applications to bond and currency options. *Review of Financial Studies*, 6, 327-343.

Heston, S. and Nandi, S. (1998). Preference-free option pricing with path-dependent volatility: A closed-form approach. Discussion Paper. Atlanta: Federal Reserve Bank of Atlanta.

Hobson, D. and Rogers, L. (1998). Complete models with stochastic volatility. *Mathematical Finance*, 8, 1, 27-48.

Hull, J. (2000). *Options, Futures and Other Derivatives*. New Jersey: Prentice Hall, 4th edition.

Hull, J., and White, A. (1987). The pricing of options on assets with stochastic volatilities. *Journal of Finance*, 42, 281-300.

Javaheri, A. (2005). *Inside Volatility Arbitrage*. New York: Wiley/ Finance.

Johnson, H. and Shanno, D. (1985). Option pricing when the variance is changing. Working Paper 85-07, University of California, Davis, CA: Graduate School of Administration.

Kazmerchuk, Y., Swishchuk, A. and Wu, J. (2005). A continuous-time GARCH model for stochastic volatility with delay. *Canadian Applied Mathematics Quarterly*, 13, 2, 123-149.

Mandelbrot, B. (1963). The variation of certain speculative prices. *Journal of Business*, 36, 394-419.

Merton, R. (1973). Theory of rational option pricing. *Bell Journal of Economic Management Science*, 4, 141-183.

Molina, G., Han, C.-H. and Fouque, J.-P. (2003). MCMC Estimation of Multiscale Stochastic Volatility Models. Preprint, Santa Barbara, CA: University of California.

Officer, R. R. (1973). The variability of the market factor of New York stock exchange. *Journal of Business*, 46, 434-453.

Ornstein, L. and Uhlenbeck, G. (1930). On the theory of Brownian motion. *Physical Review*, 36, 823-841.

Poulsen, R., Schenk-Hoppe, K.-R. and Ewald, C.-O. (2009). Risk minimization in stochastic volatility models: Model risk and empirical performance. *Quantitative Finance*, 9, 6, 693-704.

Scott, L. (1987). Option pricing when the variance changes randomly: Theory, estimation and an application. *Journal of Financial Quantitative Analysis*, 22, 419-438.

Shephard, N. (2005). *Stochastic Volatility: Selected Readings*. Oxford: Oxford University Press.

Shiryaev, A. (2008). *Essentials of Stochastic Finance: Facts, Models, Theory*. Singapore: World Scientific.

Stein, E. and Stein, J. (1991). Stock price distribution with stochastic volatility. An analytic approach. *Review of Financial Studies*, 4, 727-752.

Swishchuk, A. (2000). *Random Evolutions and Their Applications. New Trends*. Dordrecht, The Netherlands: Kluwer Academic Publishers.

Swishchuk, A. (2004). Modelling and valuing of variance and volatility swaps for financial markets with stochastic volatilites. *Wilmott Magazine*, 2, September, 64-72.

Swishchuk, A. (2005). Modeling and pricing of variance swaps for stochastic volatilities with delay. *Wilmott Magazine*, 19, September, 63-73.

Swishchuk, A. (2006). Modeling and pricing of variance swaps for multi-factor stochastic volatilities with delay. *Canadian Applied Mathematics Quarterly*, 14, 4, Winter.

Swishchuk, A. (2009a). Pricing of variance and volatility swaps with semi-Markov volatilities. *Canadian Applied Mathematics Quarterly*, 18, 4.

Swishchuk, A. (2009b). Variance swaps for local stochastic volatility with delay and jumps. Working Paper, Calgary: University of Calgary.

Swishchuk, A. and Couch, M. (2010). Volatility and variance swpas for COGARCH(1,1) model. *Wilmott Journal*, 2, 5, 231-246.

Swishchuk, A. and Manca, R. (2010). Modeling and pricing of variance swaps for local semi-Markov volatility in financnial engineering. *Mathematical Models in Engineering*, 1-17, New York: Hindawi Publishing.

Swishchuk, A. and Malenfant, K. (2010a). Pricing of variance swaps for Lévy-based stochastic volatility with delay. *International Review of Applied Financial Issues and Economics*, Paris: S.E.I.F. (accepted).

Swishchuk, A. and Li, X. (2011). Variance swaps for stochastic volatility with delay and jumps. *International Journal of Stochastic Analysis*, Volume 2011, 27 pages.

Wiggins, J. (1987). Option values under stochastic volatility. *Journal of Finanancial Economics*, 19, 351-372.

Wilmott, P., Howison, S. and Dewynne, J. (1995). *Option Pricing: Mathematical Models and Computations*. Oxford: Oxford Financial Press.

Chapter 2

Stochastic Volatility Models

2.1 Introduction

In this Chapter, we consider different types of stochastic volatilities that we use in this book. They include, in particular: Heston stochastic volatility model; stochastic volatilities with delay; multi-factor stochastic volatilities; stochastic volatilities with delay and jumps; Lévy-based stochastic volatility with delay; delayed stochastic volatility in Heston model (we call it 'delayed Heston model'); semi-Markov modulated stochastic volatilities; COGARCH(1,1) stochastic volatility; stochastic volatilities driven by fractional Brownian motion; continuous-time GARCH stochastic volatility model.

2.2 Heston Stochastic Volatility Model

Let $(\Omega, \mathcal{F}, \mathcal{F}_t, P)$ be probability space with filtration $\mathcal{F}_t, \quad t \in [0, T]$.

Assume that underlying asset S_t in the risk-neutral world and variance follow the following model, Heston (1993) model:

$$\begin{cases} \frac{dS_t}{S_t} = r_t dt + \sigma_t dw_t^1 \\ d\sigma_t^2 = k(\theta^2 - \sigma_t^2)dt + \gamma \sigma_t dw_t^2, \end{cases} \tag{2.1}$$

where r_t is deterministic interest rate, σ_0 and θ are short and long volatility, $k > 0$ is a reversion speed, $\gamma > 0$ is a volatility (of volatility) parameter, w_t^1 and w_t^2 are independent standard Wiener processes.

The Heston asset process has a variance σ_t^2 that follows Cox-Ingersoll-Ross (1985) process, described by the second equation in (2.1).

If the volatility σ_t follows Ornstein-Uhlenbeck process (see, for example, Øksendal (1998)), then Ito's lemma shows that the variance σ_t^2 follows the process described exactly by the second equation in (2.1).

This model will be studied in Chapter 6 (see also Swishchuk (2004)).

2.3   Stochastic Volatility with Delay

Let us assume that a stock (risky asset) satisfies is the stochastic process $(S(t))_{t\in[-\tau,T]}$ which satisfies the following SDDE:

$$dS(t) = \mu S(t)dt + \sigma(t, S(t-\tau))S(t)dW(t), \quad t > 0, \qquad (2.2)$$

where $\mu \in R$ is an appreciation rate, volatility $\sigma > 0$ is a continuous and bounded function and $W(t)$ is a standard Wiener process.

The initial data for (2.2) is defined by $S(t) = \varphi(t)$ is deterministic function, $t \in [-\tau, 0], \quad \tau > 0$.

Throughout the book we suppose that

$$S_t := S(t - \tau),$$

where $\tau > 0$ is a delay parameter.

We assume that the equation for the variance $\sigma^2(t, S_t)$ has the following form:

$$\frac{d\sigma^2(t, S_t)}{dt} = \gamma V + \frac{\alpha}{\tau}\left[\int_{t-\tau}^{t}\sigma(s, S_s)dW(s)\right]^2 - (\alpha+\gamma)\sigma^2(t, S_t). \qquad (2.3)$$

Here, all the parameters α, γ, τ, V are positive constants and $0 < \alpha + \gamma < 1$. The Wiener process $W(t)$ is the same as in (2.2).

This model will be studied in Chapter 7 (see also Swishchuk (2005)).

2.4   Multi-Factor Stochastic Volatility Models

Here, we define and study four multi-factor stochastic volatility models with delay, three two-factor models and one three-factor model, to model and to price variance swaps.

1. *Two-factor stochastic volatility model with delay and with geomaetric Brownian motion mean-reversion* is defined in the following way:

$$\begin{cases} \frac{d\sigma^2(t,S_t)}{dt} = \gamma V_t + \frac{\alpha}{\tau}\left[\int_{t-\tau}^{t}\sigma(s, S_s)dW(s)\right]^2 - (\alpha+\gamma)\sigma^2(t, S_t), \\ dV_t/V_t = \xi dt + \beta dW_1(t), \end{cases} \qquad (2.4)$$

where S_t is defined as $S_t := S(t - \tau,)$

$$dS(t) = \mu S(t)dt + \sigma(t, S_t)dW(t)$$

and Wiener processes $W(t)$ and $W_1(t)$ may be correlated.

2. *Two-factor stochastic volatility model with delay and with Ornstein-Uhlenbeck mean-reversion* is defined in the following way:

$$\begin{cases} \frac{d\sigma^2(t,S_t)}{dt} = \gamma V_t + \frac{\alpha}{\tau}\left[\int_{t-\tau}^{t}\sigma(s, S_s)dW(s)\right]^2 - (\alpha+\gamma)\sigma^2(t, S_t), \\ dV_t = \xi(L - V_t)dt + \beta dW_1(t), \end{cases} \qquad (2.5)$$

where Wiener processes $W(t)$ and $W^1(t)$ may be correlated, S_t is defined as $S_t := S(t - \tau)$, and

$$dS(t) = \mu S(t)dt + \sigma(t, S_t)dW(t).$$

3. *Two-factor stochastic volatility model with delay and with Pilipovich one-factor mean-reversion* is defined in the following way:

$$\begin{cases} \frac{d\sigma^2(t, S_t)}{dt} = \gamma V_t + \frac{\alpha}{\tau} \left[\int_{t-\tau}^t \sigma(s, S_s)dW(s) \right]^2 - (\alpha + \gamma)\sigma^2(t, S_t), \\ dV_t = \xi(L - V_t)dt + \beta V_t dW_1(t), \end{cases} \quad (2.6)$$

where Wiener processes $W(t)$ and $W_1(t)$ may be correlated, S_t is defined as $S_t := S(t - \tau)$,

$$dS(t) = \mu S(t)dt + \sigma(t, S_t)dW(t).$$

4. *Three-factor stochastic volatility model with delay and with Pilipovich mean-reversion* is defined in the following way:

$$\begin{cases} \frac{d\sigma^2(t, S_t)}{dt} = \gamma V_t + \frac{\alpha}{\tau} \left[\int_{t-\tau}^t \sigma(s, S_s)dW(s) \right]^2 - (\alpha + \gamma)\sigma^2(t, S_t), \\ dV_t = \xi(L_t - V_t)dt + \beta V_t dW_1(t), \\ dL_t = \beta_1 L_t dt + \eta L_t dW_2(t), \end{cases} \quad (2.7)$$

where the Wiener processes $W(t), W^1(t)$ and $W^2(t)$ may be correlated, S_t is defined as $S_t := S(t - \tau)$,

$$dS(t) = \mu S(t)dt + \sigma(t, S_t)dW(t).$$

These models have been studied in Swishchuk (2006). See also Chapter 8.

2.5 Stochastic Volatility Models with Delay and Jumps

We represent jumps in the stochastic volatility model with delay by general compound Poisson processes, and write the stochastic volatility in the following form:

$$\frac{d\sigma^2(t, S_t)}{dt} = \gamma V + \frac{\alpha}{\tau} \left[\int_{t-\tau}^t \sigma(s, S_s)dW(s) + \int_{t-\tau}^t y_s dN(s) \right]^2 - (\alpha + \gamma)\sigma^2(t, S_t)$$

$$(2.8)$$

where $W(t)$ is a Brownian motion, $N(t)$ is a Poisson process with intensity λ and y_t is the jump size at time t. We assume that $E[y_t] = A(t)$, $E[y_s y_t] = C(s, t), s < t$, $E[y_t^2] = B(t) = C(t, t)$ and $A(t)$, $B(t)$, $C(s, t)$ are all deterministic functions. Our purpose is to valuate variance swaps when the stochastic volatility satisfies this general equation.

In order to get and check the results, we first consider two simple cases which is easier to model and implement but fundamental and still capture some characteristics of the market.

We discuss the case that the jump size y_t always equals to constant one, that is, the jump part is represented by $\int_{t-\tau}^{t} dN(s)$, just simple Poisson processes. Then, we consider the case when the jump part is still compound Poisson processes denoted as $\int_{t-\tau}^{t} y_s dN(s)$, but the jump size y_t is assumed to be identically independent distributed random variable with mean value ξ and variance η.

The general case is discussed then as well. Finally, we show that the model for stochastic volatility with delay and jumps keeps those good features of the model in Swishchuk (2005). See Chapter 9 for details (see also Swishchuk and Li (2011)).

2.6   Lévy-Based Stochastic Volatility with Delay

The stock price $S(t)$ satisfies the following equation

$$dS(t) = \mu S(t)dt + \sigma(t, S_t)S(t)dW(t), \quad t > 0,$$

where $\mu \in R$ is the mean rate of return, the volatility term $\sigma > 0$ is a bounded function and $W(t)$ is a Brownian motion on a probability space $(\Omega, \mathcal{F}, P)$ with a filtration $\mathcal{F}_t$. We also let $r > 0$ be the risk-free rate of return of the market. We denote $S_t = S(t - \tau), \quad t > 0$ and the initial data of $S(t)$ is defined by $S(t) = \varphi(t)$, where $\varphi(t)$ is a deterministic function with $t \in [-\tau, 0], \quad \tau > 0$. The volatility $\sigma(t, S_t)$ satisfies the following equation:

$$\frac{d\sigma^2(t, S_t)}{dt} = \gamma V + \frac{\alpha}{\tau}\left[\int_{t-\tau}^{t} \sigma(u, S_u)dL(u)\right]^2 - (\alpha + \gamma)\sigma^2(t, S_t) \qquad (2.9)$$

where $L(t)$ is a Lévy process independent of $W(t)$ with Lévy triplet (a, γ, ν). Here, $V > 0$ is a mean-reverting level (or long-term equilibrium of $\sigma^2(t, S_t)$), $\alpha, \gamma > 0$, and $\alpha + \gamma < 1$.

Our model of stochastic volatility exhibits jumps and also past-dependence: the behavior of a stock price right after a given time t not only depends on the situation at t, but also on the whole past (history) of the process $S(t)$ up to time t. This draws some similarities with fractional Brownian motion models (see Mandelbrot (1997)) due to a long-range dependence property. Another advantage of this model is mean-reversion. This model is also a continuous-time version of GARCH(1,1) model (see Bollerslev (1986)) with jumps. See Chapter 10 for more details (see also Swishchuk and Malenfant (2011)).

2.7   Delayed Heston Model

We assume $Q-$ stock price dynamics (Z_t^Q and W_t^Q being two correlated standard Brownian motions):

$$dS_t = (r - q)S_t dt + S_t \sqrt{V_t}dZ_t^Q$$
$$dV_t = \left[\gamma(\theta^2 - V_t) + \epsilon_\tau(t)\right]dt + \delta\sqrt{V_t}dW_t^Q, \qquad (2.10)$$

where $\tau > 0$ is the delay, θ^2 (resp. γ) can be interpreted as the value of the long-range variance (resp. variance mean-reversion speed) when the delay tends to 0, δ the volatility of the variance and c the brownian correlation coefficient ($\langle W^Q, Z^Q \rangle_t = ct$).

The variance drift adjustment $\epsilon_\tau(t)$ and the adjusted long-range variance θ_τ^2 being respectively given by:

$$\epsilon_\tau(t) = \alpha\tau(\mu - r)^2 + (V_0 - \theta_\tau^2)(\gamma - \gamma_\tau)e^{-\gamma_\tau t}$$

$$\theta_\tau^2 := \theta^2 + \frac{\alpha\tau(\mu - r)^2}{\gamma}.$$

α is a continuous-time equivalent of the variance ARCH(1,1) autoregressive coefficient, and can also be seen as the amplitude of the pure delay adjustment $\epsilon_\tau(t)$.[1]

The adjusted variance mean-reversion speed γ_τ is the unique positive solution to:

$$\gamma_\tau = \alpha + \gamma + \frac{\alpha}{\gamma_\tau \tau}(1 - e^{\gamma_\tau \tau}) \qquad (0 < \gamma_\tau < \gamma).$$

Furthermore, $v_t := EQ(V_t)$ is given by:

$$v_t = \theta_\tau^2 + (V_0 - \theta_\tau^2)e^{-\gamma_\tau t}.$$

See Chapters 11-12 for more details (see also Swishchuk and Vadori (2012)).

2.8 Semi-Markov-Modulated Stochastic Volatility

Let x_t be a semi-Markov process in measurable phase space $(X, \mathcal{X})$.

We suppose that the stock price S_t satisfies the following stochastic differential equation

$$dS_t = S_t(rdt + \sigma(x_t, \gamma(t))dw_t) \qquad (2.11)$$

with the volatility $\sigma := \sigma(x_t, \gamma(t))$ depending on the process x_t, which is independent on standard Wiener process w_t, and the current life $\gamma(t) = t - \tau_{\nu(t)}$, $\mu \in R$. We call the volatility $\sigma(x_t, \gamma(t))$ the *current life semi-Markov volatility*.

Remark. We note that process (x_t, γ_t) is a Markov process on (X, R_+) with infinitesimal operator

$$Qf(x,t) = \frac{df(x,t)}{dt} + \frac{g_x(t)}{\bar{G}_x(t)} \int_X P(x, dy)[f(y,t) - f(x,t)].$$

For more details see Chapter 13 (see also Swishchuk (2010)).

---

[1]Recall that the adjustment is defined to be $\epsilon_\tau(t) := \alpha\left[\tau(\mu - r)^2 + \frac{1}{\tau}\int_{t-\tau}^t v_s ds - v_t\right]$, where $v_t := E^Q(V_t)$.

2.9   COGARCH(1,1) Stochastic Volatility Model

The COGARCH(1,1) equations (as described in Kluppleberg *et al.* (2004)) have the following form:

$$dG_t = \sigma_{t-}dL_t$$
$$d\sigma_{t-}^2 = (\beta - \eta\sigma_{t-}^2)dt + \phi\sigma_{t-}^2 d[L,L]_t, \tag{2.12}$$

where L_t is the driving Lévy process and $[L,L]_t$ is the quadratic variation of the driving Lévy process.

See Chapter 15 for more details (see also Klüppelberg *et al.* (2004) and Swishchuk and Couch (2010)).

2.10   Stochastic Volatility Driven by Fractional Brownian Motion

2.10.1   *Stochastic Volatility Driven by Fractional Ornstein-Uhlenbeck Process*

Let the stock price S_t satisfy the following equation in risk-neutral world:

$$dS_t = rS_t dt + \sigma_t S_t dW_t, \tag{2.13}$$

where $r > 0$ is an interest rate, W_t is a standard Brownian motion and volatility σ_t satisfies the following equation:

$$d\sigma_t = -a\sigma_t dt + \gamma dB_t^H, \tag{2.14}$$

where $a > 0$ is a mean-reverting speed, $\gamma > 0$ is a volatility coefficient of this stochastic volatility, B_t^H is a fractional Brownian motion with Hurst index $H > 1/2$, independent of W_t.

Note that the solution of the equation (16.7) has the following form:

$$\sigma_t = \sigma_0 e^{-at} + \gamma e^{-at} \int_0^t e^{as}dB_s^H. \tag{2.15}$$

Evidently, the Wiener integral w.r.t. fBm exists since the function $f(s) = e^{as}$ satisfies the condition (16.5).

Moreover, σ_t is the continuous Gaussian process with the second moment $E\sigma_t^2$, which is bounded on any finite interval (we present all the calculations in the next two sections), therefore the unique solution of the equation (16.7) has a form

$$S_t = S_0 \exp\left[rt - \frac{1}{2}\int_0^t \sigma^2(s)ds + \int_0^t \sigma(s)dW(s)\right],$$

where the stochastic integral $\int_0^t \sigma(s)dW(s)$ exists and is a square-integrable martingale. The process S_t itself is locally square-integrable martingale. This situation will repeat and we will not mention this again in what follows.

2.10.2 *Stochastic Volatility Driven by Fractional Vasićek Process*

Let the stock price S_t satisfy equation (16.7) (see Chapter 16) and the volatility σ_t satisfy the following equation:

$$d\sigma_t = a(b - \sigma_t)dt + \gamma dB_t^H, \tag{2.16}$$

where $a > 0$ is a mean-reverting speed, $b \geq 0$ is an equilibrium (or mean-reverting) level, $\gamma > 0$ is a volatility coefficient of this stochastic volatility, B_t^H is a fractional Brownian motion with Hurst index $H > 1/2$ independent of W_t. Since the limit case $b = 0$ corresponds the fractional Ornstein-Uhlenbeck process, we suppose that for fractional Vasićek process the parameter b is positive, $b > 0$. In this sense the model (16.10) is the generalization of the model (16.7) (see Chapter 16).

Note that the solution of the equation (10) has the following form:

$$\sigma_t = \sigma_0 e^{-at} + b(1 - e^{-at}) + \gamma e^{-at} \int_0^t e^{as} dB_s^H. \tag{2.17}$$

Evidently, this Wiener integral w.r.t. fBm exists, and fractional Vasićek process is Gaussian. As we have mentioned above, both fractional Ornstein-Uhlenbeck and Vasićek volatilities get both positive and negative values.

2.10.3 *Markets with Stochastic Volatility Driven by Geometric Fractional Brownian Motion*

Let the stock price S_t satisfy equation (16.7) and the square σ_t^2 of volatility σ_t satisfy the following equation:

$$d\sigma_t^2 = a\sigma_t^2 dt + \gamma \sigma_t^2 dB_t^H, \tag{2.18}$$

where $a > 0$ is a drift, $\gamma > 0$ is a volatility of σ_t^2, B_t^H is a fractional Brownian motion with Hurst index $H > 1/2$, independent of W_t.

Note that the solution of the equation (2.18) has the following form:

$$\sigma_t^2 = \sigma_0^2 e^{at + \gamma B_t^H}, \tag{2.19}$$

and it is evidently the positive process, so, we can put $\sigma_t = \sigma_0 \exp\{\frac{a}{2}t + \frac{1}{2}\gamma B_t^H\}$.

2.10.4 *Stochastic Volatility Driven by Fractional Continuous-Time GARCH Process*

Let the stock price S_t satisfy equation (16.7) (see Chapter 16) and the square σ_t^2 of volatility σ_t satisfy the following equation:

$$d\sigma_t^2 = a(b - \sigma_t^2)dt + \gamma \sigma_t^2 dB_t^H, \tag{2.20}$$

where $a > 0$ is a mean-reverting speed, b is a mean-reverting level, $\gamma > 0$ is a volatility of σ_t^2, B_t^H is a fractional Brownian motion with Hurst index $H > 1/2$, independent of W_t. Note that the solution of the equation (2.20) has the following form:

$$\sigma_t^2 = \sigma_0^2 e^{-at+\gamma B_t^H} + abe^{-at+\gamma B_t^H} \int_0^t e^{as-\gamma B_s^H}\, ds. \tag{2.21}$$

Note also that fractional GARCH process is a generalization of fractional geometric Brownian motion in the same sense that fractional Vasićek process is a generalization of fractional Ornstein-Uhlenbeck process.

For more details see Chapter 16 (see also Mishura and Swishchuk (2010)).

Remark. We study in this book the pricing of variance and volatility swaps for stochastic models in finance driven by fractional Brownian motion (fBM). Of course, the model with only fractional Brownian motion (fBM) is not arbitrage-free. However, we could add a Brownian motion to the fBM noise and make it arbitrage-free (see, e.g., Mishura, Y. *Stochastic Calculus for fBM and Related Processes*, Springer, 2008; the latter book contain application of fBM in finance for so-called mixed Brownian fractional Brownian models, see Chapter 5 from the book). The pricing results for financial models with only Brownian motion are available. In this way, someone may consider mixed fractional Brownian models to get further results in this direction for arbitrage-free models.

2.11 Mean-Reverting Stochastic Volatility Model (Continuous-Time GARCH Model) in Energy Markets

We consider a risky asset in energy market with stochastic volatility following a mean-reverting stochastic process that satisfies the following stochastic differential equation (we call it continuous-time GARCH model):

$$d\sigma^2(t) = a(L - \sigma^2(t))dt + \gamma\sigma^2(t)dW_t, \tag{2.22}$$

where $a > 0$ is a speed (or 'strength') of mean reversion, $L > 0$ is the mean reverting level (or equilibrium level, or long-term mean), $\gamma > 0$ is the volatility of volatility $\sigma(t)$, W_t is a standard Wiener process.

Let λ be 'market price of risk' and defined by the following constants:

$$a^* := a + \lambda\sigma, \quad L^* := aL/a^*.$$

Then, in the risk-neutral world, the drift paramater in (2.22) has the following form:

$$a^*(L^* - \sigma^2(t)) = a(L - \sigma^2(t)) - \lambda\gamma\sigma^2(t). \tag{2.23}$$

If we define the following process (W_t^*) by

$$W_t^* := W_t + \lambda t, \tag{2.24}$$

where W_t is a standard Brownian motion, then the risk-neutral stochastic volatility model has the following form

$$d\sigma^2(t) = (aL - (a + \lambda\gamma)\sigma^2(t))dt + \gamma\sigma^2(t)dW_t^*,$$

or, equivalently,

$$d\sigma^2(t) = a^*(L^* - \sigma^2(t))dt + \gamma\sigma^2(t)dW_t^*,$$

where

$$a^* := a + \lambda\gamma, \quad L^* := \frac{aL}{a + \lambda\gamma},$$

and W_t^* is defined in (2.24).

For more details, see Chapter 17 (see also Swishchuk (2012)).

2.12 Summary

- In this Chapter, we considered different types of stochastic volatilities that we use in this book.
- They include, in particular: Heston stochastic volatility model; stochastic volatilities with delay; multi-factor stochastic volatilities; stochastic volatilities with delay and jumps; Lévy-based stochastic volatility with delay; delayed stochastic volatility in Heston model (we call it 'delayed Heston model'); semi-Markov modulated stochastic volatilities; COGARCH(1,1) stochastic volatility; stochastic volatilities driven by fractional Brownian motion; continuous-time GARCH stochastic volatility model.

Bibliography

Bollerslev, T. (1986). Generalized autogressive conditional heteroscedasticity. *Journal of Econometrics*, 31, 307-327.

Heston, S. (1993). A closed-form solution for options with stochastic volatility with applications to bond and currency options. *Review of Financial Studies*, 6, 327-343.

Klüppelberg, C., Linder, A. and Maller, R. (2004). A continuous-time GARCH process driven by a lévy process: Stationarity and second-order behaviour, *Journal of Applied Probability*, 41, 3, 601-622.

Mandelbrot, B. (1997). *Fractals and Scaling in Finance, Discontinuity, Concentration, Risk*. New York: Springer.

Mishura Y. and Swishchuk A. (2010). Modeling and pricing of variance and volatility swaps for stochastic volatilities driven by fractional Brownian motion. *Applied Statistics, Actuarial and Financial Mathematics*, No. 1-2, 52-67.

Øksendal, B. (1998). *Stochastic Differential Equations: An Introduction with Applications*. New York: Springer.

Swishchuk, A. (2004). Modeling of variance and volatility swaps for financial markets with stochastic volatility. *Wilmott Magazine*, 2004, September Issue, Technical Article No. 2, 64-72,

Swishchuk, A. (2005). Modeling and pricing of variance swaps for stochastic volatilities with delay. *Wilmott Magazine*, Issue 19, September, 63-73,

Swishchuk, A. (2006). Modeling and pricing of variance swaps for multi-factor stochastic volatilities with delay. *Canadian Applied Mathematics Quaterly*, 14, 4, Winter.

Swishchuk, A. (2010). Pricing of variance and volatility swaps with semi-Markov volatilities. *Canadian Applied Mathematics Quarterly*, 18, 4.

Swishchuk, A. and Couch, M. (2010). Volatility and variance swaps for COGARCH(1,1) model. *Wilmott Journal*, 2, 5, 231-246.

Swishchuk, A. and Li, X. (2011). Pricing of variance and volatility swaps for stochastic volatilities with delay and jumps. *International Journal of Stochastic Analysis*, Volume 2011, 27 pages.

Swishchuk, A. and Malenfant, K. (2011). Variance swaps for local Lévy-based stochastic volatility with delay. *International Review of Applied Financial Issues and Economics*, 3, 2, 432-441.

Swishchuk, A. and Vadori, N. (2012). Delayed Heston model: Improvement of the volatility surface fitting. *Wilmott Magazine* (under review).

Swishchuk, A. (2012). Variance and volatility swaps in energy markets. *Journal of Energy Markets*, forthcoming.

Chapter 3

Swaps

3.1 Introduction

Swaps are useful for volatility hedging and speculation. Volatility swaps are forward contracts on future realized stock volatility and variance swaps are similar contracts on variance, the square of future volatility. Covariance and correlation swaps are covariance and correlation forward contracts, respectively, of the underlying two assets. Using change of time method, one can model and price variance, volatility, covariance and correlation swaps.

Variance, volatility, covariance and correlation swaps are relatively recent financial products that market participants can use for volatility hedging and speculation. The market for these types of swaps has been growing with many investment banks and other financial institutions are now actively quoting volatility swaps on various assets: stock indexes, currencies, and commodities.

A stock's volatility is the simplest measure of its riskiness or uncertainty. In this entry we describe, model and price variance, volatility, covariance, and correlation swaps.

3.2 Definitions of Swaps

We begin with descriptions of the different kinds of swaps that we will be discussing in this Chapter: variance swaps, volatility swaps, covariance swaps, correlation swaps and pseudo-swaps.

3.2.1 *Variance and Volatility Swaps*

A stock's volatility is the simplest measure of its risk less or uncertainty. Formally, the volatility σ_R is the annualized standard deviation of the stock's returns during the period of interest, where the subscript R denotes the observed or 'realized' volatility.

Why trade volatility or variance swaps? As mentioned in Demeterfi *et al.* (1999, p. 9), "just as stock investors think they know something about the direction of the stock market so we may think we have insight into the level of future volatility. If we think current volatility is low, for the right price we might want to take a position that profits if volatility increases."

The easiest way to trade volatility is to use volatility swaps, sometimes called realized volatility forward contracts, because they provide only exposure to volatility and not other risk.

Variance swaps are similar contracts on variance, the square of the future volatility. As noted by Carr and Madan (1998), both types of swaps provide an easy way for investors to gain exposure to the future level of volatility.

A stock volatility swap's payoff at expiration is equal to

$$N(\sigma_R(S) - K_{vol}),$$

where $\sigma_R(S)$ is the realized stock volatility (quoted in annual terms) over the life of contract,

$$\sigma_R(S) = \sqrt{\frac{1}{T} \int_0^T \sigma_s^2 ds},$$

σ_t is a stochastic stock volatility, K_{vol} is the annualized volatility delivery price, and N is the notional amount of the swap in dollar per annualized volatility point.

Although options market participants talk of volatility, it is variance, or volatility squared, that has more fundamental significance.[1] A variance swap is a forward contract on annualized variance, the square of the realized volatility. Its payoff at expiration is equal to

$$N(\sigma_R^2(S) - K_{var}),$$

where $\sigma_R^2(S)$ is the realized stock variance (quoted in annual terms) over the life of the contract; that is,

$$\sigma_R^2(S) = \frac{1}{T} \int_0^T \sigma_s^2 ds,$$

K_{var} is the delivery price for variance, and N is the notional amount of the swap in dollars per annualized volatility point squared. The holder of variance swap at expiration receives N dollars for every point by which the stock's realized variance $\sigma_R^2(S)$ has exceeded the variance delivery price K_{var}. Therefore, pricing the variance swap reduces to calculating the square of the realized volatility.

Valuing a variance forward contract or swap is no different from valuing any other derivative security. The value of a forward contract P on future realized variance with strike price K_{var} is the expected present value of the future payoff in the risk-neutral world:

$$P_{var} = E\{e^{-rT}(\sigma_R^2(S) - K_{var})\},$$

---

[1]See Demeterfi, Derman, Kamal, and Zou (1999).

where r is the risk-free interest rate corresponding to the expiration date T, and E denotes the expectation. Thus, for calculating variance swaps we need to know only $E\{\sigma_R^2(S)\}$, namely the mean value of the underlying variance.

To calculate volatility swaps we need more. Using Brockhaus and Long (2000) approximation (which the second-order Taylor expansion for function $\sqrt{x}$) we have[2]

$$E\{\sqrt{\sigma_R^2(S)}\} \approx \sqrt{E\{V\}} - \frac{Var\{V\}}{8E\{V\}^{3/2}},$$

where $V = \sigma_R^2(S)$ and $\frac{Var\{V\}}{8E\{V\}^{3/2}}$ is the convexity adjustment.

Thus, to calculate the value of volatility swaps

$$P_{vol} = \{e^{-rT}(E\{\sigma_R(S)\} - K_{vol})\}$$

we need both $E\{V\}$ and $Var\{V\}$.

3.2.2   *Covariance and Correlation Swaps*

Options dependent on exchange rate movements, such as those paying in a currency different from the underlying currency, have an exposure to movements of the correlation between the asset and the exchange rate. This risk can be eliminated by using a covariance swap.

A *covariance swap* is a covariance forward contract of the underlying rates S^1 and S^2 which have a payoff at expiration that is equal to

$$N(Cov_R(S^1, S^2) - K_{cov}),$$

where K_{cov} is a strike price, N is the notional amount, and $Cov_R(S^1, S^2)$ is a covariance between two assets S^1 and S^2.

Logically, a *correlation swap* is a correlation forward contract of two underlying rates S^1 and S^2 which payoff at expiration is the following

$$N(Corr_R(S^1, S^2) - K_{corr}),$$

where $Corr_R(S^1, S^2)$ is a realized correlation of two underlying assets S^1 and S^2, K_{corr} is a strike price, N is the notional amount.

Pricing covariance swap, from a theoretical point of view, is similar to pricing variance swaps, since

$$Cov_R(S^1, S^2) = 1/4\{\sigma_R^2(S^1 S^2) - \sigma_R^2(S^1/S^2)\}$$

where S^1 and S^2 are two underlying assets, $\sigma_R^2(S)$ is a variance swap for the underlying assets, and $Cov_R(S^1, S^2)$ is a realized covariance of the two underlying assets S^1 and S^2.

Thus, we need to know the variances for $S^1 S^2$ and for S^1/S^2. Correlation $Corr_R(S^1, S^2)$ is defined as follows:

$$Corr_R(S^1, S^2) = \frac{Cov_R(S^1, S^2)}{\sqrt{\sigma_R^2(S^1)}\sqrt{\sigma_R^2(S^2)}},$$

---

[2]See also Javaheri *et al.* (2002), p. 16.

where $Cov_R(S^1, S^2)$ is defined as above and $\sigma_R^2(S^1)$ is the realized variance for S^1.

Given two assets S_t^1 and S_t^2 with $t \in [0, T]$, sampled on days $t_0 = 0 < t_1 < t_2 < ... < t_n = T$ between today and maturity T, the log-return of each asset is

$$R_i^j = \log\left(\frac{S_{t_i}^j}{S_{t_{i-1}}^j}\right), \quad i = 1, 2, ..., n, \quad j = 1, 2.$$

Covariance and correlation can be approximated by

$$Cov_n(S^1, S^2) = \frac{n}{(n-1)T} \sum_{i=1}^{n} R_i^1 R_i^2$$

and

$$Corr_n(S^1, S^2) = \frac{Cov_n(S^1, S^2)}{\sqrt{Var_n(S^1)}\sqrt{Var_n(S^2)}},$$

respectively.

3.2.3   *Pseudo-Swaps*

The market we consider consists of the strictly positive stochastic assets $S_t^{(1)}$ and $S_t^{(2)}$ satisfying the stochastic differential equation (SDE)

$$\frac{dS_t^{(i)}}{S_t^{(i)}} = \mu_t^{(i)} \, dt + \sigma_t^{(i)} \, dW_t^i, \quad t > 0, \quad i = 1, 2$$

and a numeraire N_t which is a zero-coupon bond $df(t, T)$. Here $dW_t^{(i)}$ are standard Wiener processes (zero mean, unit variance per unit time) with correlation $\rho_t dt$. In the cases of the variance and volatility swaps, we drop the superscripts so for example we write $S_t = S_t^{(1)}$ and $\mu_t = \mu_t^{(1)}$, etc. We assume that the market is complete, so there exists a unique martingale measure Q with respect to N_t. First, let us define the following continuously realized (measured) statistics over an observation period $[T_s, T_e]$:

$$\Sigma_{(S)}^2(T_s, T_e) = \frac{1}{T_e - T_s} \int_{T_s}^{T_e} \sigma_\tau^2 d\tau, \tag{3.1}$$

$$\Sigma_{(S^{(1)}, S^{(2)})}^2(T_s, T_e) = \frac{1}{T_e - T_s} \int_{T_s}^{T_e} \sigma_\tau^{(1)} \sigma_\tau^{(2)} \rho_\tau d\tau, \tag{3.2}$$

$$\sigma_{(S)}(T_s, T_e) = \sqrt{\frac{1}{T_e - T_s} \int_{T_s}^{T_e} \sigma_\tau^2 d\tau}, \tag{3.3}$$

$$\rho_{(S^{(1)}, S^{(2)})}(T_s, T_e) = \frac{\int_{T_s}^{T_e} \sigma_\tau^{(1)} \sigma_\tau^{(2)} \rho_\tau d\tau}{\sqrt{\int_{T_s}^{T_e} {\sigma_\tau^{(1)}}^2 d\tau} \sqrt{\int_{T_s}^{T_e} {\sigma_\tau^{(2)}}^2 d\tau}}, \tag{3.4}$$

which are realized volatility-square, realized volatility-cross, realized volatility, realized correlation, respectively.

Note that when $S_t^{(1)} \equiv S_t^{(2)}$, the correlation is equal to one and the volatility-cross coincides with the volatility-square. Here we only consider the simpler case where we approximate the above quantities by ones that are discretely sampled. Thus, let $T_s = t_0 < t_1 < ... < t_n = T_e$ be the sampling dates, and we define the log-return[3] for the underlying rate S as

$$X_i^{(k)} = \log \left(\frac{S_{t_i}^{(k)}}{S_{t_{i-1}}^{(k)}} \right), \qquad k = 1, 2, \qquad i = 1, 2, ..., n$$

and also denote the arithmetic mean by

$$\bar{X}_n^{(k)} = \frac{1}{n} \sum_{i=1}^{n} X_i^{(k)}, \qquad k = 1, 2.$$

Now, we can define the following realized pseudo-statistics:

$$Var = \frac{n}{T_e - T_s} \left(\frac{1}{n-1} \sum_{i=1}^{n} (X_i - \bar{X}_n)^2 \right), \tag{3.5}$$

$$Cov = \frac{n}{T_e - T_s} \left(\frac{1}{n-1} \sum_{i=1}^{n} \prod_{k=1}^{2} (X_i^{(k)} - \bar{X}_n^{(k)}) \right), \tag{3.6}$$

$$Vol = \sqrt{\frac{n}{T_e - T_s} \left(\frac{1}{n-1} \sum_{i=1}^{n} \left(X_i - \bar{X}_n \right)^2 \right)}, \tag{3.7}$$

$$Corr = \frac{\sum_{i=1}^{n} \prod_{k=1}^{2} (X_i^{(k)} - \bar{X}_n^{(k)})}{\prod_{k=1}^{2} \sqrt{\sum_{i=1}^{n} \left(X_i^{(k)} - \bar{X}_n^{(k)} \right)^2}}, \tag{3.8}$$

which are realized pseudo-volatility-cross, realized pseudo-volatility-square, realized pseudo-volatility, realized pseudo-correlation, respectively. Based on these pseudo-statistics, we can define the swaps, which are really forward contracts, by their payoffs at maturity date $T \geq T_e$:

$$V_{var}(T) = \alpha_{var} \cdot I \cdot \left[Var - \Sigma_K^2 \right], \tag{3.9}$$

$$V_{cov}(T) = \alpha_{var} \cdot I \cdot \left[Cov - \Sigma_K^2 \right], \tag{3.10}$$

$$V_{vol}(T) = \alpha_{vol} \cdot I \cdot \left[Vol - \sigma_K \right], \tag{3.11}$$

$$V_{corr}(T) = \alpha_{corr} \cdot I \cdot \left[Corr - \rho_K \right], \tag{3.12}$$

which are pseudo-variance swap, pseudo-covariance swap, pseudo-volatility swap, pseudo-correlation swap, respectively.

---

[3]Here log denotes the natural logarithm.

In the above, α_i are the converting parameters, $I = \pm 1$ is a long-short index, and Σ_K^2, σ_K and ρ_K are the strikes.

The general pricing formula is the mathematical expectation of the N_t-discounted claim under the unique martingale measure Q:

$$
\begin{aligned}
V_i(0) &= E\left[\frac{N_0}{N_T}V_i(T)\right] \\
&= E\left[\mathrm{df}(0,T)V_i(T)\right] \\
&= df(0,T)E\left[V_i(T)\right],
\end{aligned} \tag{3.13}
$$

for $i = $ var, cov, vol, and corr.

Note that the payoffs are path-dependent. Our goal is to find analytic expressions for the above pricing problems.

3.3 Summary

- Variance, volatility, covariance, and correlation swaps are useful for volatility hedging and speculation.
- Volatility swaps are forward contracts on future realized stock volatility.
- Variance swaps are similar contracts on variance, the square of the future volatility.
- Covariance and correlation swaps are covariance and correlation forward contracts, respectively, of the underlying two assets.

Bibliography

Brockhaus, O. and Long, D. (2000). Volatility swaps made simple. *Risk Magazine*, January, 92-96.

Carr, P. and Madan, D. (1998). Towards a theory of volatility trading. In Jarrow, R. (ed.), *Volatility*, Risk Book Publications.

Cheng, R., Lawi, S., Swishchuk, A., Badescu, A., Ben Mekki, H., Gashaw, A., Hua, Y., Molyboga, M., Neocleous, T. and Petrachenko, Y. (2002). Price Pseudo-Variance, Pseudo-Covariance, Pseudo-Volatility, and Pseudo-Correlation Swaps-In Analytical Closed-Forms, *Proceedings of the Sixth PIMS Industrial Problems Solving Workshop*, PIMS IPSW 6, University of British Columbia, Vancouver, Canada, May 24-31, 2002, 45-55. Editor: J. Macki, University of Alberta, June.

Chernov, R., Gallant, E., Ghysels, E. and Tauchen, G. (2003). Alternative models for stock price dynamics. *Journal of Econometrics*, 116, 225-257.

Demeterfi, K., Derman, E., Kamal, M. and Zou, J. (1999). A guide to volatility and variance swaps. *The Journal of Derivatives*, Summer, 9-32.

Elliott, R. and Swishchuk, A. (2007). Pricing options and variance swaps in Markov-modulated Brownian markets. In Elliot, R. and Mamon, R. (eds.), *Hidden Markov Models in Finance*. New York: Springer.

Javaheri, A., Wilmott, P. and Haug, E. (2002). GARCH and volatility swaps. *Wilmott Magazine*, January, 17 pages.

Kallsen, J. and Shiryaev, A. (2002). Time change representation of stochastic integrals. *Theory Probability and Its Applications*, 46, 3, 522-528.

Swishchuk, A. (2011). Varinace and volatility swaps in energy markets. *Journal of Energy Markets*, forthcoming.

Swishchuk, A. and Malenfant, K. (2011). Pricing of variance swaps for Lévy-based stochastic volatility with delay. *International Review of Applied Financial Issues and Economics,* forthcoming.

Swishchuk, A. and Li, X. (2011). Variance swaps for stochastic volatility with delay and jumps. *International Journal of Stochastic Analysis*, Volume 2011, 27 pages.

Swishchuk, A. and Couch, M. (2010). Volatility and variance swaps for COGARCH(1,1) model. *Wilmott Magazine*, 2, 5, 231-246.

Swishchuk, A. and Manca, R. (2010). Modeling and pricing of variance swaps for local semi-Markov volatility in financial engineering. *Mathematical Models in Engineering*, Volume 2010, 1-17.

Swishchuk, A. (2009a). Pricing of variance and volatility swaps with semi-Markov volatilities. *Canadian Applied Mathematics Quaterly*, 18, 4.

Swishchuk, A. (2009b). Variance swaps for local stochastic volatility with delay and jumps. Working Paper, Calgary: University of Calgary.

Swishchuk, A. (2007). Change of time method in mathematical finance. *Canadian Applied Mathematics Quarterly*, 15, 3, 299-336.

Swishchuk, A. (2006). Modeling and pricing of variance swaps for multi-factor stochastic volatilities with delay. *Canadian Applied Mathematics Quarterly*, 14, 4.

Swishchuk, A. (2005). Modeling and pricing of variance swaps for stochastic volatilities with delay. *Wilmott Magazine*, 19, September, 63-73.

Swishchuk, A. (2004). Modeling and valuing of variance and volatility swaps for financial markets with stochastic volatilites. *Wilmott Magazine,* 2, September, 64-72.

Théoret, R., Zabré, L., and Rostan, P. (2002). Pricing volatility swaps: Empirical testing with Canadian data. Working Paper. Centre de Recherche en Gestion, Université du Québec á Montréal, Document 17-2002, July 2002, 20 pages.

Chapter 4

Change of Time Methods

4.1 Introduction

Change of time can be used to introduce stochastic volatility or solve many stochastic differential equations. The main idea of change of time method is to change time from t to a non-negative process $T(t)$ with non-decreasing sample paths (e.g., subordinator). Many Lévy processes may be written as time-changed Brownian motion. Lévy processes can also be used as a time change for other Lévy processes (subordinators). Using change of time, we can get option pricing formula for an asset following geometric Brownian motion, e.g., Black-Scholes formula, and also we can get explicit option pricing formula for an asset following mean-reverting process, e.g., continous-time GARCH proccess.

In this chapter, we provide an overview on change of time methods (CTM), and show how to solve many stochastic differential equations (SDEs) in finance (geometric Brownian motion (GBM), Ornstein-Uhlenbeck (OU), Vasiček, continuous-time GARCH, etc.) using change of time method. As applications of CTM we present two different models: geometric Brownian motion (GBM) and mean-reverting models. The solutions of these two models are different. But the nice thing is that they can be solved by CTM as many other models mentioned in this chapter. And moreover, we can use these solutions to find easy option pricing formulae: one is classic-Black-Scholes and another one is new-for a mean-reverting asset. These formulae can be used in practice (for example, in energy markets) because they all are explicit.[1]

This includes: CTM in martingale and semimartingale setting; CTM in SDEs setting; subordination as a change of time.

4.2 Descriptions of the Change of Time Methods

The main idea of change of time method is to change time from t to a non-negative process $T(t)$ with non-decreasing sample paths. One of the examples is

---

[1]Swishchuk (2007) and Swishchuk (2008c).

subordinator: if $X(t)$ and $T(t) > 0$ are some processes, then $X(T(t))$ is subordinated to $X(t)$; $T(t)$ is a change of time. Another example is time-changed Brownian motion: $M(t) = B(T(t))$, where $B(t)$ is a Brownian motion and $T(t)$ is a subordinator (e.g., variance-gamma process[2] $V(t) = B(T(t)$, where $T(t)$ is a gamma process).

Bochner (1949) introduced the notion of change of time (time-changed Brownian motion). Clark (1973) introduced Bochner's change of time into financial economics. Feller (1966) introduced subordinated process $X(T(t))$ with Markov process $X(t)$ and $T(t)$ as a process with independent increments ($T(t)$ was called 'randomized operational time'). Johnson (1979) introduced time-changed stochastic volatility model (SVM) in continuous time. Johnson and Shanno (1987) studied the pricing of options using time-changed stochastic volatility (SV) model. Ikeda and Watanabe (1981) introduced and studied change of time for the solution of SDEs. Barndorff-Nielsen, Nicolato and Shephard (2002) studied the relationship between subordination and SVM using change of time ($T(t)$-'chronometer'). Carr, Geman, Madan and Yor (2003) used subordinated processes to construct SV for Lévy processes ($T(t)$-'business time').

The change of time method is closely associated with the embedding problem: to embed a process $X(t)$ in Brownian motion is to find a Wiener process process $W(t)$ and an increasing family of stopping times $T(t)$ such that $W(T(t))$ has the same joint distribution as $X(t)$. Skorokhod (1965) first treated the embedding problem, showing that the sum of any sequence of independent random variables (r.v.) with mean zero and finite variation could be embedded in Brownian motion using stopping times. Dambis (1965) and Dubins and Schwartz (1965) independently showed that every continuous martingale could be embedded in Brownian motion. Knight (1971) discovered the multi-variate extension of Dambis (1965) and Dubins and Schwartz (1965) result. Huff (1969) showed that every process of pathwise bounded variation could be embedded in Brownian motion. Monroe (1972) proved that every right continuous martingale could be embedded in a Brownian motion. Monroe (1978): proved that a process can be embedded in Brownian motion if and only if this process is a local semimartingale. Meyer (1971), Papangelou (1972) independently discovered Knight's (1971) result for point processes.

Rosiński and Woyczyński (1986) considered time changes for integrals over a stable Lévy processes. Kallenberg (1992) considered time change representations for stable integrals.

Lévy processes can also be used as a time change for other Lévy processes (subordinators). Madan and Seneta (1990) introduced variance gamma (VG) process (Brownian motion with drift time changed by a gamma process). Geman, Madan and Yor (2001) considered time changes for Lévy processes ('business time'). Carr, Geman, Madan and Yor (2003) used change of time to introduce stochastic volatility into a Lévy model to achieve leverage effect and a long-term skew. Kallsen and

---

[2]Madan *et al.* (1990).

Shiryaev (2002) showed that Rosiński-Woyczyński-Kallenberg statement cannot be extended to any other Lévy process but symmetric α-stable. Swishchuk (2004, 2007) applied change of time method for options and swaps pricing for Gaussian models.[3]

4.2.1 *The General Theory of Time Changes*

The general theory of change of time for martingale and semimartingale theories[4] is well known. In this chapter we give a brief description of the change of time method in the following settings: martingales and stochastic differential equations.

4.2.1.1 *Martingale and Semimartingale Settings of Change of Time*

Let $(\Omega, \mathcal{F}, P)$ be a given probability space with a right continuous filtration $(\mathcal{F}_t)_{t \geq 0}$. Suppose M_t is a square integrable local continuous martingale such that $\lim_{t \to +\infty} \langle M \rangle(t) = +\infty$ almost sure (a.s.), where $\tau_t := \inf\{u : \langle M \rangle(u) > t\}$ and $\tilde{\mathcal{F}}_t = \mathcal{F}_{\tau_t}$. Then the time-changed process $B(t) := M(\tau_t)$ is an $\tilde{\mathcal{F}}_t$-Brownian motion. Also, $M(t) = B(< M > (t))$. Here, $\langle \cdot \rangle$ defines predictable quadratic variation.

If ϕ_t is a change of time process (i.e., any continuous $\mathcal{F}_t$-adapted process such that $\phi_0 = 0$, $t \to \phi_t$ is strictly increasing and $\lim_{t \to +\infty} \phi_t = +\infty$ a.s.) and if X_t is an $\mathcal{F}_t$-adapted semimartingale, then the process $\tilde{X}_t := X_{\tau_t}$ is an $\tilde{\mathcal{F}}_t$-adapted semimartingale, where $\tau_t := \inf\{u : \phi_u > t\}$, and $\tilde{\mathcal{F}}_t := \mathcal{F}_{\tau_t}$. $\tilde{X}_t$ is called the time change of X_t by ϕ_t.

Geman, Madan and Yor (2001) consider pure jump Lévy processes (which are semimartingales) of finite variation with an infinite arrival rate of jumps as models for the logarithm of asset prices. These processes may be also written as time-changed Brownian motion. Their paper exhibits the explicit time change for each of a wide class of Lévy processes and shows that the time change is a weighted price move measure of time.

4.2.1.2 *Stochastic Differential Equations Setting of Change of Time*

The change of time method is used to solve the following SDE:

$$dX_t = \alpha(t, X_t)dB(t)$$

with $B(t)$ being a Brownian motion and $\alpha(t, x)$ being a 'good' function of $t \geq 0$ and $x \in R$. Having solved the equation we can also solve the general SDE

$$dX_t = \beta(t, X_t)dt + \gamma(t, X_t)dB(t)$$

with drift $\beta(t, X_t)$ using the method of transformation of drift (the Girsanov transformation).[5]

---

[3] Barndorff-Nielsen and Shiryaev (2009) state the main ideas and results of the stochastic theory of change of time and change of measure.
[4] Ikeda and Watanabe (1981).
[5] Ikeda and Watanabe (1981), Chapter IV, Section 4.

4.2.2   Subordinators as Time Changes

4.2.2.1   Subordinators

Feller (1966) introduced a subordinated process X_{τ_t} for a Markov process X_t and τ_t a process with independent increments. τ_t was called a 'randomized operational time'. Increasing Lévy processes can also be used as a time change for other Lévy processes.[6] Lévy processes of this kind are called subordinators. They are very important ingredients for building Lévy-based models in finance.[7] If S_t is a subordinator, then its trajectories are almost surely increasing, and S_t can be interpreted as a 'time deformation' and used to 'time change' other Lévy processes. Roughly, if $(X_t)_{t \geq 0}$ is a Lévy process and $(S_t)_{t \geq 0}$ is a subordinator independent of X_t, then the process $(Y_t)_{t \geq 0}$ defined by $Y_t := X_{S_t}$ is a Lévy process.[8] This time scale has the financial interpretation of business time,[9] that is, the integrated rate of information arrival.

4.2.2.2   Subordinators and Stochastic Volatility

The time change method was used to introduce stochastic volatility into a Lévy model to achieve the leverage effect and a long-term skew.[10] In the Bates (1996) model the leverage effect and long-term skew were achieved using correlated sources of randomness in the price process and the instantaneous volatility. The sources of randomness are thus required to be Brownian motions. In the Barndorff-Nielsen *et al.* (2001, 2002) model the leverage effect and long-term skew are generated using the same jumps in the price and volatility without a requirement for the sources of randomness to be Brownian motions. Another way to achieve the leverage effect and long-term skew is to make the volatility govern the time scale of the Lévy process driving jumps in the price. Carr *et al.* (2003) suggested the introduction of stochastic volatility into an exponential-Lévy model via a time change. The generic model here is $S_t = \exp(X_t) = \exp(Y_{v_t})$, where $v_t := \int_0^t \sigma_s^2 ds$. The volatility process should be positive and mean-reverting (i.e., Ornstein-Uhlenbeck or Cox-Ingersoll-Ross processes). Barndorff-Nielsen *et al.* (2003) reviewed and placed in the context some of their recent work on stochastic volatility models including the relationship between subordination and stochastic volatility.

The main difference between the change of time method and the subordinator method is that in the former case the change of time process ϕ_t depends on the process X_t, but in the latter case, the subordinator S_t and Lévy process X_t are independent.

---

[6]Applebaum (2003), Barndorf-Nielsen *et al.* (2001), Barndorf-Nielsen *et al.* (2003), Bertoin (1996), Cont *et al.* (2004) and Schoutens (2003).

[7]Cont *et al.* (2004) and Schoutens (2003).

[8]Cont *et al.* (2004).

[9]Geman *et al.* (2002).

[10]Carr *et al.* (2003).

4.3 Applications of Change of Time Method

The change of time method may be applied to get Black-Scholes formula for GBM, explicit option pricing formula for a mean-reverting asset, pricing formula for swaps in financial models with stochastic volatility in classical Heston model and pricing formula for volatility swap in delayed Heston model.

4.3.1 *Black-Scholes by Change of Time Method*

In the early 1970s, Black *et al.* (1973) made a major breakthrough by deriving a pricing formula for vanilla option written on a stock. Their model and its extensions assume that the probability distribution of the underlying cash flow at any given future time is lognormal. There are many proofs of their result, including partial differential equation and martingale approach.[11] See Chapter 5 for details.

4.3.2 *An Option Pricing Formula for a Mean-Reverting Asset Model Using a Change of Time Method*

Some commodity prices, like oil and gas, exhibit mean reversion. This means that they tend over time to return to some long-term mean. This mean-reverting model is a one-factor version of the two-factor model made popular in the context of energy modeling by Pilipovic (1997). Black's model (1976) and Schwartz's model (1997) have become standard tools to price options on commodities. These models have the advantage that they give rise to closed-form solutions for some types of option. We note, that the recent book by Geman (2005) discusses hard and soft commodities, (that is, energy, agriculture and metals) and also presents an analysis of economic and geopolitical issues in commodities markets. Here, we show how to get explicit option pricing formula for a continuous-time GARCH asset price model using change of time. See Chapter 18 for details.

4.3.3 *Swaps by Change of Time Method in Classical Heston Model*

One of the applications of change of time method is to value variance, volatility, covariance and correlation swaps for Heston (1993) model. Change of time method for pricing of different types of swaps for Heston model and pricing of options has been considered in Swishchuk (2004, 2007, 2008c). See Chapter 6 for details. Applications of change of time method to Lévy-based stochastic volatility models, interest rates and energy derivatives have been considered in Swishchuk (2008a, 2008b, 2010c).

---

[11]Wilmott *et al.* (1995) and Elliott *et al.* (2007).

4.3.4 *Swaps by Change of Time Method in Delayed Heston Model*

We present a variance drift adjusted version of the Heston model which leads to significant improvement of the market volatility surface fitting (compared to Heston). The numerical example we performed with recent market data shows a significant (44%) reduction of the average absolute calibration error (calibration on 30th September 2011 for underlying EURUSD). Our model has two additional parameters compared to the Heston model, can be implemented very easily and was initially introduced for volatility derivatives pricing purpose. The main idea behind our model is to take into account some past history of the variance process in its (risk-neutral) diffusion. Using change of time method for continuous local martingales, we derive a closed formula for the Brockhaus and Long approximation of the volatility swap price. The model we consider is a variance drift adjusted Heston model — the delayed Heston model (see Swishchuk and Vadori (2012)). The main idea behind this model is to take into account some past history of the variance process in its (risk-neutral) diffusion. We also consider dynamic hedging of volatility swaps using a portfolio of variance swaps. See Chapters 11-12 for details.

4.4 Different Settings of the Change of Time Method

In this Section we give a brief description of the *change of time method* for the martingales and stochastic differential equations. Throughout in this chapter we consider $(\Omega, \mathcal{F}, \mathcal{F}_t, P)$ to be a probability space with a right continuous filtration $(\mathcal{F}_t)_{t \geq 0}$.

4.4.0.1 *Change of Time Method in Martingale Setting*

In this section, we describe the change of time method for a martingale $M(t) \in \mathcal{M}_2^{c,loc}$, the space of local square integrable continuous martingales.[12]

If $M(t) \in \mathcal{M}_2^{c,loc}$, $\lim_{t \to +\infty} < M > (t) = +\infty$ a.s., $\tau_t := \inf\{u :< M > (u) > t\}$ and $\tilde{\mathcal{F}}_t := \mathcal{F}_{\tau_t}$, then the following process with changed time

$$W(t) := M(\tau_t)$$

is an $\tilde{\mathcal{F}}_t$-Brownian motion (or standard Wiener process).

Consequently, we can express a local martingale $M(t)$ using an $\tilde{\mathcal{F}}_t$-Brownian motion $W(t)$ and an $\tilde{\mathcal{F}}_t$-stopping time, (since $\{< M > (t) \leq u\} = \{\tau_u \geq t\} \in \mathcal{F}_{\tau_u} = \tilde{\mathcal{F}}_u$)

$$M(t) = W(< M > (t)).$$

---

[12]Ikeda and Watanabe (1981), Theorem 7.2, Chapter 2.

4.4.0.2 *Change of Time Method in Stochastic Differential Equation Setting*

We consider the following generalization of the previous results to a SDE of the following form (without a drift)

$$dX(t) = \alpha(t, X(t))dW(t),$$

where $W(t)$ is a Brownian motion and $\alpha(t, X)$ is a continuous and measurable by t and X function on $[0, +\infty) \times R$.

The reason we consider this equation is if we solve the equation, then we can solve more general equations with a drift $\beta(t, X)$ using the *Girsanov transformation*.[13] The following result is used frequently to find a solution of a SDE using change of time method. The following theorem is due to Ikeda and Watanabe (1981).[14]

Let $\tilde{W}(t)$ be an one-dimensional $\mathcal{F}_t$-Wiener process with $\tilde{W}(0) = 0$, given on a probability space $(\Omega, \mathcal{F}, (\mathcal{F}_t)_{t \geq 0}, P)$ and let $X(0)$ be an $\mathcal{F}_0$-adopted random variable. Define a continuous process $V = V(t)$ by

$$V(t) = X(0) + \tilde{W}(t).$$

Let ϕ_t be the change of time process:

$$\phi_t = \int_0^t \alpha^{-2}(\phi_s, X(0) + \tilde{W}(s))ds.$$

If

$$X(t) := V(\phi_t^{-1}) = X(0) + \tilde{W}(\phi_t^{-1})$$

and $\tilde{\mathcal{F}}_t := \mathcal{F}_{\phi_t^{-1}}$, then there exists $\tilde{\mathcal{F}}_t$-adopted Wiener process $W = W(t)$ such that $(X(t), W(t))$ is a solution of the initial equation on probability space $(\Omega, \mathcal{F}, \tilde{\mathcal{F}}_t, P)$.[15]

We note that the solution of the following SDE

$$dX(t) = a(X(t))dW(t)$$

may be presented in the following form (which follows from the previous Theorem)

$$X(t) = X(0) + \tilde{W}(\phi_t^{-1}),$$

where $a(X)$ is a continuous measurable function, $\tilde{W}(t)$ is an one-dimensional $\mathcal{F}_t$-Wiener process with $\tilde{W}(0) = 0$, given on a probability space $(\Omega, \mathcal{F}, (\mathcal{F}_t)_{t \geq 0}, P)$ and $X(0)$ is an $\mathcal{F}_0$-adopted random variable. In this case[16]

$$\phi_t = \int_0^t a^{-2}(X(0) + \tilde{W}(s))ds,$$

and

$$\phi_t^{-1} = \int_0^t a^2(X(0) + \tilde{W}(\phi_s^{-1}))ds.$$

---

[13]Ikeda and Watanabe (1981), Chapter 4, Section 4.
[14]Chapter IV, Theorem 4.3.
[15]The proof of this Theorem may be found in Ikeda and Watanabe (1981), Chapter IV, Theorem 4.3.
[16]Ikeda and Watanabe (1981), Chapter IV, Example 4.2.

4.4.0.3   Examples: Solutions of Some SDEs[17]

1. *Solution for Ornshtein-Uhlenbeck (OU) Process Using Change of Time*
   Let S_t satisfy the following SDE:

$$dS_t = -\alpha S_t dt + \sigma dW_t.$$

Then S_t may be presented in the following form using the change of time method:

$$S_t = e^{-\alpha t}[S_0 + \tilde{W}(\phi_t^{-1})],$$

where ϕ_t^{-1} satisfies

$$\phi_t^{-1} = \sigma^2 \int_0^t (e^{\alpha s}(S_0 + \tilde{W}(\phi_s^{-1})))^2 ds.$$

2. *Solution for Vasićek Process Using Change of Time*
   Let S_t satisfy the following SDE:

$$dS_t = \alpha(b - S_t)dt + \sigma dW_t.$$

Then S_t may be presented in the following form using the change of time method

$$S_t = e^{-\alpha t}[S_0 - b + \tilde{W}(\phi_t^{-1})],$$

where ϕ_t^{-1} satisfies

$$\phi_t^{-1} = \sigma^2 \int_0^t (e^{\alpha s}(S_0 - b + \tilde{W}(\phi_s^{-1})) + b)^2 ds.$$

Above theorem may also be applied to solve Cox-Ingersoll-Ross (1985) equation, mean-reversion equation for commodity price (Pilipovic (1997) model), geometric Brownian motion equation (Black-Scholes (1973))[18] and delayed Heston model.

4.5   Summary

- The main idea of change of time method is to change time from t to a non-negative process $T(t)$ with non-decreasing sample paths (e.g., subordinator).
- Many Lévy processes may be written as time-changed Brownian motion.
- Lévy processes can also be used as a time change for other Lévy processes (subordinators).
- Change of time can be used to introduce stochastic volatility or solve many stochastic differential equations.
- Using change of time, we can get option pricing formula for an asset following geometric Brownian motion, e.g., Black-Scholes formula.
- Using change of time, we can get explicit option pricing formula for an asset following mean-reverting process, e.g., continous-time GARCH proccess.
- Using change of time, we can get pricing formulas for different swap in calssical Heston model.
- Using change of time, we can get pricing formula for variance and volatility swaps in delayed Heston model.

---

[17]Swishchuk (2007).
[18]Swishchuk (2007).

Bibliography

Applebaum, D. (2003). *Levy Processes and Stochastic Calculus*. Cambridge: Cambridge University Press.

Barndorff-Nielsen, O.E. and Shiryaev, A.N. (2009) *Change of Time and Change of Measures*. Singapore: World Scientific.

Barndorff-Nielsen, O.E. and Shephard, N. (2001). Non-Gaussian Ornstein-Uhlenbeck-based models and some of their uses in financial economics. *Journal of the Royal Statistical Society: Series B (Statistical Methodology)*, 63, 2, 167-241.

Barndorff-Nielsen, O.E. and Shephard, N. (2003). Integrated OU processes and non-Gaussian OU-based stochastic volatility models. *Scandinavian Journal of Statistics*, 30, 2, 277-295.

Barndorff-Nielsen, O.E., Nicolato, E. and Shephard, N. (2002). Some recent development in stochastic volatility modeling. *Quantitative Finance*, 2, 11-23.

Bates, D. (1996). Jumps and stochastic volatility: The exchange rate processes implicit in Deutschemark options. *Review Finance Studies*, 9, 69-107.

Black, F. and Scholes, M. (1973). The pricing of options and corporate liabilities. *Journal of Political Economy*, 81, 637-54.

Black, F. (1976). The pricing of commodity contracts. *Journal of Financial Economics*, 3, 167-179.

Bochner, S. (1949). Diffusion equation and stochastic processes. *Proceedings of National Academy of Sciences*, 85, 369-370.

Carr, P., Geman, H., Madan, D. and Yor, M. (2003). Stochastic volatility for Lévy processes. *Mathematical Finance*, 13, 3, 345-382.

Carr, P. and Wu, L. (2003). The finite moment logstable process and option pricing. *The Journal of Finance*, 43(2), 753-777, April.

Carr, P. and Wu, L. (2004). Time-changed Lévy processes and option pricing. *Journal of Financial Economics*, 71, 113-141.

Clark, P. (1973). A subordinated stochastic process model with fixed variance for speculative prices, *Econometrica*, 41, 135-156.

Cont, R. and Tankov, P. (2004). *Financial Modeling with Jump Processes*. New Jersey: Chapman & Hall/CRC Fin. Math. Series.

Cox, J., Ingersoll, J. and Ross, S. (1985). A theory of the term structure of interest rate. *Econometrics*, 53, 385-407.

Dambis, K.E. (1965). On the decomposition of continuous martingales. *Theory of Probability and Its Applications*, 10, 401-410.

Dubins, K.E. and Schwartz, E. (1965). On continuous martingales. *Proceedings of the National Academy Sciences*, 53, 913-916.

Elliott, R. and Swishchuk, A. (2007). Pricing options and variance swaps in Markov-modulated Brownian markets. *Hidden Markov Models in Finance*. New York: Springer.

Feller, W. (1966) *An Introduction to Probability Theory and Its Applications*. New York: John Wiley & Sons.

Geman, H. (2005). *Commodities and Commodity Derivatives: Modelling and Pricing for Agricaltural, Metals and Energy*. New York: Wiley & Sons.

Geman, H., Madan, D. and Yor, M. (2002). Time changes for Lévy processes. *Mathematical Finance*, 11, 79-96.

Heston, S. (1993). A closed-form solution for options with stochastic volatility with applications to bond and currency options. *Review of Financial Studies*, 6, 327-343.

Huff, B. (2000). The loose subordination of differential processes to Brownian motion. *Annals of Mathematical Statistics*, 40, 1603-1609.

Ikeda, N. and Watanabe, S. (1981). *Stochastic Differential Equations and Diffusion Processes*. Tokyo: North-Holland/Kodansha Ltd.

Johnson, H. (1979). Option pricing when the variance rate is changing. Working Paper, Los Angeles: University of California.

Johnson, H. and Shanno, D. (1987). Option pricing when the variance is changing. *Journal of Financial Quantitative Analalysis*, 22, 143-152.

Kallsen, J. and Shiryaev, A. (2002). Time change representation of stochastic integrals. *Theory Probababilbity and Its Applications*, 46, 3, 522-528.

Knight, F. (1971). A reduction of continuous, square-integrable martingales to Brownian motion. *Martingales*, Berlin: Springer, Lecture Notes in Mathematics, 190, 19-31.

Madan, D. and Seneta, E. (1990). The variance gamma (VG) model for share market returns. *Journal of Business*, 63, 511-524.

Meyer, P. A. (1971). Demonstration simplifiee d'un theoreme de Knight. *Seminaire de Probabilites V*, Berlin: Springer, Lecture Notes in Mathematics, 191, 191-195.

Monroe, I. (1972). On embedding right continuous martingales in Brownian motion. *Annals of Mathematical Statistics*, 43, 1293-1311.

Monroe, I. (1978). Processes that can be embedded in Brownian motion. *The Annals of Probability*, 6, 1, 42-56.

Papangelou, F. (1972). Integrability of expected increments of point processes and a related random change of scale. *Transactions of American Mathematical Society*, 165, 486-506.

Pilipović, D. (1997). *Valuing and Managing Energy Derivatives*. New York: McGraw-Hill.

Schoutens, W. (2003). *Lévy Processes in Finance. Pricing Financial Derivatives*. New York: Wiley & Sons.

Schwartz, E. (1997). The stochastic behaviour of commodity prices: implications for pricing and hedging. *Journal of Finance*, 52, 923-973.

Shephard, N. (2005). *Stochastic Volatility: Selected Readings*. Oxford: Oxford University Press.

Skorokhod, A. (1965). *Studies in the Theory of Random Processes*. Reading: Addison-Wesley.

Swishchuk, A. (2010a). Variance and volatility swaps in energy markets. *Journal of Energy Markets*, forthcoming.

Swishchuk, A. (2010b). Multi-factor Lévy models for pricing financial and energy derivatives. *Candadin Applied Mathemematics Quarterly*, 17, 4, 777-806.

Swishchuk, A. (2008a). Lévy-based interest rate derivatives: Change of time and PIDEs, *Canadian Applied Mathemematics Quarterly*, 16, 2, 161-192.

Swishchuk, A. (2008b). Multi-factor Lévy models: Change of time and pricing of financial and energy derivatives. Working Paper, Calgary: University of Calgary.

Swishchuk, A. (2008c). Explicit option pricing formula for a mean-reverting asset in energy market. *Journal of Numerical Applied Mathematics*, 1, 96, 216-233.

Swishchuk, A. (2007). Change of time method in mathematical finance. *Canadian Applied Mathematics Quarterly*, 15, 3, 299-336.

Swishchuk, A. (2004). Modeling and valuing of variance and volatility swaps for financial markets with stochastic volatilites. *Wilmott Magazine*, 2, September, 64-72.

Wilmott, P., Howison, S. and Dewynne, J. (1995). *Option Pricing: Mathematical Models and Computations*. Oxford: Oxford Financial Press.

Chapter 5

Black-Scholes Formula by Change of Time Method

5.1 Introduction

In this Chapter, we consider applications of the change of time method to derive the well-known Black-Scholes formula for European call options (see Swishchuk (2007)). In the early 1970s, Black and Scholes (1973) made a major breakthrough by deriving pricing formula for vanilla option written on the stock. Their model and its extensions assume that the probability distribution of the underlying cash flow at any given future time is lognormal. We mention that there are many proofs of this result, including PDE and martingale approaches (see Wilmott *et al.* (1995), Elliott and Kopp (1999)).

5.2 Black-Scholes Formula by Change of Time Method

Let $(\Omega, \mathcal{F}, \mathcal{F}_t, P)$ be a probability space with a sample space Ω, σ-algebra of Borel sets $\mathcal{F}$ and probability P. The filtration $\mathcal{F}_t$, $t \in [0, T]$, is the natural filtration of a standard Brownian motion W_t, $t \in [0, T]$, and that $\mathcal{F}_T = \mathcal{F}$.

5.2.1 *Black-Scholes Formula*

The well-known Black-Scholes (1973) formula states that if we have (B, S)-security market consisting of riskless asset $B(t)$ with interest rate r as a constant

$$dB(t) = rB(t)dt, \quad B(0) > 0, \quad r > 0, \tag{5.1}$$

and risky asset (stock) $S(t)$

$$dS(t) = \mu S(t)dt + \sigma S(t)dW(t), \quad S(0) > 0, \tag{5.2}$$

where $\mu \in R$ is an appreciation rate, $\sigma > 0$ is a volatility, then option price formula for European call option with pay-off function $f(T) = \max(S(T) - K, 0)$ ($K > 0$ is a strike price) has the following look

$$C(T) = S(0)\Phi(y_+) - e^{-rT}K\Phi(y_-), \tag{5.3}$$

where

$$y_\pm := \frac{\ln(\frac{S(0)}{K}) + (r \pm \frac{\sigma^2}{2})T}{\sigma\sqrt{T}} \qquad (5.4)$$

and

$$\Phi(y) := \frac{1}{\sqrt{2\pi}} \int_{-\infty}^{y} e^{-\frac{x^2}{2}} dx. \qquad (5.5)$$

5.2.2   Solution of SDE for Geometric Brownian Motion using Change of Time Method

Lemma 5.1. The solution of the equation (5.2) has the following look:

$$S(t) = e^{\mu t}(S(0) + \tilde{W}(\phi_s^{-1})), \qquad (5.6)$$

where $\tilde{W}(t)$ is an one-dimensional Wiener process,

$$\phi_t^{-1} = \sigma^2 \int_0^t [S(0) + \tilde{W}(\phi_s^{-1})]^2 ds$$

and

$$\phi_t = \sigma^{-2} \int_0^t [S(0) + \tilde{W}(s)]^{-2} ds.$$

Proof.
Set

$$V(t) = e^{-\mu t}S(t), \qquad (5.7)$$

where $S(t)$ is defined in (5.2).

Applying Itô formula to $V(t)$ we obtain

$$dV(t) = \sigma V(t)dW(t). \qquad (5.8)$$

Equation (5.8) looks like equation (2.9) with

$$a(X) = \sigma X.$$

In this way, the solution of the equation (5.8) using change of time method (see Corollary 4.1, Section 4.4) is (see (4.10) and (4.11))

$$V(t) = S(0) + \tilde{W}(\phi_t^{-1}), \qquad (5.9)$$

where $\tilde{W}(t)$ is an one-dimensional Wiener process,

$$\phi_t^{-1} = \sigma^2 \int_0^t [S(0) + \tilde{W}(\phi_s^{-1})]^2 ds$$

and

$$\phi_t = \sigma^{-2} \int_0^t [S(0) + \tilde{W}(s)]^{-2} ds.$$

From (5.7) and (5.9) it follows that the solution of the equation (5.2) has the representation (5.6).

5.2.3   Properties of the Process $\tilde{W}(\phi_t^{-1})$

Lemma 5.2. Process $\tilde{W}(\phi_t^{-1})$ is a mean-zero martingale with quadratic variation

$$\langle \tilde{W}(\phi_t^{-1}) \rangle = \phi_t^{-1} = \sigma^2 \int_0^t [S(0) + \tilde{W}(\phi_s^{-1})]^2 ds$$

and has the following representation

$$\tilde{W}(\phi_t^{-1}) = S(0)(e^{\sigma W(t) - \frac{\sigma^2}{2}t} - 1). \tag{5.10}$$

Proof.

From Lemma 5.1, it follows that $\tilde{W}(\phi_t^{-1})$ is a martingale with quadratic variation

$$\langle \tilde{W}(\phi_t^{-1}) \rangle = \phi_t^{-1} = \sigma^2 \int_0^t [S(0) + \tilde{W}(\phi_s^{-1})]^2 ds.$$

and the process $W(t)$ has the following look

$$W(t) = \sigma^{-1} \int_0^t [S(0) + \tilde{W}(\phi_s^{-1})]^{-1} d\tilde{W}(\phi_s^{-1}). \tag{5.11}$$

From (5.11) we obtain the following SDE for $\tilde{W}(\phi_s^{-1})$

$$d\tilde{W}(\phi_s^{-1}) = \sigma[S(0) + \tilde{W}(\phi_s^{-1})]dW(t).$$

Solving this equation we have the explicit expression (5.10) for $\tilde{W}(\phi_s^{-1})$

$$\tilde{W}(\phi_s^{-1}) = S(0)(e^{\sigma W(t) - \frac{\sigma^2}{2}t} - 1).$$

Q.E.D

We note that $E\tilde{W}(\phi_s^{-1}) = 0$ and $E[\tilde{W}(\phi_t^{-1})]^2 = S^2(0)(e^{\sigma^2 t} - 1)$, where $E := E_P$ is an expectation under physical measure P.

Since

$$E[e^{\sigma W(t) - \frac{\sigma^2}{2}t}]^n = e^{\frac{\sigma^2}{2}n(n-1)t}, \tag{5.12}$$

we can obtain all the moments for the process $\tilde{W}(\phi_s^{-1})$:

$$E[\tilde{W}(\phi_t^{-1})]^n = S^n(0) \sum_{k=0}^{n} C_n^k e^{\frac{\sigma^2 t}{2}k(k-1)} (-1)^{n-k}, \tag{5.13}$$

where $C_n^k := \frac{n!}{k!(n-k)!}$, $n! := 1 \times 2 \times 3... \times n$.

Corollary 5.1. From Lemma 5.2 (see (5.6), (5.10) and (5.12)) it follows that we can also obtain all the moment for the asset price $S(t)$ in (5.9), since

$$E[S(t)]^n = e^{n\mu t} E[S(0) + \tilde{W}(\phi_s^{-1})]^n$$

$$= e^{n\mu t} S^n(0) E[e^{\sigma W(t) - \frac{\sigma^2 t}{2}}]^n$$

$$= e^{n\mu t} S^n(0) e^{\frac{\sigma^2}{2}n(n-1)t}. \tag{5.14}$$

For example, variance of $S(t)$ is going to be

$$VarS(t) = ES^2(t) - (ES(t))^2 = S^2(0)e^{2\mu t}(e^{\sigma^2 t} - 1),$$

where $ES(t) = S(0)e^{\mu t}$ (see (5.9)).

5.3   Black-Scholes Formula by Change of Time Method

In risk-neutral world the dynamic of stock price $S(t)$ has the following look:

$$dS(t) = rS(t)dt + \sigma S(t)dW^*(t), \tag{5.15}$$

where

$$W^*(t) := W(t) + \frac{\mu - r}{\sigma}. \tag{5.16}$$

Following Section 5.2, from (5.6) we have the solution of the equation (5.15)

$$S(t) = e^{rt}[S(0) + \tilde{W}^*(\phi_t^{-1})], \tag{5.17}$$

where

$$\tilde{W}^*(\phi_t^{-1}) = S(0)(e^{\sigma W^*(t) - \frac{\sigma^2 t}{2}} - 1) \tag{5.18}$$

and $W^*(t)$ is defined in (5.16).

Let E_{P*} be an expectation under risk-neutral measure (or martingale measure) P^* (i.e., process $e^{-rT}S(t)$ is a martingale under the measure P^*).

Then the option pricing formula for European call option with pay-off function

$$f(T) = \max[S(T) - K, 0]$$

has the following look

$$C(T) = e^{-rT}E_{P*}[f(T)] = e^{-rT}E_{P*}[\max(S(T) - K, 0)]. \tag{5.19}$$

Proposition 5.1.

$$C(T) = S(0)\Phi(y_+) - Ke^{-rT}\Phi(y_-), \tag{5.20}$$

where $y_\pm$ and $\Phi(y)$ are defined in (5.4) and (5.5).

Proof. Using change of time method we have the following representation for the process $S(t)$ (see (5.17))

$$S(t) = e^{rt}[S(0) + \tilde{W}^*(\phi_t^{-1})],$$

where $\tilde{W}^*(\phi_t^{-1})$ is defined in (5.18). From (5.17)-(5.19), after substitution $\tilde{W}^*(\phi_t^{-1})$ into (5.17) and $S(T)$ into (5.19), it follows that

$$C(T) = e^{-rT}E_{P*}[\max(S(T) - K, 0)]$$

$$= e^{-rT}E_{P*}[\max(e^{rt}[S(0) + \tilde{W}^*(\phi_t^{-1})] - K, 0)]$$

$$= e^{-rT}E_{P*}[\max(e^{rt}S(0)e^{\sigma W^*(T) - \frac{\sigma^2 T}{2}} - K, 0)]$$

$$= e^{-rT}E_{P*}[\max(S(0)e^{\sigma W^*(T) + (r - \frac{\sigma^2}{2})T} - K, 0)]$$

$$= e^{-rT}\frac{1}{\sqrt{2\pi}}\int_{-\infty}^{+\infty}\max[S(0)e^{\sigma u\sqrt{T} + (r - \frac{\sigma^2}{2})T} - K, 0]e^{-\frac{u^2}{2}}\,du. \tag{5.21}$$

Let y_0 be a solution of the following equation

$$S(0)e^{\sigma y\sqrt{T} + (r - \sigma^2/2)T} = K,$$

namely,

$$y_0 = \frac{\ln(\frac{K}{S(0)}) - (r - \sigma^2/2)T}{\sigma\sqrt{T}}.$$

Then (5.21) may be presented in the following form

$$C(T) = e^{-rT} \frac{1}{\sqrt{2\pi}} \int_{y_0}^{+\infty} (S(0)e^{\sigma u\sqrt{T} + (r - \frac{\sigma^2}{2})T} - K)e^{-\frac{u^2}{2}} du. \qquad (5.22)$$

Finally, straightforward calculation of the integral in the right-hand side of (5.22) gives us the Black-Scholes result

$$C(T) = \frac{1}{\sqrt{2\pi}} \int_{y_0}^{+\infty} S(0)e^{\sigma u\sqrt{T} - \frac{\sigma^2 T}{2}} e^{-u^2/2} du - Ke^{-rT}[1 - \Phi(y_0)]$$

$$= \frac{S(0)}{\sqrt{2\pi}} \int_{y_0 - \sigma\sqrt{T}}^{+\infty} e^{-u^2/2} du - Ke^{-rT}[1 - \Phi(y_0)]$$

$$= S(0)[1 - \Phi(y_0 - \sigma\sqrt{T})] - Ke^{-rT}[1 - \Phi(y_0)]$$

$$= S(0)\Phi(y_+) - Ke^{-rT}\Phi(y_-),$$

where $y_\pm$ and $\Phi(y)$ are defined in (5.4) and (5.5). Q.E.D.

5.4 Summary

- In this Chapter, we considered applications of the change of time method to give one more derivation of the well-known Black-Scholes formula for European call options (see Swishchuk (2007)).
- We mention that there are many proofs of this result, including PDE and martingale approaches (see Wilmott *et al.* (1995), Elliott and Kopp (1999)).

Bibliography

Black, F. and Scholes, M. (1973). The pricing of options and corporate liabilities. *Journal of Political Economy*, 81, 637-54.

Elliott, R. and Kopp, P. (1999). *Mathematics of Financial Markets*. New York: Springer-Verlag.

Swishchuk, A. (2007). Change of time method in mathematical finance. *Canadian Applied Mathematics Quarterly*, 15, 3, 199-235.

Wilmott, P., Howison, S. and Dewynne, J. (1995). *The Mathematics of Financial Derivatives*. Cambridge: Cambridge University Press.

Chapter 6

Modeling and Pricing of Swaps for Heston Model

6.1 Introduction

In this Chapter, we study variance and volatility swaps for financial markets with underlying asset and variance following the Heston (1993) model. We also study covariance and correlation swaps for the financial markets. As an application, we provide a numerical example using $S\&P60$ Canada Index to price swap on the volatility.

In the early 1970s, Black and Scholes (1973) made a major breakthrough by deriving pricing formulas for vanilla options written on the stock. The Black-Scholes model assumes that the volatility term is a constant. This assumption is not always satisfied by real-life options as the probability distribution of an equity has a fatter left tail and thinner right tail than the lognormal distribution (see Hull (2000)), and the assumption of constant volatility σ in financial model (such as the original Black-Scholes model) is incompatible with derivatives prices observed in the market.

Instead, people started to use local volatility and stochastic volatility models. Local volatility models are used because they can be calibrated to market prices using well-known Dupire (1994) formula. Once they have been calibrated, the corresponding SDE has an associated time t distribution which is consistent with the market implied distribution from option prices with expiration t. However, local volatility models are well known to generate forward skews which do not match the ones observed in the market (forward skews are skews implied from forward starting options). In this way, people came up with stochastic volatility models.

The above issues have been addressed and studied in several ways, such as:
(i) Volatility is assumed to be a deterministic function of the time: $\sigma \equiv \sigma(t)$ (see Wilmott *et al.* (1995)); Merton (1973) extended the term structure of volatility to $\sigma := \sigma_t$ (deterministic function of time), with the implied volatility for an option of maturity T given by $\hat{\sigma}_T^2 = \frac{1}{T} \int_0^T \sigma_u^2 du$;
(ii) Volatility is assumed to be a function of the time and the current level of the stock price $S(t)$ (local volatility): $\sigma \equiv \sigma(t, S(t))$ (see Dupire (1994), Derman and Kani (1994)); the dynamics of the stock price satisfies the following stochastic

45

differential equation:

$$dS(t) = \mu S(t)dt + \sigma(t, S(t))S(t)dW_1(t),$$

where $W_1(t)$ is a standard Wiener process. A time and level dependent volatility coefficient makes the arithmetic more challenging and usually precludes the existence of a closed-form solution. When local volatility models are used the price of the financial claim is computed via Monte Carlo simulations (when the number of underliers is larger than two) or via finite difference schemes applied to the underlying pricing PDE (when the number of underliers is smaller than two).

(iii) Volatility is described by stochastic differential equation with the same sourse of randomness as stock's price (see Javaheri, Wilmott and Haug (2002));

(iv) The time variation of the volatility involves an additional source of randomness, besides $W_1(t)$, represented by $W_2(t)$, and is given by

$$d\sigma(t) = a(t, \sigma(t))dt + b(t, \sigma(t))dW_2(t),$$

where $W_2(t)$ and $W_1(t)$ (the initial Wiener process that governs the price process) may be correlated (see Buff (2002), Hull and White (1987), Heston (1993)); Another approach is connected with stochastic volatility, namely, uncertain volatility scenario (see Buff (2002)). This approach is based on the uncertain volatility model developed in Avellaneda *et al.* (1995), where a concrete volatility surface is selected among a candidate set of volatility surfaces. This approach addresses the sensitivity question by computing an upper bound for the value of the portfolio under arbitrary candidate volatility, and this is achieved by choosing the local volatility $\sigma(t, S(t))$ among two extreme values $\sigma_{\min}$ and $\sigma_{\max}$ such that the value of the portfolio is maximized locally;

(v) The volatility depends on a random parameter x such as $\sigma(t) \equiv \sigma(x(t))$, where $x(t)$ is some random process (see Elliott and Swishchuk (2007), Griego and Swishchuk (2000), Swishchuk (1995), Swishchuk (2000), Swishchuk *et al.* (2000));

(vi) The volatility $\sigma(t, S_t)$ depends on $S_t := S(t + \theta)$ for $\theta \in [-\tau, 0]$, namely, stochastic volatility with delay (see Kazmerchuk, Swishchuk and Wu (2002)), which is past-dependent model of volatility;

(vii) Stochastic local volatility models in local volatility context (see Ren, Madan and Qian Qian (2007)): generalization of the Dupire (1994), and Derman and Kani (1994) equations for determination of local volatility. Ren, Madan and Qian Qian (2007) exploit the Gyöngy (1986) result to represent the marginal laws of Itô processes by Markov processes in a local volatility context. They gave quanto corrections in local volatility models.

In the approach **(i)**, the volatility coefficient is independent of the current level of the underlying stochastic process $S(t)$. This is a deterministic volatility model, and the special case where σ is a constant reduces to the well-known Black-Scholes model that suggests changes in stock prices are lognormal distributed. But the empirical test by Bollerslev (1986) seems to indicate otherwise. One explanation for this problem of a lognormal model is the possibility that the variance of $\log(S(t)/S(t-1))$

changes randomly. This motivated the work of Chesney and Scott (1989), where the prices are analyzed for European options using the modified Black-Scholes model of foreign currency options and a random variance model. In their works the results of Hull and White (1987), Scott (1987) and Wiggins (1987) were used in order to incorporate randomly changing variance rates.

In the approach (ii), several ways have been developed to derive the corresponding Black-Scholes formula: one can obtain the formula by using stochastic calculus and, in particular, the Ito's formula (see Øksendal (1998), for example).

A generalized volatility coefficient of the form $\sigma(t, S(t))$ is said to be *level-dependent*. Because volatility and asset price are perfectly correlated, we have only one source of randomness given by $W_1(t)$. A time and level-dependent volatility coefficient makes the arithmetic more challenging and usually precludes the existence of a closed-form solution. However, the *arbitrage argument* based on portfolio replication and a completeness of the market remain unchanged.

The situation becomes different if the volatility is influenced by a second "non-tradable" source of randomness. This is addressed in the approach (iii)–(vii). In this case, we usually obtains a *stochastic volatility model*, which is general enough to include the deterministic model as a special case. The concept of stochastic volatility was introduced by Hull and White (1987), and subsequent development includes the work of Wiggins (1987), Johnson and Shanno (1987), Scott (1987), Stein and Stein (1991) and Heston (1993). We also refer to Frey (1997) for an excellent survey on level-dependent and stochastic volatility models.

Hobson and Rogers (1998) suggested a new class of non-constant volatility models, which can be extended to include the aforementioned level-dependent model and share many characteristics with the stochastic volatility model. The volatility is non-constant and can be regarded as an endogenous factor in the sense that it is defined in terms of the *past behaviour* of the stock price. This is done in such a way that the price and volatility form a multi-dimensional Markov process.

Also, in Elliott and Swishchuk (2007) we found value of variance swap for financial market with Markov stochastic volatility (case (v) above).

In working paper Schoutens, Simons and Tistaert (2003), the authors shows that several advanced equity option models incorporating stochastic volatility (the Heston stochastic volatility model with jumps and without jumps in the stock price process, the Barndorff-Nielsen-Shephard model and Levy models with stochastic time) can be calibrated very nicely to a realistic option surface.

Volatility swaps are forward contracts on future realized stock volatility, variance swaps are similar contract on variance, the square of the future volatility, both these instruments provide an easy way for investors to gain exposure to the future level of volatility.

The stock volatility is the simplest measure of its risk less or uncertainty. Formally, the volatility σ_R is the annualized standard deviation of the stock's returns during the period of interest, where the subscript R denotes the observed or "realized" volatility.

The easy way to trade volatility is to use volatility swaps, sometimes called realized volatility forward contracts, because they provide pure exposure to volatility (and only to volatility).

Demeterfi, Derman, Kamal and Zou (1999) explained the properties and the theory of both variance and volatility swaps. They derived an analytical formula for theoretical fair value in the presence of realistic volatility skews, and pointed out that volatility swaps can be replicated by dynamically trading the more straightforward variance swap.

Javaheri, Wilmott and Haug (2002) discussed the valuation and hedging of a GARCH(1,1) stochastic volatility model. They used a general and exible PDE approach to determine the first two moments of the realized variance in a continuous or discrete context. Then they approximate the expected realized volatility via a convexity adjustment.

Brockhaus and Long (2000) provided an analytical approximation for the valuation of volatility swaps and analyzed other options with volatility exposure.

Working paper by Théoret, Zabré and Rostan (2002) presented an analytical solution for pricing of volatility swaps, proposed by Javaheri, Wilmott and Haug (2002). They priced the volatility swaps within framework of GARCH(1,1) stochastic volatility model and applied the analytical solution to price a swap on volatility of the $S\&P60$ Canada Index (5-year historical period: 1997–2002).

In this Chapter we study a stochastic volatility model, Heston (1993) model, to model variance and volatility swaps. The Heston asset process has a variance σ_t^2 that follows a Cox, Ingersoll and Ross (1985) process. We find some analytical close forms for expectation and variance of the realized both continuously and discrete sampled variance, which are needed for study of variance and volatility swaps, and price of pseudo-variance, pseudo-volatility, the problems proposed by He and Wang (2002) for financial markets with deterministic volatility as a function of time. This approach was first applied to the study of stochastic stability of Cox-Ingersoll-Ross process in Swishchuk and Kalemanova (2000).

The same expressions for $E[V]$ and for $Var[V]$ were obtained by Brockhaus and Long (2000) using another analytical approach. Most articles on volatility products focus on the relatively straightforward variance swaps. They take the subject further with a simple model of volatility swaps.

We also study covariance and correlation swaps for the securities markets with two underlying assets with stochastic volatilities.

As an application of our analytical solutions, we provide a numerical example using $S\&P60$ Canada Index to price swap on the volatility.

6.2   Variance and Volatility Swaps

Volatility swaps are forward contracts on future realized stock volatility, variance swaps are similar contract on variance, the square of the future volatility, both these instruments provide an easy way for investors to gain exposure to the future level of volatility.

The stock volatility is the simplest measure of its risk less or uncertainty. Formally, the volatility $\sigma_R(S)$ is the annualized standard deviation of the stock's returns during the period of interest, where the subscript R denotes the observed or "realized" volatility for the stock S.

The easy way to trade volatility is to use volatility swaps, sometimes called realized volatility forward contracts, because they provide pure exposure to volatility (and only to volatility) (see Demeterfi, Derman, Kamal and Zou (1999)).

A stock *volatility swap* is a forward contract on the annualized volatility. Its payoff at expiration is equal to

$$N(\sigma_R(S) - K_{vol}),$$

where $\sigma_R(S)$ is the realized stock volatility (quoted in annual terms) over the life of contract,

$$\sigma_R(S) := \sqrt{\frac{1}{T}\int_0^T \sigma_s^2 ds},$$

σ_t is a stochastic stock volatility, K_{vol} is the annualized volatility delivery price, and N is the notional amount of the swap in dollar per annualized volatility point. The holder of a volatility swap at expiration receives N dollars for every point by which the stock's realized volatility σ_R has exceeded the volatility delivery price K_{vol}. The holder is swapping a fixed volatility K_{vol} for the actual (floating) future volatility σ_R. We note that usually $N = \alpha I$, where α is a converting parameter such as 1 per volatility-square, and I is a long-short index ($+1$ for long and -1 for short).

Although options market participants talk of volatility, it is variance, or volatility squared, that has more fundamental significance (see Demeterfi, Derman, Kamal and Zou (1999)).

A *variance swap* is a forward contract on annualized variance, the square of the realized volatility. Its payoff at expiration is equal to

$$N(\sigma_R^2(S) - K_{var}),$$

where $\sigma_R^2(S)$ is the realized stock variance (quoted in annual terms) over the life of the contract,

$$\sigma_R^2(S) := \frac{1}{T}\int_0^T \sigma_s^2 ds,$$

K_{var} is the delivery price for variance, and N is the notional amount of the swap in dollars per annualized volatility point squared. The holder of variance swap at expiration receives N dollars for every point by which the stock's realized variance $\sigma_R^2(S)$ has exceeded the variance delivery price K_{var}.

Therefore, pricing the variance swap reduces to calculating the realized volatility square.

Valuing a variance forward contract or swap is no different from valuing any other derivative security. The value of a forward contract P on future realized variance with strike price K_{var} is the expected present value of the future payoff in the risk-neutral world:

$$P = E\{e^{-rT}(\sigma_R^2(S) - K_{var})\},$$

where r is the risk-free discount rate corresponding to the expiration date T, and E denotes the expectation.

Thus, for calculating variance swaps we need to know only $E\{\sigma_R^2(S)\}$, namely, mean value of the underlying variance.

To calculate volatility swaps we need more. From Brockhaus-Long (2000) approximation (which is used the second order Taylor expansion for function $\sqrt{x}$) we have (see also Javaheri *et al.* (2002), p.16):

$$E\{\sqrt{\sigma_R^2(S)}\} \approx \sqrt{E\{V\}} - \frac{Var\{V\}}{8E\{V\}^{3/2}},$$

where $V := \sigma_R^2(S)$ and $\frac{Var\{V\}}{8E\{V\}^{3/2}}$ is the convexity adjustment.

Thus, to calculate volatility swaps we need both $E\{V\}$ and $Var\{V\}$.

The realised continuously sampled variance is defined in the following way:

$$V := Var(S) := \frac{1}{T}\int_0^T \sigma_t^2 dt.$$

The realised discrete sampled variance is defined as follows:

$$Var_n(S) := \frac{n}{(n-1)T}\sum_{i=1}^{n} \log^2 \frac{S_{t_i}}{S_{t_{i-1}}},$$

where we neglected by $\frac{1}{n}\sum_{i=1}^{n}\log\frac{S_{t_i}}{S_{t_{i-1}}}$ since we assume that the mean of the returns is of the order $\frac{1}{n}$ and can be neglected. The scaling by $\frac{n}{T}$ ensures that these quantities annualized (daily) if the maturity T is expressed in years (days).

$Var_n(S)$ is unbiased variance estimation for σ_t. It can be shown that (see Brockhaus and Long (2000))

$$V := Var(S) = \lim_{n\to+\infty} Var_n(S).$$

Realised discrete sampled volatility is given by:

$$Vol_n(S) := \sqrt{Var_n(S)}.$$

Realised continuously sampled volatility is defined as follows:

$$Vol(S) := \sqrt{Var(S)} = \sqrt{V}.$$

The expressions for V, $Var_n(S)$ and $Vol(S)$ are used for calculation of variance and volatility swaps.

6.2.1 *Variance and Volatility Swaps for Heston Model*

6.2.1.1 *Stochastic Volatility Model*

Let $(\Omega, \mathcal{F}, \mathcal{F}_t, P)$ be probability space with filtration $\mathcal{F}_t, \quad t \in [0, T]$.

Assume that underlying asset S_t in the risk-neutral world and variance follow the following model, Heston (1993) model:

$$\begin{cases} \frac{dS_t}{S_t} = r_t dt + \sigma_t dw_t^1 \\ d\sigma_t^2 = k(\theta^2 - \sigma_t^2)dt + \gamma \sigma_t dw_t^2, \end{cases} \tag{6.1}$$

where r_t is deterministic interest rate, σ_0 and θ are short and long volatility, $k > 0$ is a reversion speed, $\gamma > 0$ is a volatility (of volatility) parameter, w_t^1 and w_t^2 are independent standard Wiener processes.

The Heston asset process has a variance σ_t^2 that follows Cox-Ingersoll-Ross (1985) process, described by the second equation in (6.1).

If the volatility σ_t follows Ornstein-Uhlenbeck process (see, for example, Øksendal (1998)), then Ito's lemma shows that the variance σ_t^2 follows the process described exactly by the second equation in (5.1).

6.2.1.2 *Explicit Expression for σ_t^2*

In this section we propose a new probabilistic approach to solve the equation for variance σ_t^2 in (5.1) explicitly, using change of time method (see Ikeda and Watanabe (1981)).

Define the following process:

$$v_t := e^{kt}(\sigma_t^2 - \theta^2). \tag{6.2}$$

Then, using Ito formula (see Øksendal (1995)) we obtain the equation for v_t :

$$dv_t = \gamma e^{kt} \sqrt{e^{-kt} v_t + \theta^2} dw_t^2. \tag{6.3}$$

Using change of time approach to the general equation (see Ikeda and Watanabe (1981))

$$dX_t = \alpha(t, X_t) dw_t^2,$$

we obtain the following solution of the equation (6.3):

$$v_t = \sigma_0^2 - \theta^2 + \tilde{w}^2(\phi_t^{-1}),$$

or (see (6.2)),

$$\sigma_t^2 = e^{-kt}(\sigma_0^2 - \theta^2 + \tilde{w}^2(\phi_t^{-1})) + \theta^2, \tag{6.4}$$

where $\tilde{w}^2(t)$ is an $\mathcal{F}_t$-measurable one-dimensional Wiener process, ϕ_t^{-1} is an inverse function to ϕ_t:

$$\phi_t = \gamma^{-2} \int_0^t \{e^{k\phi_s}(\sigma_0^2 - \theta^2 + \tilde{w}^2(t)) + \theta^2 e^{2k\phi_s}\}^{-1} ds.$$

6.2.1.3   *Properties of Processes $\tilde{w}^2(\phi_t^{-1})$ and σ_t^2*

The properties of $\tilde{w}^2(\phi_t^{-1}) := b(t)$ are the following:

$$Eb(t) = 0; \tag{6.5}$$

$$E(b(t))^2 = \gamma^2 \left\{ \frac{e^{kt} - 1}{k}(\sigma_0^2 - \theta^2) + \frac{e^{2kt} - 1}{2k}\theta^2 \right\}; \tag{6.6}$$

$$Eb(t)b(s) = \gamma^2 \left\{ \frac{e^{k(t\wedge s)} - 1}{k}(\sigma_0^2 - \theta^2) + \frac{e^{2k(t\wedge s)} - 1}{2k}\theta^2 \right\}, \tag{6.7}$$

where $t \wedge s := \min(t, s)$.

Using representation (6.4) and properties (6.5)–(6.7) of $b(t)$ we obtain the properties of σ_t^2. Straightforward calculations give us the following results:

$$E\sigma_t^2 = e^{-kt}(\sigma_0^2 - \theta^2) + \theta^2;$$

$$E\sigma_t^2\sigma_s^2 = \gamma^2 e^{-k(t+s)} \left\{ \frac{e^{k(t\wedge s)} - 1}{k}(\sigma_0^2 - \theta^2) + \frac{e^{2k(t\wedge s)} - 1}{2k}\theta^2 \right\}$$

$$+ e^{-k(t+s)}(\sigma_0^2 - \theta^2)^2 + e^{-kt}(\sigma_0^2 - \theta^2)\theta^2$$

$$+ e^{-ks}(\sigma_0^2 - \theta^2)\theta^2 + \theta^4. \tag{6.8}$$

6.2.1.4   *Valuing Variance and Volatility Swaps*

From formula (6.8) we obtain mean value for V :

$$E\{V\} = \frac{1}{T}\int_0^T E\sigma_t^2 dt$$

$$= \frac{1}{T}\int_0^T \{e^{-kt}(\sigma_0^2 - \theta^2) + \theta^2\}dt$$

$$= \frac{1 - e^{-kT}}{kT}(\sigma_0^2 - \theta^2) + \theta^2. \tag{6.9}$$

The same expression for $E[V]$ may be found in Brockhaus and Long (2000). Substituting $E[V]$ from (6.9) into formula

$$P = e^{-rT}(E\{\sigma_R^2(S)\} - K_{var}) \tag{6.10}$$

we obtain the value of the variance swap.

Variance for V equals to:

$$Var(V) = EV^2 - (EV)^2.$$

From (9) we have:

$$(EV)^2 = \frac{1 - 2e^{-kT} + e^{-2kT}}{k^2 T^2}(\sigma_0^2 - \theta^2)^2 + \frac{2(1 - e^{-kT})}{kT}(\sigma_0^2 - \theta^2)\theta^2 + \theta^4. \tag{6.11}$$

Second moment may be found as follows using formula (6.8):

$$EV^2 = \frac{1}{T^2} \int_0^T \int_0^T E\sigma_t^2 \sigma_s^2 \, dt ds$$

$$= \frac{\gamma^2}{T^2} \int_0^T \int_0^T e^{-k(t+s)} \left\{ \frac{e^{k(t \wedge s)} - 1}{k} (\sigma_0^2 - \theta^2) + \frac{e^{2k(t \wedge s)} - 1}{2k} \theta^2 \right\} dt ds$$

$$+ \frac{1 - 2e^{-kT} + e^{-2kT}}{k^2 T^2} (\sigma_0^2 - \theta^2)^2 + \frac{2(1 - e^{-kT})}{kT} (\sigma_0^2 - \theta^2)\theta^2 + \theta^4. \quad (6.12)$$

Taking into account (6.11) and (6.12) we obtain:

$$Var(V) = EV^2 - (EV)^2$$

$$= \frac{\gamma^2}{T^2} \int_0^T \int_0^T e^{-k(t+s)} \left\{ \frac{e^{k(t \wedge s)} - 1}{k} (\sigma_0^2 - \theta^2) + \frac{e^{2k(t \wedge s)} - 1}{2k} \theta^2 \right\} dt ds.$$

After calculations the last expression we obtain is the following expression for variance of V:

$$Var(V) = \frac{\gamma^2 e^{-2kT}}{2k^3 T^2} [(2e^{2kT} - 4e^{kT}kT - 2)(\sigma_0^2 - \theta^2)$$

$$+ (2e^{2kT}kT - 3e^{2kT} + 4e^{kT} - 1)\theta^2]. \quad (6.13)$$

Similar expression for $Var[V]$ may be found in Brockhaus and Long (2000). Substituting EV from (9) and $Var(V)$ from (6.13) into formula

$$P = \{e^{-rT}(E\{\sigma_R(S)\} - K_{var})\} \quad (6.14)$$

with

$$E\{\sigma_R(S)\} = E\{\sqrt{\sigma_R^2(S)}\} \approx \sqrt{E\{V\}} - \frac{Var\{V\}}{8E\{V\}^{3/2}},$$

we obtain the value of volatility swap.

6.2.1.5 *Calculation of $E\{V\}$ in Discrete Case*

The realised discrete sampled variance:

$$Var_n(S) := \frac{n}{(n-1)T} \sum_{i=1}^{n} \log^2 \frac{S_{t_i}}{S_{t_{i-1}}},$$

where we neglected by $\frac{1}{n} \sum_{i=1}^{n} \log \frac{S_{t_i}}{S_{t_{i-1}}}$ for simplicity reason only. We note that

$$\log \frac{S_{t_i}}{S_{t_{i-1}}} = \int_{t_{i-1}}^{t_i} (r_t - \sigma_t^2/2) dt + \int_{t_{i-1}}^{t_i} \sigma_t dw_t^1.$$

$$E\{Var_n(S)\} = \frac{n}{(n-1)T} \sum_{i=1}^{n} E\left\{ \log^2 \frac{S_{t_i}}{S_{t_{i-1}}} \right\}.$$

$$E\left\{\log^2\frac{S_{t_i}}{S_{t_{i-1}}}\right\} = \left(\int_{t_{i-1}}^{t_i} r_t dt\right)^2 - \int_{t_{i-1}}^{t_i} r_t dt \int_{t_{i-1}}^{t_i} E\sigma_t^2 dt$$

$$+\frac{1}{4}\int_{t_{i-1}}^{t_i}\int_{t_{i-1}}^{t_i} E\sigma_t^2\sigma_s^2 dt ds$$

$$-E\left(\int_{t_{i-1}}^{t_i}\sigma_t^2 dt \int_{t_{i-1}}^{t_i}\sigma_t dw_t^1\right) + \int_{t_{i-1}}^{t_i} E\sigma_t^2 dt.$$

We know the expressions for $E\sigma_t^2$ and for $E\sigma_t^2\sigma_s^2$, and the fourth expression is equal to zero. Hence, we can easily calculate all the above expressions and, hence, $E\{Var_n(S)\}$ and variance swap in this case.

Remark 6.1. Some expressions for price of the realised discrete sampled variance $Var_n(S) := \frac{n}{(n-1)T}\sum_{i=1}^n \log^2\frac{S_{t_i}}{S_{t_{i-1}}}$, (or pseudo-variance) were obtained in the Proceedings of the 6th PIMS Industrial Problems Solving Workshop, PIMS IPSW 6, UBC, Vancouver, Canada, May 27-31, 2002. Editor: J. Macki, University of Alberta, Canada, June, 2002, pp. 45-55.

6.3 Covariance and Correlation Swaps for Two Assets with Stochastic Volatilities

6.3.1 *Definitions of Covariance and Correlation Swaps*

Option dependent on exchange rate movements, such as those paying in a currency different from the underlying currency, have an exposure to movements of the correlation between the asset and the exchange rate, this risk may be eliminated by using covariance swap.

A *covariance swap* is a covariance forward contact of the underlying rates S^1 and S^2 which payoff at expiration is equal to

$$N(Cov_R(S^1, S^2) - K_{cov}),$$

where K_{cov} is a strike price, N is the notional amount, $Cov_R(S^1, S^2)$ is a covariance between two assets S^1 and S^2.

Logically, a *correlation swap* is a correlation forward contract of two underlying rates S^1 and S^2 which payoff at expiration is equal to:

$$N(Corr_R(S^1, S^2) - K_{corr}),$$

where $Corr(S^1, S^2)$ is a realized correlation of two underlying assets S^1 and S^2, K_{corr} is a strike price, N is the notional amount.

Pricing covariance swap, from a theoretical point of view, is similar to pricing variance swaps, since

$$Cov_R(S^1, S^2) = 1/4\{\sigma_R^2(S^1 S^2) - \sigma_R^2(S^1/S^2)\}$$

where S^1 and S^2 are given two assets, $\sigma_R^2(S)$ is a variance swap for underlying assets, $Cov_R(S^1, S^2)$ is a realized covariance of the two underlying assets S^1 and S^2.

Thus, we need to know variances for $S^1 S^2$ and for S^1/S^2 (see Section 4.2 for details).

Correlation $Corr_R(S^1, S^2)$ is defined as follows:

$$Corr_R(S^1, S^2) = \frac{Cov_R(S^1, S^2)}{\sqrt{\sigma_R^2(S^1)}\sqrt{\sigma_R^2(S^2)}},$$

where $Cov_R(S^1, S^2)$ is defined above and $\sigma_R^2(S^1)$ in Section 3.4.

Given two assets S_t^1 and S_t^2 with $t \in [0, T]$, sampled on days $t_0 = 0 < t_1 < t_2 < ... < t_n = T$ between today and maturity T, the log-return each asset is:

$$R_i^j := \log\left(\frac{S_{t_i}^j}{S_{t_{i-1}}^j}\right), \quad i = 1, 2, ..., n, \quad j = 1, 2.$$

Covariance and correlation can be approximated by

$$Cov_n(S^1, S^2) = \frac{n}{(n-1)T} \sum_{i=1}^{n} R_i^1 R_i^2$$

and

$$Corr_n(S^1, S^2) = \frac{Cov_n(S^1, S^2)}{\sqrt{Var_n(S^1)}\sqrt{Var_n(S^2)}},$$

respectively.

6.3.2 *Valuing of Covariance and Correlation Swaps*

To value covariance swap we need to calculate the following

$$P = e^{-rT}(ECov(S^1, S^2) - K_{cov}). \tag{6.15}$$

To calculate $ECov(S^1, S^2)$ we need to calculate $E\{\sigma_R^2(S^1 S^2) - \sigma_R^2(S^1/S^2)\}$ for a given two assets S^1 and S^2.

Let S_t^i, $i = 1, 2$, be two strictly positive Ito's processes given by the following model

$$\begin{cases} \frac{dS_t^i}{S_t^i} = \mu_t^i dt + \sigma_t^i dw_t^i, \\ d(\sigma^i)_t^2 = k^i(\theta_i^2 - (\sigma^i)_t^2)dt + \gamma^i \sigma_t^i dw_t^j, \quad i = 1, 2, \quad j = 3, 4, \end{cases} \tag{6.16}$$

where μ_t^i, $i = 1, 2$, are deterministic functions, k^i, θ^i, γ^i, $i = 1, 2$, are defined in similar way as in (1), standard Wiener processes w_t^j, $j = 3, 4$, are independent, $[w_t^1, w_t^2] = \rho_t dt$, ρ_t is deterministic function of time, $[,]$ means the quadratic covariance, and standard Wiener processes w_t^i, $i = 1, 2$, and w_t^j, $j = 3, 4$, are independent.

We note that

$$d \ln S_t^i = m_t^i dt + \sigma_t^i dw_t^i, \tag{6.17}$$

where

$$m_t^i := \left(\mu_t^i - \frac{(\sigma_t^i)^2}{2} \right), \tag{6.18}$$

and

$$Cov_R(S_T^1, S_T^2) = \frac{1}{T}[\ln S_T^1, \ln S_T^2] = \frac{1}{T}\left[\int_0^T \sigma_t^1 dw_t^1, \int_0^T \sigma_t^2 dw_t^2 \right] = \frac{1}{T}\int_0^T \rho_t \sigma_t^1 \sigma_t^2 dt. \tag{6.19}$$

Let us show that

$$[\ln S_T^1, \ln S_T^2] = \frac{1}{4}([\ln(S_T^1 S_T^2)] - [\ln(S_T^1/S_T^2)]). \tag{6.20}$$

Remark first that

$$d \ln(S_t^1 S_t^2) = (m_t^1 + m_t^2)dt + \sigma_t^+ dw_t^+, \tag{6.21}$$

and

$$d \ln(S_t^1/S_t^2) = (m_t^1 - m_t^2)dt + \sigma_t^- dw_t^-, \tag{6.22}$$

where

$$(\sigma_t^\pm)^2 := (\sigma_t^1)^2 \pm 2\rho_t \sigma_t^1 \sigma_t^2 + (\sigma_t^2)^2, \tag{6.23}$$

and

$$dw_t^\pm := \frac{1}{\sigma_t^\pm}(\sigma_t^1 dw_t^1 \pm \sigma_t^2 dw_t^2). \tag{6.24}$$

Processes $w_t^\pm$ in (6.24) are standard Wiener processes by Levi-Kunita-Watanabe theorem and $\sigma_t^\pm$ are defined in (6.23).

In this way, from (6.21) and (6.22) we obtain that

$$[\ln(S_t^1 S_t^2)] = \int_0^t (\sigma_s^+)^2 ds = \int_0^t ((\sigma_s^1)^2 + 2\rho_t \sigma_s^1 \sigma_s^2 + (\sigma_s^2)^2) ds, \tag{6.25}$$

and

$$[\ln(S_t^1/S_t^2)] = \int_0^t (\sigma_s^-)^2 ds = \int_0^t ((\sigma_s^1)^2 - 2\rho_t \sigma_s^1 \sigma_s^2 + (\sigma_s^2)^2) ds. \tag{6.26}$$

From (6.20), (6.25) and (6.26) we have directly formula (6.20):

$$[\ln S_T^1, \ln S_T^2] = \frac{1}{4}([\ln(S_T^1 S_T^2)] - [\ln(S_T^1/S_T^2)]). \tag{6.27}$$

Thus, from (6.27) we obtain that

$$Cov_R(S^1, S^2) = 1/4(\sigma_R^2(S^1 S^2) - \sigma_R^2(S^1/S^2)).$$

Returning to the valuation of the covariance swap we have

$$P = E\{e^{-rT}(Cov(S^1, S^2) - K_{cov})\} = \frac{1}{4}e^{-rT}(E\sigma_R^2(S^1S^2) - E\sigma_R^2(S^1/S^2) - 4K_{cov}).$$

The problem now has reduced to the same problem as in the Section 6.2, but instead of σ_t^2 we need to take $(\sigma_t^+)^2$ for S^1S^2 and $(\sigma_t^-)^2$ for S^1/S^2 (see (6.23)), and proceed with the similar calculations as in Section 6.2.

6.3.3 *Variance Swaps for Lévy-Based Heston Model*

Assume that in the risk-neutral world the underlying asset S_t and the variance follow the following model:

$$\begin{cases} \frac{dS_t}{S_t} = r_t dt + \sigma_t dw_t \\ d\sigma_t^2 = k(\theta^2 - \sigma_t^2)dt + \gamma\sigma_t dL_t, \end{cases}$$

where r_t is the deterministic interest rate, σ_0 and θ are the short and long volatilities, $k > 0$ is a reversion speed, $\gamma > 0$ is a volatility (of volatility) parameter, w_t and L_t are independent standard Wiener and α-stable Lévy processes ($\alpha \in (0,2]$).

The solution for the second equation has the following form (see Swishchuk (2009)):

$$\sigma^2(t) = e^{-kt}[\sigma_0^2 - \theta^2 + \hat{L}(\hat{T}_t)] + \theta^2,$$

where $\hat{T}_t = \gamma^\alpha \int_0^t [e^{k\hat{T}_s}(\sigma_0^2 - \theta^2 + \hat{L}(\hat{T}_s)) + \theta^2 e^{2k\hat{T}_s}]^{\alpha/2}ds.$

The realized variance in our case is:

$$\sigma_R^2(S) := \frac{1}{T}\int_0^T \sigma^2(s)ds = \frac{1}{T}\int_0^T \{e^{-ks}[\sigma_0^2 - \theta^2 + \hat{L}(\hat{T}_s)] + \theta^2\}ds.$$

The value of the variance swap then is:

$$P_{var} = E\{e^{-rT}(\sigma_R^2(S) - K_{var})\}$$

$$= E\left\{e^{-rT}\left(\frac{1}{T}\int_0^T \{e^{-ks}[\sigma_0^2 - \theta^2 + \hat{L}(\hat{T}_s)] + \theta^2\}ds - K_{var}\right)\right\}.$$

Thus, for calculating variance swaps we need to know only $E\{\sigma_R^2(S)\}$, namely, the mean value of the underlying variance, or $E\{\hat{L}(\hat{T}_s)\}$.

Only moments of order less than α exist for the non-Gaussian family of α-stable distributions. We suppose that $1 < \alpha < 2$ to find $E\{\hat{L}(\hat{T}_s)\}$.

The value of a variance swap for the Lévy-based Heston model is:

$$P_{var} = e^{-rT}\left[\frac{1 - e^{-kT}}{kT}(\sigma_0^2 - \theta^2) + \theta^2 + \frac{\delta T}{2} - K_{var}\right],$$

where δ is a location parameter.

If $\delta = 0$, then the value of a variance swap for the Lévy-based Heston model is:

$$P_{var} = e^{-rT} \left[\frac{1 - e^{-kT}}{kT}(\sigma_0^2 - \theta^2) + \theta^2 - K_{var} \right],$$

which coincides with the well-known above result (see Brockhaus and Long (2000) and Swishchuk (2004)).

6.4   Numerical Example: $S\&P60$ Canada Index

In this Section, we apply the analytical solutions from Section 3 to price a swap on the volatility of the $S\&P60$ Canada index for five years (January 1997–February 2002).

These data were kindly provided to the author by Raymond Théoret (Université du Québec à Montréal, Montréal, Québec, Canada) and Pierre Rostan (Analyst at the R&D Department of Bourse de Montréal and Université du Québec à Montréal, Montréal, Québec, Canada). They calibrated the GARCH parameters from five years of daily historic $S\&P60$ Canada Index (from January 1997 to February 2002) (see working paper "Pricing volatility swaps: Empirical testing with Canadian data" by Theoret, Zabre and Rostan (2002)).

At the end of February 2002, we wanted to price the fixed leg of a volatility swap based on the volatility of the $S\&P60$ Canada index. The statistics on log returns $S\&P60$ Canada Index for 5 years (January 1997–February 2002) are presented in Table 6.1:

Table 6.1

Statistics on Log Returns $S\&P60$ Canada Index	
Series:	LOG RETURNS $S\&P60$ CANADA INDEX
Sample:	1 1300
Observations:	1300
Mean	0.000235
Median	0.000593
Maximum	0.051983
Minimum	−0.101108
Std. Dev.	0.013567
Skewness	−0.665741
Kurtosis	7.787327

From the histogram of the $S\&P60$ Canada index log returns on a 5-year historical period (1,300 observations from January 1997 to February 2002) it may be seen leptokurtosis in the histogram. If we take a look at the graph of the $S\&P60$ Canada index log returns on a 5-year historical period we may see volatility clustering in the returns series. These facts indicate about the conditional heteroscedasticity. A GARCH(1,1) regression is applied to the series and the results is obtained as in the next Table 6.2:

Table 6.2

Estimation of the GARCH(1,1) process				
Dependent Variable: Log returns of S&P60 Canada Index Prices				
Method: ML-ARCH				
Included Observations: 1300				
Convergence achieved after 28 observations				
–	**Coefficient:**	**Std. error:**	z-**statistic:**	**Prob.**
C	0.000617	0.000338	1.824378	0.0681
Variance Equation				
C	2.58E-06	3.91E-07	6.597337	0
ARCH(1)	0.060445	0.007336	8.238968	0
GARCH(1)	0.927264	0.006554	141.4812	0
R-squared	−0.000791	Mean dependent var	–	0.000235
Adjusted R-squared	−0.003108	S.D. dependent var	–	0.013567
S.E. of regression	0.013588	Akaike info criterion	–	−5.928474
Sum squared resid	0.239283	Schwartz criterion	–	−5.912566
Log likelihood	3857.508	Durbin-Watson stat	–	1.886028

This table allows us to generate different input variables to the volatility swap model.

We use the following relationship

$$\theta = \frac{V}{dt},$$

$$k = \frac{1 - \alpha - \beta}{dt},$$

$$\gamma = \alpha\sqrt{\frac{\xi - 1}{dt}},$$

to calculate the following discrete GARCH(1,1) parameters:

ARCH(1,1) coefficient $\alpha = 0.060445$;

GARCH(1,1) coefficient $\beta = 0.927264$;

the Pearson kurtosis (fourth moment of the drift-adjusted stock return) $\xi = 7.787327$;

long volatility $\theta = 0.05289724$;

$k = 3.09733$;

$\gamma = 2.499827486$;

a short volatility σ_0 equals to 0.01;

Parameter V may be found from the expression $V = \frac{C}{1-\alpha-\beta}$, where $C = 2.58 \times 10^{-6}$ is defined in Table 6.2. Thus, $V = 0.00020991$;

$dt = 1/252 = 0.003968254$.

Now, applying the analytical solutions (6.9) and (6.13) for a swap maturity T of 0.91 year, we find the following values:

$$E\{V\} = \frac{1 - e^{-kT}}{kT}(\sigma_0^2 - \theta^2) + \theta^2 = .3364100835,$$

and

$$Var(V) = \frac{\gamma^2 e^{-2kT}}{2k^3 T^2}[(2e^{2kT} - 4e^{kT}kT - 2)(\sigma_0^2 - \theta^2)$$

$$+ (2e^{2kT}kT - 3e^{2kT} + 4e^{kT} - 1)\theta^2] = .0005516049969.$$

The convexity adjustment $\frac{Var\{V\}}{8E\{V\}^{3/2}}$ is equal to .0003533740855.

If the non-adjusted strike is equal to 18.7751%, then the adjusted strike is equal to

$$18.7751\% - 0.03533740855\% = 18.73976259\%.$$

This is the fixed leg of the volatility swap for a maturity $T = 0.91$.

Repeating this approach for a series of maturities up to 10 years we obtain the following plot (see Figure 6.2) of $S\&P60$ Canada Index Volatility Swap.

Figure 6.1 illustrates the non-adjusted and adjusted volatility for the same series of maturities.

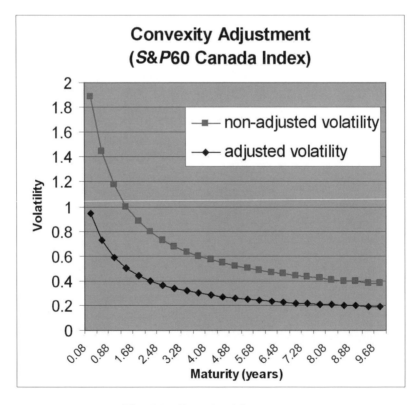

Fig. 6.1 Convexity Adjustment.

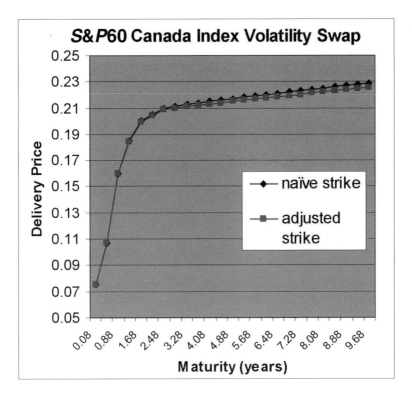

Fig. 6.2 *S&P*60 Canada Index Volatility Swap.

6.5   Summary

– In this Chapter, we priced variance and volatility swaps for financial markets with underlying asset and variance following the Heston (1993) model.
– We also priced covariance and correlation swaps for the financial markets.
– As an application, we provide a numerical example using *S&P*60 Canada Index to price swap on the volatility.

Bibliography

Avellaneda, M., Levy, A. and Paras, A. (1995). Pricing and hedging derivative securities in markets with uncertain volatility. *Applied Mathematical Finance*, 2, 73-88.

Black, F. and Scholes, M. (1973). The pricing of options and corporate liabilities. *Journal of Political Economy*, 81, 637-54.

Bollerslev, T. (1986). Generalized autoregressive conditional heteroscedasticity. *Journal of Economics*, 31, 307-27.

Brockhaus, O. and Long, D. (2000). Volatility swaps made simple. *Risk Magazine*, January, 92-96.

Buff, R. (2002). *Uncertain Volatility Model. Theory and Applications.* New York: Springer.

Carr, P. and Madan, D. (1998). Towards a theory of volatility trading. In Jarrow, R. (ed.), *Volatility*, Risk Book Publications.

Chesney, M. and Scott, L. (1989). Pricing European Currency Options: A comparison of modifeied Black-Scholes model and a random variance model. *Journal of Financial and Quantitative Analysis*, 24, 3, 267-284.

Cox, J., Ingersoll, J. and Ross, S. (1985). A theory of the term structure of interest rates. *Econometrica*, 53, 385-407.

Demeterfi, K., Derman, E., Kamal, M. and Zou, J. (1999). A guide to volatility and variance swaps. *The Journal of Derivatives*, Summer, 9-32.

Duan, J. (1995). The GARCH option pricing model. *Mathematical Finance*, 5, 13-32.

Dupire, B. (1994). Pricing with smile. *Risk Magazine,* Incisive Media (January).

Derman, E. and Kani, I. (1994). Riding on a smile. *Risk Magazine*, February, 32-39.

Elliott, R. and Swishchuk, A. (2007), Pricing Options and Variance Swaps in Markov-Modulated Brownian Markets, in Elliot, R. and Mamon, R. (eds.), *Hidden Markov Models in Finance.* New York: Springer.

Frey, R. (1997). Derivative asset analysis in models with level-dependent and stochastic volatility. *CWI Quarterly*, 10, 1-34.

Gyöngy, I. (1986). Mimicking the one-dimensional marginal distributions of processes having an Itô differential. *Probability Theory and Related Fields*, 71, 501-516.

Griego, R. and Swishchuk, A. (2000). Black-Scholes formula for a market in a Markov environment. *Theory of Probability and Mathematical Statistics* 62, 9-18.

He, R. and Wang, Y. (2002). Price Pseudo-Variance, Pseudo Covariance, Pseudo-Volatility, and Pseudo-Correlation Swaps-In Analytical Close Forms, RBC Financial Group, Query Note for 6th ISPW, PIMS, Vancouver, University of British Columbia, May 2002.

Heston, S. (1993). A closed-form solution for options with stochastic volatility with applications to bond and currency options. *Review of Financial Studies*, 6, 327-343.

Hobson, D. and Rogers, L. (1998). Complete models with stochastic volatility. *Mathematical Finance*, 8, 1, 27-48.

Hull, J. (2000). *Options, Futures and Other Derivatives* (4th edition). New Jersey: Prentice Hall.

Hull, J. and White, A. (1987). The pricing of options on assets with stochastic volatilities. *Journal of Finance*, 42, 281-300.

Ikeda, N. and Watanabe, S. (1981). *Stochastic Differential Equations and Diffusion Processes.* Tokyo: Kodansha Ltd.

Javaheri, A., Wilmott, P. and Haug, E. (2002). GARCH and volatility swaps. *Wilmott Technical Article*, January, 17 pages.

Johnson, H. and Shanno, D. (1987). Option pricing when the variance is changing. *Journal of Financial and Quantitative Analysis*, 22, 143-151.

Kazmerchuk, Y., Swishchuk, A. and Wu, J. (2002). The pricing of options for security markets with delayed response (submitted to *Mathematical Finance Journal*).

Kazmerchuk, Y., Swishchuk, A. and Wu, J. (2002). A continuous-time GARCH model for stochastic volatility with delay, 18 pages. (Submitted to *European Journal of Applied Mathematics*).

Merton, R. (1973). Theory of rational option pricing. *Bell Journal of Economic Management Science*, 4, 141-183.

Øksendal, B. (1998). *Stochastic Differential Equations: An Introduction with Applications.* New York: Springer.

Ren, Y., Madan, D. and Qian Qian, M. (2007). Calibrating and pricing with embedded loca volatility models. *Risk Magazine*, September, 138-143.

Schoutens, W., Simons, E. and Tistaert, J. (2003). A perfect calibration! Now what? *Wilmott Magazine*, November 18.

Scott, L. (1987). Option pricing when the variance changes randomly: theory, estimation and an application. *Journal of Financial Quantitative Analysis*, 22, 419-438.

Stein, E. and Stein, J. (1991). Stock price distributions with stochastic volatility: An analytic approach. *Review of Financial Studies*, 4, 727-752.

Swishchuk, A. (1997). Hedging of options under mean-square criterion and with semi-Markov volatility. *Ukrainian Mathematics Journal*, 47, 7, 1119-1127.

Swishchuk, A. (2000). *Random Evolutions and Their Applications. New Trends*. Dordrecht, The Netherlands: Kluwer Academic Publishers.

Swishchuk, A. and Kalemanova, A. (2000). Stochastic stability of interest rates with jumps. *Theory of Probability and Mathematical statistics*, TBiMC Science Publication, 61. (Preprint on-line: www.math.yorku.ca/~aswishch/sample.html)

Swishchuk, A.V., Cheng, R., Lawi, S., Badescu, A., Mekki, H.B., Gashaw, A.F., Hua, Y., Molyboga, M., Neocleous, T. and Petrachenko, Y. (2002). Price pseudo-variance, pseudo-covariance, pseudo-volatility, and pseudo-correlation swaps-in analytical closed-forms. Proceedings of the Sixth PIMS Industrial Problems Solving Workshop, PIMS IPSW 6, 45-55. May 24-31, University of British Columbia, Vancouver, Canada.

Swishchuk, A. (2009). Multi-factor Levy models for pricing financial and energy derivatives. *Canadian Applied Mathematics Quarterly*, 17, 4, 777-806.

Theoret, R., Zabre, L. and Rostan, P. (2002). Pricing volatility swaps: Empirical testing with Canadian data. Working Paper, Centre de Recherche en Gestion, Document 17-2002, July 2002.

Wiggins, J. (1987). Option values under stochastic volatility: Theory and empirical estimates. *Journal of Financial Economics*, 19, 351-372.

Wilmott, P., Howison, S. and Dewynne, J. (1995). *Option Pricing: Mathematical Models and Computations*. Oxford: Oxford Financial Press.

Chapter 7

Modeling and Pricing of Variance Swaps for Stochastic Volatilities with Delay

7.1 Introduction

Variance swaps for financial markets with underlying asset and stochastic volatilities with delay are modeled and priced in this Chapter. We find some analytical close forms for expectation and variance of the realized continuously sampled variance for stochastic volatility with delay both in stationary regime and in general case. The key features of the stochastic volatility model with delay are the following: i) continuous-time analogue of discrete-time GARCH model; ii) mean-reversion; iii) contains the same source of randomness as stock price; iv) market is complete; v) incorporates the expectation of log-return. We also present an upper bound for delay as a measure of risk. As applications, we provide two numerical examples using $S\&P60$ Canada Index (1998–2002) and $S\&P500$ Index (1990–1993) to price variance swaps with delay. Varinace swaps for stochastic volatility with delay is very similar to variance swaps for stochastic volatility in Heston model, but simplier to model and to price it. A stock variance is a square of stock volatility (or standard deviation) and the stock's volatility is the simplest measure of stock's riskless or uncertainty. Formally, the volatility σ_R is the annualized standard deviation of the stock's returns during the period of interest, where the subscript R denotes the observed or "realized" volatility, and σ_R^2 is the "'realized'" variance.

The easy way to trade variance, square of volatility, is to use variance swaps, sometimes called realized variance forward contracts (see Carr and Madan (1998)).

Variance swaps are forward contracts on future realized stock variance, the square of the future volatility. This instrument provides an easy way for investors to gain exposure to the future level of variance.

In the previous Chapter we found the values of variance and volatility swaps for financial markets with underlying asset and variance that follow the Heston (1993) model. We also studied covariance and correlation swaps for the financial markets. As an application, we provided a numerical example using $S\&P60$ Canada Index to price swap on the volatility.

In this Chapter, we are going to consider the case of stochastic volatility with delay to price variance swap.

Why Delay?   Some statistical studies of stock prices (see Sheinkman and LeBaron (1989), and Akgiray (1989)) indicate the dependence on past returns. For example, Kind, Liptser and Runggaldier (1991) obtained a diffusion approximation result for processes satisfying some equations with past-dependent coefficients, and they applied this result to a model of option pricing, in which the underlying asset price volatility depends on the past evolution to obtain a generalized (asymptotic) Black-Scholes formula. Hobson and Rogers (1998) suggested a new class of non-constant volatility models, which can be extended to include the aforementioned level-dependent model and share many characteristics with the stochastic volatility model. The volatility is nonconstant and can be regarded as an endogenous factor in the sense that it is defined in terms of the past behavior of the stock price. This is done in such a way that the price and volatility form a multi-dimensional Markov process. Chang and Yoree (1999) studied the pricing of a European contingent claim for the (B, S)-securities markets with a hereditary price structure in the sense that the rate of change of the unit price of the bond account and rate of change of the stock account S depend not only on the current unit price but also on their histor-ical prices. The price dynamics for the bank account and that of the stock account are described by a linear functional differential equation and a linear stochastic functional differential equation, respectively. They show that the rational price for a European contingent claim is independent of the mean growth rate of the stock. Later Chang and Yoree (1999) generalized the celebrated Black-Scholes formula to include the (B, S)-securities market with hereditary price structure. Clearly related to our work is the work by Mohammed, Arriojas and Pap (2001) devoted to the derivation of a delayed Black-Scholes formula for the (B, S)-securities market using PDE approach. Hobson and Rogers (1998) also observed in their past-dependent model that the resulting implied volatility is U-shaped as a function of strike price. However, they dealt with only a special case where the model can be reduced to a system of SDEs. Unfortunately, not every past-dependent model can be reduced to a system of SDEs, and a more sophisticated approach, as developed in this Chapter, is needed.

Our model of stochastic volatility exhibits past-dependence: the behavior of a stock price right after a given time t not only depends on the situation at t, but also on the whole past (history) of the process $S(t)$ up to time t. This draws some similarities with fractional Brownian motion models due to a long-range dependence property. Our work is also based on the GARCH(1,1) model (see Bollerslev (1986))

$$\sigma_n^2 = \gamma V + \alpha \ln^2(S_{n-1}/S_{n-2}) + (1 - \alpha - \gamma)\sigma_{n-1}^2$$

or, more general,

$$\sigma_n^2 = \gamma V + \frac{\alpha}{l} \ln^2(S_{n-1}/S_{n-1-l}) + (1 - \alpha - \gamma)\sigma_{n-1}^2$$

and the work of Duan (1995) where he showed that it is possible to use the GARCH model as the basis for an internally consistent option pricing model. If we write down the last equation in differential form we can get the continuous-time GARCH with expectation of log-returns of zero:

$$\frac{d\sigma^2(t)}{dt} = \gamma V + \frac{\alpha}{\tau} \ln^2(\frac{S(t)}{S(t-\tau)}) - (\alpha + \gamma)\sigma^2(t).$$

If we incorporate non-zero expectation of log-return (using Itô Lemma for $\ln \frac{S(t)}{S(t-\tau)}$) then we arrive to our continuous-time GARCH model for stochastic volatility with delay:

$$\frac{d\sigma^2(t, S_t)}{dt} = \gamma V + \frac{\alpha}{\tau} \left[\int_{t-\tau}^{t} \sigma(s, S_s)dW(s) \right]^2 - (\alpha + \gamma)\sigma^2(t, S_t).$$

We should mention that in the work of Kind *et al.* (1991), a past-dependent model was defined by diffusion approximation. In their model, the volatility depends on the quadratic variation of the process, while our model deals with more general dependence of the volatility on the history of the process over a finite interval.

In Kazmerchuk, Swishchuk and Wu (2002a) we found the Black-Scholes formula for security markets with delayed response and in Kazmerchuk, Swishchuk and Wu (2002b) we proposed and studied the continuous-time GARCH model for stochastic volatility with delay.

We note that the work by Mohammed, Arriojas and Pap (2001) is devoted to the derivation of a delayed Black-Scholes formula for the (B, S)-securities market using PDE approach. In their paper, the stock price satisfies the following equation:

$$dS(t) = \mu S(t-a)S(t)dt + \sigma(S(t-b))S(t)dW(t),$$

where a and b are positive constants and σ is a continuous function, and the price of the option at time t has the form $F(t, S(t))$. They found Black-Scholes formula works for this model.

7.2 Variance Swaps

As indicated in Chapter 3, variance swaps are forward contracts on future realized stock variance, the square of the future volatility.

The easy way to trade variance is to use variance swaps, sometimes called realized variance forward contracts (see Carr and Madan (1998)).

Although options market participants talk of volatility, it is variance, or volatility squared, that has more fundamental significance (see Demeterfi, Derman, Kamal and Zou (1999)).

A *variance swap* is a forward contract on annualized variance, the square of the realized volatility. Its payoff at expiration is equal to

$$N(\sigma_R^2(S) - K_{var}),$$

where $\sigma_R^2(S)$ is the realized stock variance (quoted in annual terms) over the life of the contract,

$$\sigma_R^2(S) := \frac{1}{T} \int_0^T \sigma^2(s)ds,$$

K_{var} is the delivery price for variance, and N is the notional amount of the swap in dollars per annualized volatility point squared. The holder of variance swap at expiration receives N dollars for every point by which the stock's realized variance $\sigma_R^2(S)$ has exceeded the variance delivery price K_{var}. We note that usually $N = \alpha I$, where α is a converting parameter such as 1 per volatility-square, and I is a long-short index (+1 for long and −1 for short).

Valuing a variance forward contract or swap is no different from valuing any other derivative security. The value of a forward contract P on future realized variance with strike price K_{var} is the expected present value of the future payoff in the risk-neutral world:

$$P^* = E_{P^*}\{e^{-rT}(\sigma_R^2(S) - K_{var})\},$$

where r is the risk-free discount rate corresponding to the expiration date T, and E_{P^*} denotes the expectation under the risk-neutral measure P^*.

Thus, for calculating variance swaps we need to know only $E\{\sigma_R^2(S)\}$, namely, mean value of the underlying variance.

In this Chapter we are interested in the valuing of variance swap for security markets with stochastic volatility $\sigma(t, S_t)$ with delay, where $S_t := S(t-\tau)$, $\tau > 0$, and $S(t)$ is a stock price at time $t \in [0, T]$.

In this way, *a variance swap for stochastic volatility with delay* is a forward contract on annualized variance $\sigma_R^2(t, S_t)$. Its payoff at expiration equals to

$$N(\sigma_R^2(S) - K_{var}),$$

where $\sigma_R^2(S)$ is the realized stock variance(quoted in annual terms) over the life of the contract,

$$\sigma_R^2(S) := \frac{1}{T}\int_0^T \sigma^2(u, S(u-\tau))du, \quad \tau > 0.$$

7.2.1 *Modeling of Financial Markets with Stochastic Volatility with Delay*

In this Section, we recall some notions and facts from the paper Kazmerchuk, Swishchuk and Wu (2002b).

7.2.1.1 *Model of Financial Markets with Delay*

The *bond* (riskless asset) is represented by the price function $B(t)$ such that

$$B(t) = B_0 e^{rt}, \ t \in [0, T], \tag{7.1}$$

where $r > 0$ is the risk-free rate of return.

The *stock* (risky asset) in our model is the stochastic process $(S(t))_{t\in[-\tau,T]}$ which satisfies the following SDDE:

$$dS(t) = \mu S(t)dt + \sigma(t, S(t-\tau))S(t)dW(t), \quad t > 0, \tag{7.2}$$

where $\mu \in R$ is an appreciation rate, volatility $\sigma > 0$ is a continuous and bounded function and $W(t)$ is a standard Wiener process.

The *initial data* for (7.1) is defined by $S(t) = \varphi(t)$ is deterministic function, $t \in [-\tau, 0], \quad \tau > 0$.

Throughout the paper we note

$$S_t := S(t - \tau).$$

The *discounted stock price* is defined by

$$Z(t) := \frac{S(t)}{B(t)}. \tag{7.3}$$

Using Girsanov's theorem, we obtain the following result concerning the change of probability measure in above market. Under the assumption $\int_0^T \left(\frac{r-\mu}{\sigma(t,S_t)} \right)^2 dt < \infty$, a.s. the following holds:

1) There is a probability measure P^* equivalent to P such that

$$\frac{dP^*}{dP} := \exp\left\{ \int_0^T \frac{r-\mu}{\sigma(s,S_s)} dW(s) - \frac{1}{2} \int_0^T \left(\frac{r-\mu}{\sigma(s,S_s)} \right)^2 ds \right\} \tag{7.4}$$

is its Radon-Nikodym density.

2) The discounted stock price $Z(t)$ is a positive local martingale with respect to P^*, and it is given by

$$Z(t) = Z_0 \exp\left\{ -\frac{1}{2} \int_0^t \sigma^2(s,S_s) ds + \int_0^t \sigma(s,S_s) dW^*(s) \right\},$$

where

$$W^*(t) := \int_0^t \frac{\mu - r}{\sigma(s,S_s)} ds + W(t) \tag{7.5}$$

is a standard Wiener process with respect to P^*.

Remarks: 1. Another form, the process $Z(t)$ can be written in, is

$$dZ(t) = Z(t)\sigma(t,S_t)dW^*(t),$$

and for $\ln Z(t)$ we obtain the following equation

$$d\ln Z(t) = -\frac{1}{2}\sigma^2(t,S_t)dt + \sigma(t,S_t)dW^*(t).$$

2. A sufficient condition for the right-hand side of (7.4) to be martingale with t in place of T is

$$E \exp\left\{ \frac{1}{2} \int_0^T \left(\frac{r-\mu}{\sigma(t,S_t)} \right)^2 dt \right\} < \infty.$$

In this way, the only source of randomness in our model for the market consisting of one stock $S(t)$ and the bond $B(t)$ is a standard Wiener process $W(t)$, $t \in [0,T]$, with T denoting the terminal time. This Wiener process generates the filtration $\mathcal{F}_t := \sigma\{W(s) : 0 \le s \le t\}$.

From an intuitive point of view, the filtration generated by S (or Z), rather than by W, is more natural one, since S is the observed process. The following lemma holds (see Kallsen and Taqqu (1995)).

Lemma 7.1. *The P^*-completed filtrations generated by either W, W^*, S or Z all coincide.*

Since the initial process φ is deterministic, we must not worry about this.

Theorem 7.1. *Suppose that $Z(t)$ is a martingale under P^*. Then the model is complete. The price at time 0 for a given integrable contingent claim C is given by*

$$\phi_0 = E_{P^*}(e^{-rT}C).$$

For instance, the European call option with expiration date T and strike price K is defined by $C = (S(T) - K)^+$.

Let us introduce a *premium per unit of risk* process

$$\lambda(t) := \frac{\mu - r}{\sigma(t, S_t)}$$

for $t \geq 0$. Changing probability measure in equation (2) for stock price and using Ito's lemma lead to the following:

$$\ln S(t) = \int_0^t (r - \frac{1}{2}\sigma^2(u, S_u))du + \int_0^t \sigma(u, S_u)dW^*(u) \tag{7.6}$$

or, equivalently,

$$\ln \frac{S(t)}{S(t - \tau)} = r\tau - \frac{1}{2}\int_{t-\tau}^t \sigma^2(u, S_u)du + \int_{t-\tau}^t \sigma(u, S_u)dW^*(u),$$

where $W^*(t) = \int_0^t \lambda(s)ds + W(t)$. The expression (7.6), as well as the following in terms of physical measure P, will be needed later for deriving a continuous-time analogue of GARCH(1,1)-model for stochastic volatility:

$$\ln \frac{S(t)}{S(t - \tau)} = r\tau + \int_{t-\tau}^t \left[\lambda(u)\sigma(u, S_u) - \frac{1}{2}\sigma^2(u, S_u)\right]du + \int_{t-\tau}^t \sigma(u, S_u)dW(u). \tag{7.7}$$

For discounted stock price $Z(t)$ we obtain:

$$\ln \frac{Z(t)}{Z(t - \tau)} = \int_{t-\tau}^t \left[\lambda(u)\sigma(u, S_u) - \frac{1}{2}\sigma^2(u, S_u)\right]du + \int_{t-\tau}^t \sigma(u, S_u)dW(u)$$

or

$$\ln \frac{Z(t)}{Z(t - \tau)} = -\frac{1}{2}\int_{t-\tau}^t \sigma^2(u, S_u)du + \int_{t-\tau}^t \sigma(u, S_u)dW^*(u).$$

We note that $Z(t)$ has the representation:

$$Z(t) = Z(0)\exp\left\{\int_0^t \sigma(u, S_u)dW^*(u) - \frac{1}{2}\int_0^t \sigma^2(u, S_u)du\right\}.$$

The process $Z(t)$ is a martingale under risk-neutral probability measure P^*. Combining this result with the Theorem 7.1 we have the following theorem.

Theorem 7.2. *(Completeness) The equation (7.2) for stock price $S(t)$ is complete and the initial price of any integrable claim C is given by*

$$\phi_0 = E_{P^*}(e^{-rT}C),$$

and the price of the claim at any time $0 \leq t \leq T$ is given by

$$\phi_t = e^{rt}E_{P^*}(e^{-rT}C|\mathcal{F}_t).$$

Let us show that $S(t) > 0$ a.s. for all $t \in [0, T]$, when $\varphi(0) > 0$ a.s. Define the following process:

$$N(t) := \mu t + \int_0^t \sigma(s, S_s)dW(s), \quad t \in [0, T]$$

This is a semimartingale with quadratic variation $< N > (t) = \int_0^t \sigma^2(s, S_s)ds$. Then, from equation (7.2) we get:

$$dS(t) = S(t)dN(t), \quad S(0) = \varphi(0).$$

This equation has a solution:

$$S(t) = \varphi(0) \exp \left\{ N(t) - \frac{1}{2} < N > (t) \right\}$$

$$= \varphi(0) \exp \left\{ \mu t + \int_0^t \sigma(u, S_u)dW(u) - \frac{1}{2}\int_0^t \sigma^2(u, S_u)du \right\},$$

From this we see that if $\varphi(0) > 0$ a.s., then $S(t) > 0$ a.s. for all $t \in [0, T]$.

7.2.1.2 *Continuous-time GARCH model for Stochastic Volatility with Delay*

As we have seen, in the risk-neutral world the stock price $S(t)$ has the dynamics:

$$dS(t) = rS(t)dt + \sigma(t, S_t)dW^*(t), \tag{7.8}$$

where $W^*(t)$ was defined in (7.5). Let us consider the following equation for the variance $\sigma^2(t, S_t)$:

$$\frac{d\sigma^2(t, S_t)}{dt} = \gamma V + \frac{\alpha}{\tau} \left[\int_{t-\tau}^t \sigma(s, S_s)dW(s) \right]^2 - (\alpha + \gamma)\sigma^2(t, S_t). \tag{7.9}$$

Here, all the parameters α, γ, τ, V are positive constants and $0 < \alpha + \gamma < 1$. The Wiener process $W(t)$ is the same as in (7.2).

Taking into account (7.7), the equation (7.9) is equivalent to the following:

$$\frac{d\sigma^2(t, S_t)}{dt} = \gamma V + \frac{\alpha}{\tau} \left(\ln \frac{S(t)}{S(t-\tau)} - r\tau - \int_{t-\tau}^t (\lambda(u)\sigma(u, S_u) - \frac{1}{2}\sigma^2(u, S_u))du \right)^2$$

$$- (\alpha + \gamma)\sigma^2(t, S_t). \tag{7.10}$$

Our first attempt (see Kazmerchuk. Swishchuk and Wu (2002a)) was to introduce a continuous version of GARCH in the following way:

$$\frac{d\sigma^2(t)}{dt} = \gamma V + \frac{\alpha}{\tau} \ln^2 \left(\frac{S(t)}{S(t-\tau)} \right) - (\alpha + \gamma)\sigma^2(t), \qquad (7.11)$$

where all the parameters were inherited from its discrete-time analogue:

$$\sigma_n^2 = \gamma V + \frac{\alpha}{l} \ln^2(S_{n-1}/S_{n-1-l}) + (1 - \alpha - \gamma)\sigma_{n-1}^2, \quad l = \frac{\tau}{\Delta},$$

which, in the special case $l = 1$, is a well-known GARCH(1,1) model for stochastic volatility without conditional mean of log-return (see Bolerslev (1986)).

J.-C. Duan remarked that it is important to incorporate the expectation of log-return $\ln(S(t)/S(t-\tau))$ into (7.11), which is explicitly shown in (7.10). Therefore, the stochastic delay differential equation (7.9) is a continuous-time analogue of GARCH(1,1)-model with incorporating of conditional mean of log-return.

Using risk-neutral measure argument, we obtain from (7.9):

$$\frac{d\sigma^2(t, S_t)}{dt} = \gamma V + \frac{\alpha}{\tau} \left[\int_{t-\tau}^{t} \sigma(s, S_s)dW^*(s) - \int_{t-\tau}^{t} \lambda(u)\sigma(u, S_u)du \right]^2$$

$$- (\alpha + \gamma)\sigma^2(t, S_t)$$

$$= \gamma V + \frac{\alpha}{\tau} \left[\int_{t-\tau}^{t} \sigma(s, S_s)dW^*(s) + (\mu - r)\tau \right]^2$$

$$- (\alpha + \gamma)\sigma^2(t, S_t). \qquad (7.12)$$

7.2.2   *Variance Swaps for Stochastic Volatility with Delay*

7.2.2.1   *Key Features of Stochastic Volatility Model with Delay*

We assume that the underlying asset $S(t)$ follows the process

$$dS(t) = \mu S(t)dt + \sigma(t, S_t)dW(t) \qquad (7.14)$$

and the asset volatility is defined as the solution of the following equation:

$$\frac{d\sigma^2(t, S_t)}{dt} = \gamma V + \frac{\alpha}{\tau} \left[\int_{t-\tau}^{t} \sigma(s, S_s)dW(s) \right]^2 - (\alpha + \gamma)\sigma^2(t, S_t). \qquad (7.15)$$

The key features of the *stochastic volatility model with delay* in (7.15) are the following:

i) *continuous-time analogue of discrete-time GARCH model;*

ii) *mean-reversion;*

iii) *does not contain another Wiener process;*

iv) *market is complete;*

v) *incorporates the expectation of log-return.*

In the risk-neutral world, the underlying asset $S(t)$ follows the process

$$dS(t) = rS(t)dt + \sigma(t, S_t)dW^*(t)$$

and the asset volatility is defined then as follows or

$$\frac{d\sigma^2(t, S_t)}{dt} = \gamma V + \frac{\alpha}{\tau} \left[\int_{t-\tau}^{t} \sigma(s, S_s)dW^*(s) - \int_{t-\tau}^{t} \lambda(u)\sigma(u, S_u)du \right]^2$$

$$- (\alpha + \gamma)\sigma^2(t, S_t)$$

$$= \gamma V + \frac{\alpha}{\tau} \left[\int_{t-\tau}^{t} \sigma(s, S_s)dW^*(s) + (\mu - r)\tau \right]^2$$

$$- (\alpha + \gamma)\sigma^2(t, S_t). \tag{7.16}$$

where $W^*(t)$ is defined in (7.5).

Let us take the expectations under risk-neutral measure $\mathcal{P}^*$ on both sides of the equation above. Denoting $v(t) := E_{P^*}[\sigma^2(t, S_t)]$, we obtain the following deterministic delay differential equation:

$$\frac{dv(t)}{dt} = \gamma V + \alpha\tau(\mu - r)^2 + \frac{\alpha}{\tau} \int_{t-\tau}^{t} v(s)ds - (\alpha + \gamma)v(t). \tag{7.17}$$

Notice that (7.17) has a stationary solution

$$v(t) \equiv X = V + \alpha\tau(\mu - r)^2/\gamma. \tag{7.18}$$

7.2.2.2 Valuing of Variance Swaps with Delay in Stationary Regime under Risk-Neutral Measure

In this case of risk-neutral measure P^* we have

$$v(t) = E_{P^*}[\sigma^2(t, S_t)] = V + \alpha\tau(\mu - r)^2/\gamma. \tag{7.19}$$

Hence,

$$E_{P^*}[Var(S)] = \frac{1}{T} \int_{0}^{T} E_{P^*}[\sigma^2(t, S_t)]dt$$

$$= V + \alpha\tau(\mu - r)^2/\gamma. \tag{7.20}$$

Therefore, from (7.19) and (7.20) it follows that the *price $\mathcal{P}^*$ of variance swap for stochastic volatility with delay in stationaty regime under risk-netral measure P^* equals* to

$$\mathcal{P}^* = e^{-rT}[V - K + \alpha\tau(\mu - r)^2/\gamma].$$

It is interesting to note that (7.19) contains parameter μ even after risk-neutral valuation. This is because of the delay τ: if $\tau = 0$, then

$$E_{P^*}[Var(S)] = V$$

and

$$\mathcal{P}^* = e^{-rT}[V - K].$$

7.2.2.3   Valuing of Variance Swaps with Delay in General Case

There is no way to write a solution in explicit form for arbitrarily given initial data. But we can understand an approximate behavior of solutions of (7.17) by looking at its eigenvalues. Let us substitute $v(t) = X + Ce^{\rho t}$ into (7.17), wher X is defined in (7.18). Then, the characteristic equation for ρ is:

$$\rho = \frac{\alpha}{\rho\tau}(1 - e^{-\rho\tau}) - (\alpha + \gamma), \qquad (7.21)$$

which is equivalent to (when $\rho \neq 0$):

$$\rho^2 = \frac{\alpha}{\tau} - \frac{\alpha}{\tau}e^{-\rho\tau} - (\alpha + \gamma)\rho.$$

The only solution to this equation (7.21) is $\rho \approx -\gamma$, assuming that γ is sufficiently small.

Then, the behavior of any solution is stable near X, and

$$v(t) \approx X + Ce^{-\gamma t} \qquad (7.22)$$

for large values of t.

Let find variance swap for stochastic volatility with delay in the case

$$v(t) \approx X + Ce^{-\gamma t} = V + \alpha\tau(\mu - r)^2/\gamma + Ce^{-\gamma t}. \qquad (7.23)$$

Since

$$v(0) = \sigma(0, S(0 - \tau)) = \sigma(0, \phi(-\tau)) := \sigma_0,$$

we can find the value of C from (7.22)

$$C = v(0) - X = \sigma_0^2 - V - \alpha\tau(\mu - r)^2/\gamma. \qquad (7.24)$$

In this way, from (7.23) and (7.24) we obtain

$$v(t) = E_{P*}[\sigma^2(t, S_t)] \approx V + \alpha\tau(\mu - r)^2/\gamma + (\sigma_0^2 - V - \alpha\tau(\mu - r)^2/\gamma)e^{-\gamma t}. \qquad (7.25)$$

Hence,

$$E_{P*}[Var(S)] = \frac{1}{T}\int_0^T E_{P*}[\sigma^2(t, S_t)]dt$$

$$\approx \frac{1}{T}\int_0^T [V + \alpha\tau(\mu - r)^2/\gamma + (\sigma_0^2 - V - \alpha\tau(\mu - r)^2/\gamma)e^{-\gamma t}]dt$$

$$= V + \alpha\tau(\mu - r)^2/\gamma + (\sigma_0^2 - V - \alpha\tau(\mu - r)^2/\gamma)\frac{1 - e^{-\gamma T}}{T\gamma}. \qquad (7.26)$$

Therefore, from (7.25) and (7.26) it follows that the price $\mathcal{P}^*$ of variance swap for stochastic volatility with delay in the case in risk-netral measure P^* equals to

$$\mathcal{P}^* = e^{-rT}\left[V - K + \alpha\tau(\mu - r)^2/\gamma + (\sigma_0^2 - V - \alpha\tau(\mu - r)^2/\gamma)\frac{1 - e^{-\gamma T}}{T\gamma}\right].$$

7.2.3 Delay as A Measure of Risk

As we could see from the previous Section (see (7.26)), in risk-neutral world we have the following expression for $E_{P^*}[Var(S)]$:

$$E_{P^*}[Var(S)] = \frac{1}{T}\int_0^T E_{P^*}[\sigma^2(t, S_t)]dt$$

$$\approx \frac{1}{T}\int_0^T [V + \alpha\tau(\mu - r)^2/\gamma + (\sigma_0^2 - V - \alpha\tau(\mu - r)^2/\gamma)e^{-\gamma t}]dt$$

$$= V + \alpha\tau(\mu - r)^2/\gamma + (\sigma_0^2 - V - \alpha\tau(\mu - r)^2/\gamma)\frac{1 - e^{-\gamma T}}{T\gamma}. \quad (7.27)$$

This expression contains all the information about our model, since it contains all the initial parameters. We note that $\sigma_0^2 := \sigma^2(0, \phi(-\tau))$.

The sign of the second term in (27) depends on the relationship between σ_0^2 and $V + \alpha\tau(\mu - r)^2/\gamma$. If $\sigma_0^2 > V + \alpha\tau(\mu - r)^2/\gamma$), then the second term in (7.27) is positive and $E_{P^*}[Var(S)]$ stays above $V + \alpha\tau(\mu - r)^2/\gamma$. *It means that variance is high and, hence, risk is high.* If $\sigma_0^2 < V + \alpha\tau(\mu - r)^2/\gamma$, then the second term in (7.27) is negative and $E_{P^*}[Var(S)]$ stays below $V + \alpha\tau(\mu - r)^2/\gamma$. *It means that variance is low and, hence, risk is low.* In this way, the relationship

$$\sigma^2(0, \phi(-\tau)) = V + \alpha\tau(\mu - r)^2/\gamma$$

defines the *measure of risk in the stochastic volatility model with delay.*

To reduce the risk we need to take into account the following relationship with respect to the delay τ (which follows from (7.27)):

$$\tau < \frac{(V - \sigma^2(0, \phi(-\tau)))\gamma}{\alpha(\mu - r)^2}.$$

7.2.4 Comparison of Stochastic Volatility in Heston Model and Stochastic Volatility with Delay

In the paper Swishchuk (2004) we studied variance and volatility swaps for financial markets with underlying asset and variance that follow the Heston (1993) model.

The underlying asset S_t in the risk-neutral world and variance follow the following model, Heston (1993) model:

$$\begin{cases} \frac{dS_t}{S_t} = r_t dt + \sigma_t dw_t^1 \\ d\sigma_t^2 = k(\theta^2 - \sigma_t^2)dt + \gamma\sigma_t dw_t^2, \end{cases}$$

where r_t is deterministic interest rate, σ_0 and θ are short and long volatility, $k > 0$ is a reversion speed, $\gamma > 0$ is a volatility (of volatility) parameter, w_t^1 and w_t^2 are independent standard Wiener processes.

The Heston asset process has a variance σ_t^2 that follows Cox-Ingersoll-Ross (1985) process, described by the second equation above.

76   *Modeling and Pricing of Swaps for Financial and Energy Markets*

It was found that mean value for realized variance in Heston model has the following expression

$$E\{V\} := E_{P^*}[Var(S)] = \frac{1}{T}\int_0^T E\sigma_t^2 dt$$

$$= \frac{1}{T}\int_0^T \{e^{-kt}(\sigma_0^2 - \theta^2) + \theta^2\}dt$$

$$= \frac{1 - e^{-kT}}{kT}(\sigma_0^2 - \theta^2) + \theta^2,$$

where σ_0^2 is a short variance (initial value), θ^2 is a long variance (long-term variance), k is a reversion speed.

From (7.27) we have the following expression for realized varinace with delay

$$E\{V\} := E_{P^*}[Var(S)] = \frac{1}{T}\int_0^T E_{P^*}[\sigma^2(t, S_t)]dt$$

$$\approx \frac{1}{T}\int_0^T [V + \alpha\tau(\mu - r)^2/\gamma + (\sigma_0^2 - V - \alpha\tau(\mu - r)^2/\gamma)e^{-\gamma t}]dt$$

$$= \frac{1 - e^{-\gamma T}}{\gamma T}(\sigma_0^2 - V - \alpha\tau(\mu - r)^2/\gamma) + [V + \alpha\tau(\mu - r)^2/\gamma],$$

where $\sigma_0^2 := \sigma^2(0, \phi(-\tau))$ is a short variance (initial value), V is a long variance (long-term variance) and $0 < \gamma < 1$ is a constant, weight of V.

If we compare these two models we can see that they are very similar, especially when $\tau = 0$: for realized variance in Heston model we have

$$E\{V\} = \frac{1 - e^{-kT}}{kT}(\sigma_0^2 - \theta^2) + \theta^2, \tag{7.28}$$

and for realized variance for stochastic volatility with delay we have

$$E\{V\} \approx \frac{1 - e^{-\gamma T}}{\gamma T}(\sigma^2(0, \phi(-\tau)) - V - \alpha\tau(\mu - r)^2/\gamma) + [V + \alpha\tau(\mu - r)^2/\gamma] \tag{7.29}$$

and for $\tau = 0$ we obtain

$$E\{V\} = \frac{1 - e^{-\gamma T}}{\gamma T}(\sigma^2(0, \phi(-\tau)) - V) + V.$$

The parameter $\gamma > 0$ in (7.29), the weight of V, plays the role of a reversion speed in stochastic volatility model with delay, as well as paramater k in Heston model (see (7.28)).

Therefore, we can use our continuous-time GARCH model for stochastic volatility with delay in a row with stochastic volatility in Heston model to price the variance swaps.

7.3 Numerical Example 1: *S&P*60 Canada Index

In this Section, we apply the analytical solutions from Section 7.2 to price the variance swap of the *S&P*60 Canada index for five years (January 1998–February 2002) (see Theoret, Zabre and Rostan (2002)).

At the end of February 2002, we wanted to price the fixed leg of a variance swap based on the *S&P*60 Canada index. The statistics on log returns *S&P*60 Canada Index for 5 year (January 1997–February 2002) is presented in Table 7.1:

Table 7.1

Statistics on Log Returns *S&P*60 Canada Index	
Series:	LOG RETURNS *S&P*60 CANADA INDEX
Sample:	1 1300
Observations:	1300
Mean	0.000235
Median	0.000593
Maximum	0.051983
Minimum	−0.101108
Std. Dev.	0.013567
Skewness	−0.665741
Kurtosis	7.787327

From the histogram of the *S&P*60 Canada index log returns on a 5-year historical period (1,300 observations from January 1998 to February 2002) it may be seen leptokurtosis in the histogram. If we take a look at the graph of the *S&P*60 Canada index log returns on a 5-year historical period we may see volatility clustering in the returns series. These facts indicate about the conditional heteroscedasticity.

A GARCH(1,1) regression is applied to the series and the results are obtained as in the next Table 7.2:

Table 7.2

Estimation of the GARCH(1,1) process				
Dependent Variable: Log returns of S&P60 Canada Index Prices				
Method: ML-ARCH				
Included Observations: 1300				
Convergence achieved after 28 observations				
–	**Coefficient:**	**Std. error:**	**z-statistic:**	**Prob.**
C	0.000617	0.000338	1.824378	0.0681
Variance Equation				
C	2.58E-06	3.91E-07	6.597337	0
ARCH(1)	0.060445	0.007336	8.238968	0
GARCH(1)	0.927264	0.006554	141.4812	0
R-squared	−0.000791	Mean dependent var	–	0.000235
Adjusted R-squared	−0.003108	S.D. dependent var	–	0.013567
S.E. of regression	0.013588	Akaike info criterion	–	−5.928474
Sum squared resid	0.239283	Schwartz criterion	–	−5.912566
Log likelihood4	3857.508	Durbin-Watson stat	–	1.886028

This table allows to generate different input variables to the volatility swap model.

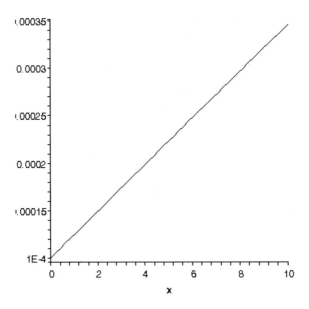

Fig. 7.1 Dependence of Variance Swap with Delay on Delay (*S&P*60 Canada Index).

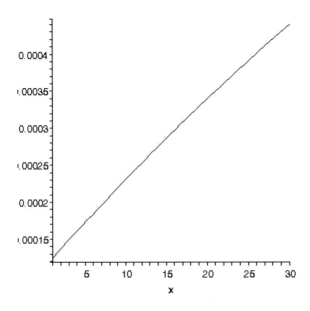

Fig. 7.2 Dependence of Variance Swap with Delay on Maturity (*S&P*60 Canada Index).

Dependence of Var(S) Swap on Delay and Maturity

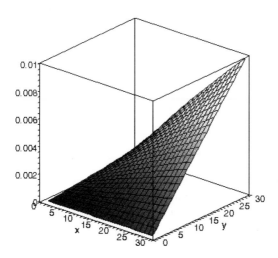

Fig. 7.3   Variance Swap with Delay for $S\&P60$ Canada Index.

We use the following relationship

$$\theta = \frac{V}{dt},$$

$$k = \frac{1 - \alpha - \beta}{dt},$$

to calculate the following discrete GARCH(1,1) parameters:

ARCH(1,1) coefficient $\alpha = 0.060445$;

GARCH(1,1) coefficient $\beta = 0.927264$;

GARCH(1,1) coefficient $\gamma = 0.012391$;

the Pearson kurtosis (fourth moment of the drift-adjusted stock return) $\xi = 7.787327$;

long volatility $\theta = 0.05289724$;

$k = 3.09733$;

a short volatility σ_0 equals to 0.01;

$\mu = 0.000235$,

$r = 0.02$ and $\tau = 1(day)$.

Parameter V may be found from the expression $V = \frac{C}{1-\alpha-\beta}$, where $C = 2.58 \times 10^{-6}$ is defined in Table 7.2. Thus, $V = 0.00020991$;

$dt = 1/252 = 0.003968254$.

Now, applying the analytical solutions (7.26) for a variance swap maturity T of

1 year, we find the following value:

$$E_{P^*}\{Var(S)\} = V + \frac{\alpha\tau(\mu-r)^2}{\gamma} + (\sigma_0^2 - V - \alpha\tau(\mu-r)^2/\gamma)\frac{1-e^{-\gamma T}}{T\gamma}$$

$$= 0.0002 + (0.0604 \times (0.0002 - 0.02)^2/0.0124)$$

$$+ (0.0001 - 0.0002 - 0.0604 \times (0.0002 - 0.02)^2/0.0124)$$

$$\times \frac{1-e^{-0.0124}}{0.0124}$$

$$= 0.000125803041.$$

Repeating this approach for a series of maturities up to 30 years and series of delays up to 30 days we obtain the following plot (see Figure 7.3) of $S\&P60$ Canada Index Variance Swap. Figures 7.1 and 7.2 depicts the dependence of the variance swap with delay on delay and maturity, respectively.

7.4   Numerical Example 2: $S\&P500$ Index

In this section, we apply the analytical solutions from Section 7.3 to price a swap on the variance of the $S\&P500$ index for four years (1990–1993)(see Kazmerchuk, Swishchuk and Wu (2002b)).

The statistics on log returns $S\&P500$ Index for 4 year (1990–1993) is presented in Table 7.3:

Table 7.3

Statistics on Log Returns $S\&P500$ Index	
Series:	LOG RETURNS $S\&P500$ INDEX
Sample:	1 1006
Observations:	1006
Mean	0.000263014
Median	8.84424E-05
Maximum	0.034025839
Minimum	−0.045371484
Std. Dev.	0.00796645
Sample Variance	6.34643E-05
Skewness	−0.178481359
Kurtosis	3.296144083

Using maximum likehood method we obtain the following parameters required:
  $\alpha = 0.3828$;
  $\beta = 0.1062$;
  $\gamma = 0.511$;
the Pearson kurtosis (fourth moment of the drift-adjusted stock return) $\xi = 3.296144083$;

Dependence of Var(S) Swap on Delay

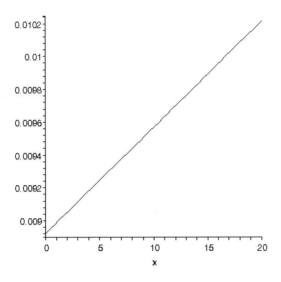

Fig. 7.4 Dependence of Variance Swap with Delay on Delay ($S\&P500$ Index).

long volatility $\theta = 0.04038144$;
a short volatility σ_0 equals to 0.00796645;
parameter $V = 0.04038144$;
$\mu = 0.000263$,
$r = 0.02$ and $\tau = 14(days)$.
Now, applying the analytical solutions (7.26) for a swap maturity T of 1 year,
we find the following value:

$$E_{P^*}\{Var(S)\} = V + \frac{\alpha\tau(\mu - r)^2}{\gamma} + (\sigma_0^2 - V - \alpha\tau(\mu - r)^2/\gamma)\frac{1 - e^{-\gamma T}}{T\gamma}$$

$$= 0.004038144 + (0.3828 \times 14 \times (0.000263 - 0.02)^2/0.511)$$

$$+ (0.000063 - 0.04038144 - 0.3828 \times 14$$

$$\times (0.000263 - 0.02)^2/0.511)$$

$$\times \frac{1 - e^{-0.511}}{0.511}$$

$$= 0.00988086882.$$

Repeating this approach for a series of maturities up to 30 years and series of
delays up to 30 days we obtain the following plot (see Figure 7.6) of $S\&P500$ Index
Variance Swap. Figures 7.4 and 7.5 depicts the dependence of the variance swap
with delay on delay and maturity, respectively.

Dependence of Var(S) Swap on Maturity

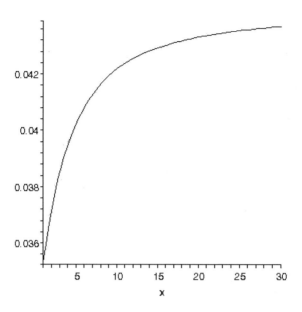

Fig. 7.5 Dependence of Variance Swap with Delay on Maturity (*S&P*500 Index).

Dependence of Var(S) Swap on Delay and Maturity

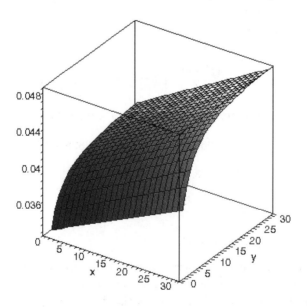

Fig. 7.6 Variance Swap with Delay for *S&P*500 Index.

Figure 7.1 depicts the dependence of Variance Swap with Delay on delay for *S&P*60 Canada Index (1998–2002).

Figure 7.2 depicts the dependence of Variance Swap with Delay on maturity for *S&P*60 Canada Index (1998–2002).

Figure 7.3 depicts the Variance Swap with Delay on maturity and delay for *S&P*60 Canada Index (1998–2002).

Figure 7.4 depicts the dependence of Variance Swap with Delay on delay for *S&P*500 Index (1990–1993).

Figure 7.5 depicts the dependence of Variance Swap with Delay on maturity for *S&P*500 Index (1990–1993).

Figure 7.6 depicts the Variance Swap with Delay on maturity and delay for *S&P*500 Index (1990–1993).

7.5 Summary

– Variance swaps for financial markets with underlying asset and stochastic volatilities with delay have been modeled and priced in this Chapter. We found some analytical close forms for expectation and variance of the realized continuously sampled variance for stochastic volatility with delay both in stationary regime and in general case.
– The key features of the stochastic volatility model with delay are the following: i) continuous-time analogue of discrete-time GARCH model; ii) mean-reversion; iii) contains the same source of randomness as stock price; iv) market is complete; v) incorporates the expectation of log-return. We also present an upper bound for delay as a measure of risk.
– As applications, we provide two numerical examples using *S&P*60 Canada Index (1998–2002) and *S&P*500 Index (1990–1993) to price variance swaps with delay.
– Varinace swaps for stochastic volatility with delay is very similar to variance swaps for stochastic volatility in Heston model, but simpler to model and to price it.

Bibliography

Akrigay, V. (1989). The finite moment logstable process and option pricing. *The Journal of Finance*, 58, 2, 753-777.

Avellaneda, M., Levy, A. and Paras, A. (1995). Pricing and hedging derivative securities in markets with uncertain volatility. *Applied Mathematical Finance*, 2, 73-88.

Black, F. and Scholes, M. (1973). The pricing of options and corporate liabilities. *Journal of Political Economy*, 81, 637-54.

Bollerslev, T. (1986). Generalized autoregressive conditional heteroscedasticity. *Journal of Economics*, 31, 307-27.

Brockhaus, O. and Long, D. (2000). Volatility swaps made simple. *Risk Magazine*, January, 92-96.

Buff, R. (2002). *Uncertain Volatility Model. Theory and Applications.* New York: Springer.

Carr, P. and Madan, D. (1998). Towards a theory of volatility trading. In Jarrow, R. (ed.), *Volatility*, Risk Book Publications.

Chang, M.H. and Yoree, R.K. (1999). The European option with hereditary price structure: Basic theory. *Applied Mathematics and Computation*, 102, 279-296.

Demeterfi, K., Derman, E., Kamal, M. and Zou, J. (1999). A guide to volatility and variance swaps. *The Journal of Derivatives*, Summer, 9-32.

Duan, J. (1995). The GARCH option pricing model, *Mathematical Finance*, 5, 13-32.

Elliott, R. and Swishchuk, A. (2004). Pricing Options and Variance Swaps in Brownian and Fractional Brownian Markets. Working Paper, University of Calgary, Calgary, AB, Canada, 20 pages.

Fouque, J.-P., Papanicolaou, G. and Sircar, K. R. (2000). *Derivatives in Financial Markets with Stochastic Volatilities.* Berlin-Heidelberg: Springer-Verlag.

Griego, R. and Swishchuk, A. (2000). Black-Scholes Formula for a Market in a Markov Environment. *Theory of Probability and Mathematical Statistics*, 62, 9-18.

Heston, S. (1993). A closed-form solution for options with stochastic volatility with applications to bond and currency options. *Review of Financial Studies*, 6, 327-343

Hobson, D. and L.C. Rogers (1998). Complete models with stochastic volatility. *Mathematical Finance*, 8, 1, 27-48.

Hull, J. (2000). *Options, Futures and Other Derivatives* (4th edition). New Jersey: Prentice Hall.

Hull, J. and White, A. (1987). The pricing of options on assets with stochastic volatilities. *Journal of Finance*, 42, 281-300.

Javaheri, A., Wilmott, P. and Haug, E. (2002). GARCH and volatility swaps. *Wilmott Technical Article*, January, 17 pages.

Kallsen, J. and Taqqu, M. (1995). Option Pricing in ARCH-type Models: With Detailed Proofs. Technical Report No. 10, University of Freiburg.

Kazmerchuk, Y., Swishchuk, A., Wu, J. (2002a). The pricing of options for security markets with delayed response (submitted to *Mathematical Finance Journal*).

Kazmerchuk, Y., Swishchuk, A. and Wu, J. (2002b). A continuous-time GARCH model for stochastic volatility with delay, 18 pages (Submitted to *European Journal of Applied Mathematics*).

Kind, P., Liptser, R. and Runggaldier, W. (1991). Diffusion approximation in past-dependent models and applications to option pricing. *Annals of Probability*, 1, 3, 379-405.

Mandelbrot, B. (1997). *Fractals and Scaling in Finance, Discontinuity, Concentration, Risk.* New York: Springer.

Merton, R. (1973). Theory of rational option pricing. *Bell Journal of Economic Management Science*, 4, 141-183.

Mohammed, S.E.A, Arriojas, M. and Pap, Y. (2001). A delayed Black and Scholes formula. Preprint, Southern Illinois University.

Mohammed, S.E.A. (1998). Stochastic differential systems with memory: Theory, examples and applications. *Stochastic Analysis and Related Topics*, 6, 1-77.

Sheinkman, J. and LeBaron, B. (1989). Nonlinear dynamics and stock returns. *Journal of Business*, 62, 311-337.

Swishchuk, A. (2005). Modeling and pricing of variance swaps for stochastic volatilities with delay. *Wilmott Magazine*, September Issue, Technical article No. 19, 63-73.

Swishchuk, A. (2004). Modeling of variance and volatility swaps for financial markets with stochastic volatilities. *Wilmott Magazine*, September Issue, Technical article No. 2, 64-72.

Swishchuk, A. (2000). *Random Evolutions and Their Applications. New Trends.* Dordrecht, The Netherlands: Kluwer Academic Publishers.

Swishchuk, A. (1995). Hedging of options under mean-square criterion and with semi-Markov volatility. *Ukrainian Mathematics Journal*, 47, 7, 1119-1127.

Theoret, R., Zabre, L. and Rostan, P. (2002). Pricing volatility swaps: Empirical testing with Canadian data. Working Paper, Centre de Recherche en Gestion, Document 17-2002, July 2002.

Wilmott, P. (2000). *Quantitative Finance: Volumes I and II.* Chichester, West Sussex, England: Wiley.

Wilmott, P. (1999). *Derivatives: The Theory and Practice of Financial Engineering.* Chichester, West Sussex, England: Wiley.

Wilmott, P., Howison, S. and Dewynne, J. (1995). *Option Pricing: Mathematical Models and Computations.* Oxford: Oxford Financial Press.

Chapter 8

Modeling and Pricing of Variance Swaps for Multi-Factor Stochastic Volatilities with Delay

8.1 Introduction

Variance swaps for financial markets with underlying asset and multi-factor, i.e., two- and three-factors, stochastic volatilities with delay are modelled and priced in this Chapter. We found some analytical close forms for expectation and variance of the realized continuously sampled variances for multi-factor stochastic volatilities with delay. As applications, we provide a numerical examples using $S\&P60$ Canada Index (1998–2002) to price variance swaps with delay for all these models.

8.1.1 *Variance Swaps*

A stock's variance is a square of stock's volatility (or standard deviation) and the stock's volatility is the simplest measure of stock's riskless or uncertainty. Formally, the volatility σ_R is the annualized standard deviation of the stock's returns during the period of interest, where the subscript R denotes the observed or "realized" volatility, and σ_R^2 is the " 'realized' " variance.

The easy way to trade variance, square of volatility, is to use variance swaps, sometimes called realized variance forward contracts (see Carr and Madan (1998)).

Variance swaps are forward contracts on future realized stock variance, the square of the future volatility. This instrument provides an easy way for investors to gain exposure to the future level of variance.

Demeterfi, Derman, Kamal and Zou (1999) explained the properties and the theory of both variance and volatility swaps. They derived an analytical formula for theoretical fair value in the presence of realistic vilatility skews, and pointed out that volatility swaps can be replicated by dynamically trading the more straightforward variance swap.

Javaheri, Wilmott and Haug (2002) discussed the valuation and hedging of a GARCH(1,1) stochastic volatility model. They used a general and exible PDE approach to determine the first two moments of the realized variance in a continuous or discrete context. Then they approximate the expected realized volatility via a convexity adjustment.

Working paper by Théoret, Zabré and Rostan (2002) presented an analytical solution for pricing of volatility swaps, proposed by Javaheri, Wilmott and Haug (2002). They priced the volatility swaps within framework of GARCH(1,1) stochastic volatility model and applied the analytical solution to price a swap on volatility of the $S\&P60$ Canada Index (5-year historical period: 1997–2002).

Brockhaus and Long (2000) provided an analytical approximation for the valuation of volatility swaps and analyzed other options with volatility exposure.

In the paper Swishchuk (2004) we found the values of variance and volatility swaps for financial markets with underlying asset and variance that follow the Heston (1993) model. We also studied covariance and correlation swaps for the financial markets. As an application, we provided a numerical example using $S\&P60$ Canada Index to price swap on the volatility.

Working paper Elliott and Swishchuk (2004) contains the value of variance swap for financial market with Markov stochastic volatility.

Paper Swishchuk (2005) studied the modeling and pricing of variance swaps for one-factor stochastic volatility with delay. As an application, we provided a numerical example using $S\&P60$ Canada Index to price variance swap.

8.1.2   *Volatility*

It is known that the probability distribution of an equity has a fatter left tail and thinner right tail than the lognormal distribution (see Wilmott (1999, 2000), Hull (2000)), and the assumption of constant volatility σ in financial model (such as the original Black-Scholes model (see Balck-Scholes (1973)) is incompatible with derivatives prices observed in the market.

The latest issue has been addressed and studied in several ways, such as:

(i) Volatility is assumed to be a deterministic function of the time: $\sigma \equiv \sigma(t)$ (see Wilmott *et al.* (1995)); Merton (1973) extended the term structure of volatility to $\sigma := \sigma_t$ (deterministic function of time), with the implied volatility for an option of maturity T given by $\hat{\sigma}_T^2 = \frac{1}{T} \int_0^T \sigma_u^2 du$;

(ii) Volatility is assumed to be a function of the time and the current level of the stock price $S(t)$: $\sigma \equiv \sigma(t, S(t))$ (see Demeterfi, Derman, Kamal and Zou (1999));

(iii) Volatility is described by stochastic differential equation with the same source of randomness as stock's price (see Javaheri, Wilmott and Haug (2002));

(iv) The time variation of the volatility involves an additional source of randomness, besides $W_1(t)$, represented by $W_2(t)$, and is given by

$$d\sigma(t) = a(t, \sigma(t))dt + b(t, \sigma(t))dW_2(t),$$

where $W_2(t)$ and $W_1(t)$ (the initial Wiener process that governs the price process) may be correlated (see Buff (2002), Hull and White (1987), Heston (1993), Fouque, Papanicolaou and Sircar (2000));

(v) The volatility depends on a random parameter x such as $\sigma(t) \equiv \sigma(x(t))$, where $x(t)$ is some random process (see Elliott and Swishchuk (2004), Griego and Swishchuk (2000), Swishchuk (2000), Swishchuk (1995));

(vi) Another approach is connected with so-called uncertain volatility scenario (see Avellaneda, Levy and Paras (1995), Buff (2002));

(vii) The volatility $\sigma(t, S_t)$ depends on $S_t := S(t + \theta)$ for $\theta \in [-\tau, 0]$, namely, stochastic volatility with delay (see Kazmerchuk, Swishchuk and Wu (2005)).

8.2 Multi-Factor Models

Eydeland and Geman (1998) proposed extending the Heston (1993) stochastic volatility model to gas or electricity prices by introducing mean-reversion in the spot price and leaving CIR model for the variance, resulting in two-state variable model for commodity prices:

$$\begin{cases} dS_t = k(a - \ln S_t)S_t dt + \sigma(t)S_t dW^1(t) \\ d\sigma^2(t) = b(c - \sigma^2(t))dt + e\sigma(t)dW^2(t), \end{cases}$$

where k, a, b, c, e are all positive, $dW^1(t)W^2(t) = \rho dt$, the correlation coefficient ρ is, in general, negative, since, in contrast to stock price, the volatility of commodity prices tends to increase with prices-the *inverse leverage effect*, which leads in option prices to a volatility smile 'skewed' to the right.

Geman (2000) proposed a three-state variable for oil prices by introducing mean-reversion in the spot price, geometric Brownian motion in the equilibrium (or mean-reverting) price and CIR model in the variance.

$$\begin{cases} dS_t = k(L_t - \ln S_t)S_t dt + \sigma(t)S_t dW^1(t) \\ \frac{L_t}{L_t} = \mu dt + \sigma_2 dW^2(t) \\ d\sigma^2(t) = b(c - \sigma^2(t))dt + e\sigma(t)dW^3(t), \end{cases}$$

where $dW^1(t)W^2(t) = \rho dt$, independence between W^1 and W^2, and between W^2 and W^3, may be assumed or not. Here, a positive drift μ in the second equation would translate to a rise, on average, of the value L_t toward which the commodity spot price S_t tends to revert, while this spot price itself may fluctuate around L_t depending on the arrival of positive or negative news about the situation of world- and company-specific reserves.

Gibson and Schwartz (1990) note that the convenience yield $y(t)$ has been shown to be a key factor driving the relationship between spot and futures prices, and they proposed the two-state variable model for oil-contingent claim pricing

$$\begin{cases} \frac{dS_t}{S_t} = (\mu - y(t))dt + \sigma_1 dW^1(t) \\ dy(t) = k(\alpha - y(t))dt + \sigma_2 dW^2(t) \\ dW^1(t)dW^2(t) = \rho dt, \end{cases}$$

where the second equation defines the Ornstein-Uhlenbeck process driving the convenience yield and leading to positive and negative values.

To go to the more complex level of two factor models, keeping the same mean L_t, the SDEs for S_t and L_t can be generalized by either allowing the long run mean L_t or the volatility σ to be governed by an SDE. This leads to two distinct two factor models, with different dynamics. The first model assumes a stochastic long run mean and was introduced by Pilipovic (1997):

$$\begin{cases} dS_t = \alpha(L_t - S_t)dt + \sigma S_t^\gamma dZ_t \\ dL_t = \mu L_t dt + \xi L_t^\delta dW_t, \end{cases}$$

where γ and δ are either 0 or 1. The second generalization is the two-factor model where volatility is allowed to be stochastic (see Lari-Lavassani, Sadeghi and Ware (2001)):

$$\begin{cases} dS_t = \alpha(L_t - S_t)dt + \sqrt{\sigma_t} S_t^\gamma dZ_t \\ d\sigma_t = \mu(\sigma_0 - \sigma_t)dt + \xi \sigma_t^\gamma dW_t, \end{cases}$$

where γ is either 0 or 1, $\alpha, L, \mu, \xi, \sigma_0$ are constants. We mention, that from dynamic point of view, one common feature of these two two-factor models is that their determinsitic parts, giving the mean of the systems, are linear hyperbolic systems.

Pilipovic (1997) describes a two-factor mean-reverting model where spot prices revert to a long term equilibrium level which is itself a random variable. Pilipovic derives a closed-form solution for forward prices to her model when the spot and long term prices are uncorrelated, but however does not discuss option pricing in her two-factor model.

Gibson and Schwartz (1990), Schwartz (1997) and Hilliard and Reis (1998) all analyze versions of the same two-factor model that allows for a stochastic convenience yield and permits a high level of analytical tractability. The first factor is the spot price process which is assumed to follow the geometric Brownian motion (GBM) and the second factor is the instanteneous convenience yield of the spot energy and is assumed to follow the mean reverting process.

Schwartz (1997) extends his two-factor model to include stochastic interest rates. In this three-factor model the short term rate is assumed to follow the Vasiček (1977) mean-reverting process.

We now exhibit two different types of three-factor mean-reverting models (see Lari-Lavassani, Sadeghi and Ware (2001)). The first is a three-factor mean-reverting model with an intermediate level of stochastic mean L_t, governing growth or decay of S_t, and a second level of stochastic mean-reversion M_t, inducing higher local shocks to S_t via L_t. This system may be a suitable for modeling the spikes of electricity prices:

$$\begin{cases} dS_t = \alpha(L_t - S_t)dt + \sigma S_t^\gamma dZ_t \\ dL_t = \mu(M_t - L_t)dt + \xi L_t^\gamma dW_t \\ dM_t = \beta M_t dt + \lambda M_t^\gamma dV_t, \end{cases}$$

where $\alpha, \sigma, \mu, \xi, \beta, \gamma$ are constants and γ is either 0 or 1.

A different three-factor mean-reverting model with both stochastic mean and volatility is (the second one):

$$\begin{cases} dS_t = \alpha(L_t - S_t)dt + \sqrt{\sigma_t}S_t^\gamma dZ_t \\ dL_t = \mu L_t dt + \xi L_t^\gamma dW_t \\ d\sigma_t = \beta(\sigma_0 - \sigma_t)dt + \lambda\sigma_t^\gamma dW_t, \quad \gamma = 0 \quad or \quad 1, \end{cases}$$

where $\alpha, \sigma_0, \mu, \xi, \beta, \lambda$ are constants. Local shocks to S_t are triggered by combined nonlinear effects of L_t and σ_t.

Fouque, Papanicolaou and Sircar (2000) considered the following multi-factor stochastic volatility model

$$\begin{cases} dS_t = \mu S_t dt + \sigma_t S_t dW_t \\ \sigma_t = f(Y_t, Z_t) \\ dY_t = ac_1(Y_t)dt + \sqrt{a}g_1(Z_t)dW_t^1 \\ dZ_t = bc_2(Z_t)dt + \sqrt{b}g_2(Z_t)dW_t^2, \end{cases}$$

where S_t is the underlying asset price with a constant rate of return μ and a stochastic volatility σ_t driven by the stochastic processes Y_t and Z_t varying on the respective time scales $1/a$ and $1/b$. The standard Brownian motions (W_t, W_t^1, W_t^2) are correlated.

Fouque and Han (2003) found that two-factor SV models provide a better fit to the term structure of implied volatility than one factor SV models by capturing the behavior at short and long maturities.

Chernov *et al.* (2003) used two-factor SV family models to obtain comparable empirical goodness-of-fit.

Molina *et al.* (2003) found a strong evidence of two-factor SV models with well-separated time scales in foreign exchange data.

Multi-factor SV models do not admit in genaral explicit solutions for option prices, such as a class, for example, jump-diffusion models (see Bates (1996)), but have direct implications on hedges.

8.3 Multi-Factor Stochastic Volatility Models with Delay

In this Chapter, we are going to incorporate the case (vii) (see Introduction) above to price variance swap for multi-factor stochastic volatility with delay, namely, for three two-factor and one three-factors stochastic volatility models with delay.

In Kazmerchuk, Swishchuk and Wu (2002) we found the Black-Scholes formula for secrity markets with delayed response.

In Kazmerchuk, Swishchuk and Wu (2005) we proposed and studied the continuous-time GARCH model for one-factor stochastic volatility with delay:

$$\begin{cases} dS(t) = \mu S(t)dt + \sigma(t, S_t)dW(t) \\ \frac{d\sigma^2(t, S_t)}{dt} = \gamma V + \frac{\alpha}{\tau}\left[\int_{t-\tau}^t \sigma(s, S_s)dW(s)\right]^2 - (\alpha + \gamma)\sigma^2(t, S_t), \end{cases}$$

where $S_t := S(t - \tau)$, $\tau > 0$ is a delay. Here, all the parameters α, γ, τ, V are positive constants and $0 < \alpha + \gamma < 1$.

Paper Swishchuk (2005) modeled and priced of variance swaps for this one-factor stochastic volatilities with delay.

In this Chapter, we define and study four multi-factor stochastic volatility models with delay, three two-factor models and one three-factor model, to model and to price variance swaps.

1. *Two-factor stochastic volatility model with delay and with geomaetric Brownian motion mean-reversion* is defined in the following way:

$$
\begin{cases}
\frac{d\sigma^2(t,S_t)}{dt} = \gamma V_t + \frac{\alpha}{\tau} \left[\int_{t-\tau}^t \sigma(s, S_s) dW(s) \right]^2 - (\alpha + \gamma)\sigma^2(t, S_t) \\
dV_t/V_t = \xi dt + \beta dW_1(t),
\end{cases}
\tag{8.1}
$$

where S_t is defined as $S_t := S(t - \tau)$,

$$
dS(t) = \mu S(t) dt + \sigma(t, S_t) dW(t)
$$

and Wiener processes $W(t)$ and $W_1(t)$ may be correlated.

2. *Two-factor stochastic volatility model with delay and with Ornstein-Uhlenbeck mean-reversion* is defined in the following way:

$$
\begin{cases}
\frac{d\sigma^2(t,S_t)}{dt} = \gamma V_t + \frac{\alpha}{\tau} \left[\int_{t-\tau}^t \sigma(s, S_s) dW(s) \right]^2 - (\alpha + \gamma)\sigma^2(t, S_t) \\
dV_t = \xi(L - V_t) dt + \beta dW_1(t),
\end{cases}
\tag{8.2}
$$

where Wiener processes $W(t)$ and $W^1(t)$ may be correlated, S_t is defined as $S_t := S(t - \tau)$, and

$$
dS(t) = \mu S(t) dt + \sigma(t, S_t) dW(t).
$$

3. *Two-factor stochastic volatility model with delay and with Pilipovic one-factor mean-reversion* is defined in the following way:

$$
\begin{cases}
\frac{d\sigma^2(t,S_t)}{dt} = \gamma V_t + \frac{\alpha}{\tau} \left[\int_{t-\tau}^t \sigma(s, S_s) dW(s) \right]^2 - (\alpha + \gamma)\sigma^2(t, S_t) \\
dV_t = \xi(L - V_t) dt + \beta V_t dW_1(t),
\end{cases}
\tag{8.3}
$$

where Wiener processes $W(t)$ and $W_1(t)$ may be correlated, S_t is defined as $S_t := S(t - \tau)$,

$$
dS(t) = \mu S(t) dt + \sigma(t, S_t) dW(t).
$$

4. *Three-factor stochastic volatility model with delay and with Pilipovic mean-reversion* is defined in the following way:

$$
\begin{cases}
\frac{d\sigma^2(t,S_t)}{dt} = \gamma V_t + \frac{\alpha}{\tau} \left[\int_{t-\tau}^t \sigma(s, S_s) dW(s) \right]^2 - (\alpha + \gamma)\sigma^2(t, S_t) \\
dV_t = \xi(L_t - V_t) dt + \beta V_t dW_1(t), \\
dL_t = \beta_1 L_t dt + \eta L_t dW_2(t),
\end{cases}
\tag{8.4}
$$

where the Wiener processes $W(t), W^1(t)$ and $W^2(t)$ may be correlated, S_t is defined as $S_t := S(t - \tau)$,

$$dS(t) = \mu S(t)dt + \sigma(t, S_t)dW(t).$$

We find some analytical close forms for expectation and variance of the realized continuously sampled variance for these stochastic volatility models with delay. As applications of our analytical solutions, we provide two numerical examples using $S\&P60$ Canada Index (1998–2002) and $S\&P500$ Index (1990–1993) to price variance swaps with delay.

8.4   Pricing of Variance Swaps for Multi-Factor Stochastic Volatility Models with Delay

In this Section, we calculate the variance swaps for four multi-factor stochastic volatility models with delay defined in the previous section.

8.4.1   *Pricing of Variance Swap for Two-Factor Stochastic Volatility Model with Delay and with Geometric Brownian Motion Mean-Reversion*

Two-Factor Stochastic Volatility Model with Delay is defined in the following way (see (8.1)):

$$\begin{cases} \frac{d\sigma^2(t,S_t)}{dt} = \gamma V_t + \frac{\alpha}{\tau} \left[\int_{t-\tau}^t \sigma(s, S_s)dW(s) \right]^2 - (\alpha + \gamma)\sigma^2(t, S_t) \\ dV_t/V_t = \xi dt + \beta dW_1(t), \end{cases} \tag{8.5}$$

where S_t is defined as $S_t := S(t - \tau)$,

$$dS(t) = \mu S(t)dt + \sigma(t, S_t)dW(t).$$

and Wiener processes $W(t)$ and $W_1(t)$ may be correlated.

In order to incorporate a correlation between the Brownian motions $(W(t), W_1(t))$, we set

$$W(t) = \tilde{W}(t),$$
$$W_1(t) = \rho\tilde{W}(t) + \sqrt{1 - \rho^2}\tilde{W}_1(t),$$

where $(\tilde{W}(t), \tilde{W}_1(t))$ are independent standard Brownian motions, and the correlation coefficient ρ satisfies $|\rho| < 1$.

We note, that the market is incomplete. The second, beside P^* in equation (7.4) (see Chapter 7), martingale (or risk-neutral) measure $\tilde{P}$ may be defined, for example, in the following way (see, for example, Kallianpur and Karandikar (2000)):

$$\frac{d\tilde{P}}{dP} := \eta(T)\tilde{\eta}(T),$$

where

$$\tilde{\eta}(T) := \exp\left(\tilde{W}_1(T) - \frac{1}{2}T\right),$$

and $\eta(T)$ is defined in Chapter 7.

The measure P^* is the minimal martingale probability measure associated with P (see Kallianpur and Karandikar (2000)).

Under the risk-neutral probability measure P^*, a family of two-factor stochastic volatility model with delay and with GBM mean-reversion can be described as follows

$$\begin{cases} \frac{d\sigma^2(t,S_t)}{dt} = \gamma V_t + \frac{\alpha}{\tau}\left[\int_{t-\tau}^t \sigma(s,S_s)dW^*(s) + (\mu - r)\tau\right]^2 \\ \qquad - (\alpha + \gamma)\sigma^2(t,S_t) \\ dV_t/V_t = (\xi - \lambda\beta)dt + \beta(\rho dW^*(t) + \sqrt{1-\rho^2}dW_1^*(t)), \end{cases} \tag{8.6}$$

where $(W^*(t), W_1^*(t))$ are independent standard Brownian motions,

$$dS(t) = rS(t)dt + \sigma(t, S_t)dW^*(t),$$

$W^*(t)$ is defined in Chapter 7.

The constant λ is the market price of volatility risk.

Since $E_{P^*}V_t = V_0 e^{(\xi - \lambda\beta)t}$, the first equation takes the following form:

$$\frac{dv(t)}{dt} = \gamma V_0 e^{(\xi - \lambda\beta)t} + \alpha\tau(\mu - r)^2 + \frac{\alpha}{\tau}\int_{t-\tau}^t v(s)ds - (\alpha + \gamma)v(t),$$

where $v(t) := E_{P^*}\sigma^2(t, S_t)$.

To get variance swap we have to find a solution for the last equation, nonhomogeneous integro-differential equation with delay.

After taking the first derivative of this equation we obtain

$$v''(t) = (\xi - \lambda\beta)\gamma V_0 e^{(\xi - \lambda\beta)t} + \frac{\alpha}{\tau}[v(t) - v(t - \tau)] - (\alpha + \gamma)v'(t).$$

To solve this equation we rewrite it in vector form:

$$\vec{v}'(t) = Av(t) + B\vec{v}(t - \tau) + \vec{f}(t),$$

where

$$\vec{v}(t) := \begin{pmatrix} v_1(t) \\ v_2(t) \end{pmatrix} := \begin{pmatrix} v(t) \\ v'(t) \end{pmatrix}, \quad A := \begin{pmatrix} 0 & 0 \\ \frac{\alpha}{\tau} & -(\alpha + \gamma) \end{pmatrix},$$

$$B := \begin{pmatrix} 0 & 0 \\ 0 & -\frac{\alpha}{\tau} \end{pmatrix}, \quad \vec{f}(t) := \begin{pmatrix} 0 \\ (\xi - \lambda\beta)\gamma V_0 e^{(\xi - \lambda\beta)t} \end{pmatrix}.$$

This is a nonhomogeneous first-order differential equation with delay. Complementary homogeneous equation has the following look:

$$\vec{v}_h'(t) = Av(t) + B\vec{v}_h(t - \tau),$$

and has the following approximated solution (see Hale (1977) and Section 7.2, Chapter 7.)

$$v_h(t) \approx X + Ce^{-\gamma t},$$

where

$$X := V_0 + \alpha\tau(\mu - r)^2/\gamma, \quad C := \sigma_0^2 - V_0 - \alpha\tau(\mu - r)^2/\gamma. \tag{8.7}$$

Therefore, the nonhomogeneous eqaution has the following approximated solution (see Hale (1977))

$$v(t) \approx X + Ce^{-\gamma t} + (\xi - \lambda\beta)\gamma V_0 \int_0^t (X + Ce^{\gamma(t-s)})e^{(\xi-\lambda\beta)s}ds.$$

After calculations, we get

$$v(t) \approx X + Ce^{-\gamma t}$$

$$+ (\xi - \lambda\beta)\gamma V_0 \left[\frac{X}{\xi - \lambda\beta}(e^{(\xi-\lambda\beta)t} - 1) + \frac{C}{\xi - \lambda\beta - \gamma}(e^{(\xi-\lambda\beta)t} - e^{\gamma t})\right]. \tag{8.8}$$

We note, that

$$v(t) := E_{P^*}[\sigma^2(t, S_t)],$$

and to find swap we have to calculate

$$E_{P^*}[Var(S)] := \frac{1}{T}\int_0^T E_{P^*}[\sigma^2(t, S_t)]dt = \frac{1}{T}\int_0^T v(t)dt. \tag{8.9}$$

After substitution of $v(t)$ in (8.7) into (8.8) we obtain the price $\mathcal{P}^*$ of variance swap for two-factor stochastic volatility with delay:

$$\mathcal{P}^* = e^{-rT}\left\{\left[V - K + \alpha\tau(\mu - r)^2/\gamma + (\sigma_0^2 - V - \alpha\tau(\mu - r)^2/\gamma)\frac{1 - e^{-\gamma T}}{T\gamma}\right]\right.$$

$$+ \frac{(\xi - \lambda\beta)\gamma V_0}{T}\left[\frac{X}{(\xi - \lambda\beta)}\left(\frac{e^{(\xi-\lambda\beta)T} - 1}{(\xi - \lambda\beta)} - T\right)\right.$$

$$\left.\left. + \frac{C(e^{(\xi-\lambda\beta)T} - 1)}{(\xi - \lambda\beta)((\xi - \lambda\beta) - \gamma)} - \frac{C(e^{\gamma T} - 1)}{\gamma(\xi - \lambda\beta - \gamma)}\right]\right\}$$

$$= e^{-rT}\left\{\left[X - K + C\frac{1 - e^{-\gamma T}}{T\gamma}\right]\right.$$

$$+ \frac{(\xi - \lambda\beta)\gamma V_0}{T}\left[\frac{X}{(\xi - \lambda\beta)}\left(\frac{e^{(\xi-\lambda\beta)T} - 1}{(\xi - \lambda\beta)} - T\right)\right.$$

$$\left.\left. + \frac{C(e^{(\xi-\lambda\beta)T} - 1)}{(\xi - \lambda\beta)((\xi - \lambda\beta) - \gamma)} - \frac{C(e^{\gamma T} - 1)}{\gamma((\xi - \lambda\beta) - \gamma)}\right]\right\},$$

or, finally,

$$\mathcal{P}^* = e^{-rT}\left\{\left[X - K + C\frac{1 - e^{-\gamma T}}{T\gamma}\right] + \frac{(\xi - \lambda\beta)\gamma V_0}{T}\left[\frac{X}{(\xi - \lambda\beta)}\left(\frac{e^{(\xi - \lambda\beta)T} - 1}{(\xi - \lambda\beta)} - T\right)\right.\right.$$

$$\left.\left. + \frac{C(e^{(\xi - \lambda\beta)T} - 1)}{(\xi - \lambda\beta)(\xi - \lambda\beta - \gamma)} - \frac{C(e^{\gamma T} - 1)}{\gamma(\xi - \lambda\beta - \gamma)}\right]\right\}, \tag{8.10}$$

where constants X and C are defined in equation (8.7).

8.4.2 Pricing of Variance Swap for Two-Factor Stochastic Volatility Model with Delay and with Ornstein-Uhlenbeck Mean-Reversion

Two-Factor Stochastic Volatility Model with Delay is defined in the following way (see (8.2)):

$$\begin{cases} \frac{d\sigma^2(t,S_t)}{dt} = \gamma V_t + \frac{\alpha}{\tau}\left[\int_{t-\tau}^t \sigma(s, S_s)dW(s)\right]^2 - (\alpha + \gamma)\sigma^2(t, S_t). \\ dV_t = \xi(L - V_t)dt + \beta dW_1(t), \end{cases} \tag{8.11}$$

where Wiener processes $W(t)$ and $W^1(t)$ may be correlated, S_t is defined as $S_t := S(t - \tau)$, and

$$dS(t) = \mu S(t)dt + \sigma(t, S_t)dW(t).$$

Taking into account the same reasonings like in the previous Section 8.4.1, a family of two-factor stochastic volatility model with delay and with OU mean-reversion under the risk-neutral probability measure P^* can be described as follows

$$\begin{cases} \frac{d\sigma^2(t,S_t)}{dt} = \gamma V_t + \frac{\alpha}{\tau}\left[\int_{t-\tau}^t \sigma(s, S_s)dW^*(s) + (\mu - r)\tau\right]^2 \\ \qquad - (\alpha + \gamma)\sigma^2(t, S_t) \\ dV_t = \xi((L - \frac{\lambda\beta}{\xi}) - V_t)dt + \beta(\rho dW^*(t) + \sqrt{1 - \rho^2}dW_1^*(t)), \end{cases} \tag{8.12}$$

where $(W^*(t), W_1^*(t))$ are independent standard Brownian motion,

$$dS(t) = rS(t)dt + \sigma(t, S_t)dW^*(t),$$

$W^*(t)$ is defined in (Chapter 7, the correlation coefficient $|\rho| < 1$.

We note, that

$$E_{P^*}V_t = e^{-\xi t}\left(V_0 - \left(L - \frac{\lambda\beta}{\xi}\right)\right) + \left(L - \frac{\lambda\beta}{\xi}\right).$$

Thus, after taking the expectation E_{P^*}, the first equation finally takes the following form:

$$\frac{dv(t)}{dt} = \gamma\left(e^{-\xi t}\left(V_0 - \left(L - \frac{\lambda\beta}{\xi}\right)\right) + \left(L - \frac{\lambda\beta}{\xi}\right)\right)$$

$$+ \alpha\tau(\mu - r)^2 + \frac{\alpha}{\tau}\int_{t-\tau}^t v(s)ds - (\alpha + \gamma)v(t).$$

We are going to proceed with the same steps like in the previous Section 8.4.1.

To get variance swap we have to find a solution for the last equation, nonhomogeneous integro-differential equation with delay.

After taking the first derivative of this equation we obtain

$$v''(t) = -\xi\gamma e^{-\xi t}\left(V_0 - \left(L - \frac{\lambda\beta}{\xi}\right)\right) + \frac{\alpha}{\tau}[v(t) - v(t-\tau)] - (\alpha+\gamma)v'(t).$$

To solve this equation we rewrite it in vector form:

$$\vec{v}'(t) = Av(t) + B\vec{v}(t-\tau) + \vec{f}(t),$$

where

$$\vec{v}(t) := \begin{pmatrix} v_1(t) \\ v_2(t) \end{pmatrix} := \begin{pmatrix} v(t) \\ v'(t) \end{pmatrix}, \quad A := \begin{pmatrix} 0 & 0 \\ \frac{\alpha}{\tau} & -(\alpha+\gamma) \end{pmatrix},$$

$$B := \begin{pmatrix} 0 & 0 \\ 0 & -\frac{\alpha}{\tau} \end{pmatrix}, \quad \vec{f}(t) := \begin{pmatrix} 0 \\ -\xi\gamma(V_0 - (L - \frac{\lambda\beta}{\xi}))e^{-\xi t} \end{pmatrix}.$$

This is a nonhomogeneous first-order differential equation with delay.

Complementary homogeneous equation has the following look:

$$\vec{v}_h'(t) = Av(t) + B\vec{v}_h(t-\tau),$$

and has the following approximated solution (see Hale (1977) and Section 7.2, Chapter 7.)

$$v_h(t) \approx X + Ce^{-\gamma t},$$

where

$$X := V_0 + \alpha\tau(\mu-r)^2/\gamma, \quad C := \sigma_0^2 - V_0 - \alpha\tau(\mu-r)^2/\gamma. \qquad (8.13)$$

Therefore, the nonhomogeneous eqaution has the following approximated solution (see Hale (1977))

$$v(t) \approx X + Ce^{-\gamma t} - \xi\gamma\left(V_0 - \left(L - \frac{\lambda\beta}{\xi}\right)\right)\int_0^t (X + Ce^{\gamma(t-s)})e^{-\xi s}ds.$$

After calculations, we get

$$v(t) \approx X + Ce^{-\gamma t}$$

$$+ \xi\gamma\left(V_0 - \left(L - \frac{\lambda\beta}{\xi}\right)\right)\left[\frac{X}{\xi}(e^{-\xi t} - 1) + \frac{C}{\xi+\gamma}(e^{-\xi t} - e^{\gamma t})\right]. \qquad (8.14)$$

We note, that

$$v(t) := E_{P^*}[\sigma^2(t, S_t)],$$

and to find swap we have to calculate

$$E_{P^*}[Var(S)] := \frac{1}{T}\int_0^T E_{P^*}[\sigma^2(t, S_t)]dt = \frac{1}{T}\int_0^T v(t)dt. \qquad (8.15)$$

After substitution of $v(t)$ in (8.14) into (8.15) we obtain the price $\mathcal{P}^*$ of variance swap for two-factor stochastic volatility with delay and with OU mean-reversion:

$$\mathcal{P}^* = e^{-rT}\left\{\left[V - K + \alpha\tau(\mu - r)^2/\gamma + (\sigma_0^2 - V - \alpha\tau(\mu - r)^2/\gamma)\frac{1 - e^{-\gamma T}}{T\gamma}\right]\right.$$

$$\left. + \frac{\xi\gamma(V_0 - (L - \frac{\lambda\beta}{\xi}))}{T}\left[\frac{X}{\xi}\left(\frac{e^{-\xi T} - 1}{\xi} + T\right) + \frac{C(e^{-\xi T} - 1)}{\xi(\xi + \gamma)} + \frac{C(e^{\gamma T} - 1)}{\gamma(\gamma + \xi)}\right]\right\}$$

$$= e^{-rT}\left\{\left[X - K + C\frac{1 - e^{-\gamma T}}{T\gamma}\right]\right.$$

$$\left. + \frac{\xi\gamma(V_0 - (L - \frac{\lambda\beta}{\xi}))}{T}\left[\frac{X}{\xi}\left(\frac{e^{-\xi T} - 1}{\xi} + T\right) + \frac{C(e^{-\xi T} - 1)}{\xi(\xi + \gamma)} + \frac{C(e^{\gamma T} - 1)}{\gamma(\gamma + \xi)}\right]\right\},$$

or, finally,

$$\mathcal{P}^* = e^{-rT}\left\{\left[X - K + C\frac{1 - e^{-\gamma T}}{T\gamma}\right]\right.$$

$$\left. + \frac{\xi\gamma(V_0 - (L - \frac{\lambda\beta}{\xi}))}{T}\left[\frac{X}{\xi}\left(\frac{e^{-\xi T} - 1}{\xi} + T\right) + \frac{C(e^{-\xi T} - 1)}{\xi(\xi + \gamma)} + \frac{C(e^{\gamma T} - 1)}{\gamma(\gamma + \xi)}\right]\right\},$$

$$(8.16)$$

where constants X and C are defined in (8.13).

8.4.3 *Pricing of Variance Swap for Two-Factor Stochastic Volatility Model with Delay and with Pilipovic One-Factor Mean-Reversion*

Two-Factor Stochastic Volatility Model with Delay and with Pilipovic one-factor mean-reversion is defined in the following way (see (8.3)):

$$\begin{cases} \frac{d\sigma^2(t,S_t)}{dt} = \gamma V_t + \frac{\alpha}{\tau}\left[\int_{t-\tau}^t \sigma(s, S_s)dW(s)\right]^2 - (\alpha + \gamma)\sigma^2(t, S_t) \\ dV_t = \xi(L - V_t)dt + \beta V_t dW_1(t), \end{cases} \quad (8.17)$$

where Wiener processes $W(t)$ and $W_1(t)$ may be correlated, S_t is defined as $S_t := S(t - \tau)$,

$$dS(t) = \mu S(t)dt + \sigma(t, S_t)dW(t).$$

Taking into account the same reasonings like in the previous section, a family of two-factor stochastic volatility model with delay and with Pilipovic one-factor mean-reversion under risk-neutral probability measure P^* can be described as follows

$$\begin{cases} \frac{d\sigma^2(t,S_t)}{dt} = \gamma V_t + \frac{\alpha}{\tau}\left[\int_{t-\tau}^t \sigma(s, S_s)dW^*(s) + (\mu - r)\tau\right]^2 \\ \qquad - (\alpha + \gamma)\sigma^2(t, S_t). \\ dV_t = (\xi + \lambda\beta)(L\frac{\xi}{\xi + \lambda\beta} - V_t)dt + \beta V_t(\rho dW^*(t) + \sqrt{1 - \rho^2}dW_1^*(t)), \end{cases} \quad (8.18)$$

where $(W^*(t), W_1^*(t))$ are independent standard Brownian motion, λ is the market price of volatility risk,

$$dS(t) = rS(t)dt + \sigma(t, S_t)dW^*(t),$$

$W^*(t)$ is defined in Chapter 7, the correlation coefficient $|\rho| < 1$.

We note, that

$$E_{P^*}V_t = e^{-(\xi + \lambda\beta)t}\left(V_0 - L\frac{\xi}{\xi + \lambda\beta}\right) + L\frac{\xi}{\xi + \lambda\beta}.$$

Thus, after taking the expectation E_{P^*}, the first equation finally takes the following form:

$$\frac{dv(t)}{dt} = \gamma\left(e^{-(\xi + \lambda\beta)t}\left(V_0 - L\frac{\xi}{\xi + \lambda\beta}\right) + L\frac{\xi}{\xi + \lambda\beta}\right)$$

$$+ \alpha\tau(\mu - r)^2 + \frac{\alpha}{\tau}\int_{t-\tau}^t v(s)ds - (\alpha + \gamma)v(t).$$

Proceeding with the similar calculations like in the previous Section 4.2, we can get the following expression for function $v(t)$:

$$v(t) \approx X + Ce^{-\gamma t}$$

$$+ \frac{\gamma\xi}{\xi + \lambda\beta}\left[X\left(\frac{V_0(\xi + \lambda\beta)}{\xi} - L\right)(1 - e^{-(\xi + \lambda\beta)t}) + XLt\right.$$

$$\left. + \frac{C(\frac{V_0(\xi + \lambda\beta)}{\xi} - L)}{\xi + \lambda\beta + \gamma} + \frac{CL}{\gamma}(e^{\gamma t} - 1)\right]. \tag{8.19}$$

The price of variance swap for two-factor stochastic volatility with delay and with Pilipovic one-factor mean-reversion has the following look:

$$\mathcal{P}^* = e^{-rT}\left\{\left[X - K + C\frac{1 - e^{-\gamma T}}{T\gamma}\right]\right.$$

$$+ \frac{\gamma\xi}{(\xi + \lambda\beta)T}\left[X\left(\left(\frac{V_0(\xi + \lambda\beta)}{\xi} - L\right)\left(\frac{e^{-(\xi + \lambda\beta)} - 1}{\xi + \lambda\beta} - T\right)\right)\right.$$

$$+ \frac{XLT^2}{2} + \frac{C(\frac{V_0(\xi + \lambda\beta)}{\xi} - L)}{\xi + \lambda\beta + \gamma}\left(\frac{e^{\gamma T} - 1}{\gamma} + \frac{e^{-(\xi + \lambda\beta)T} - 1}{\xi + \lambda\beta}\right)$$

$$\left.\left. + \frac{CL}{\gamma}\left(\frac{e^{\gamma T} - 1}{\gamma} - T\right)\right]\right\}, \tag{8.20}$$

where constants X and C are defined in (8.13).

8.4.4   Variance Swap for Three-Factor Stochastic Volatility Model with Delay and with Pilipovic Mean-Reversion

Three-Factor Stochastic Volatility Model with Delay and with Pilipovic Mean-Reversion is defined in the following way (see (8.4)):

$$
\begin{cases}
\frac{d\sigma^2(t,S_t)}{dt} = \gamma V_t + \frac{\alpha}{\tau}\left[\int_{t-\tau}^t \sigma(s,S_s)dW(s)\right]^2 - (\alpha+\gamma)\sigma^2(t,S_t) \\
dV_t = \xi(L_t - V_t)dt + \beta V_t dW_1(t), \\
dL_t = \beta_1 L_t dt + \eta L_t dW_2(t),
\end{cases}
\tag{8.21}
$$

where the Wiener processes $W(t), W^1(t)$ and $W^2(t)$ may be correlated, S_t is defined as $S_t := S(t-\tau)$,

$$dS(t) = \mu S(t)dt + \sigma(t,S_t)dW(t).$$

In order to incorporate a correlation between the Brownian motions $(W(t), W_1(t)), W_2(t)$, we set

$$W(t) = \tilde{W}(t),$$

$$W_1(t) = \rho_1 \tilde{W}(t) + \sqrt{1-\rho_1^2}\tilde{W}_1(t),$$

$$W_2(t) = \rho_2 \tilde{W}(t) + \rho_{12}\tilde{W}_1(t) + \sqrt{1-\rho_2^2-\rho_{12}^2}\tilde{W}_2(t),$$

where $(\tilde{W}(t), \tilde{W}_1(t)), \tilde{W}_2(t)$ are independent standard Brownian motions, and the correlation coefficients $\rho_1, \rho_2, \rho_{12}$, satisfy $|\rho_1| < 1$ and $|\rho_2^2 + \rho_{12}^2| < 1$.

Taking into account the same reasonings like in the previous Section, a family of two-factor stochastic volatility model with delay and with OU mean-reversion under the risk-neutral probability measure P^* can be described as follows

$$
\begin{cases}
\frac{d\sigma^2(t,S_t)}{dt} = \gamma V_t + \frac{\alpha}{\tau}\left[\int_{t-\tau}^t \sigma(s,S_s)dW^*(s) + (\mu-r)\tau\right]^2 \\
\qquad\qquad - (\alpha+\gamma)\sigma^2(t,S_t). \\
dV_t = (\xi+\lambda\beta)(L_t\frac{\xi}{\xi+\lambda\beta} - V_t)dt + \beta V_t(\rho_1 dW^*(t) + \sqrt{1-\rho_1^2}dW_1^*(t)) \\
dL_t = (\beta_1 - \lambda_1\eta)L_t dt + \eta L_t(\rho_2 W^*(t) + \rho_{12}W_1^*(t) + \sqrt{1-\rho_2^2-\rho_{12}^2}W_2^*(t)),
\end{cases}
\tag{8.22}
$$

where $(W^*(t), W_1^*(t)), W_2^*(t)$ are independent standard Brownian motions, λ and λ_1 are the market prices of volatility risk.

We note, that (see Pilipovic (1997))

$$E_{P^*}V_t = e^{-(\xi+\lambda\beta)t}V_0 + \frac{\xi+\lambda\beta}{\xi+\lambda\beta+\beta_1}L_0(e^{(\beta_1-\lambda_1\eta)t} - e^{-(\xi+\lambda\beta)t}).$$

After taking the expectation E_{P^*}, the first equation finally takes the following form:

$$\frac{dv(t)}{dt} = \gamma \left(e^{-(\xi+\lambda\beta)t}V_0 + \frac{\xi+\lambda\beta}{\xi+\lambda\beta+\beta_1}L_0(e^{(\beta_1-\lambda_1\eta)t} - e^{-(\xi+\lambda\beta)t}) \right)$$

$$+ \alpha\tau(\mu - r)^2 + \frac{\alpha}{\tau}\int_{t-\tau}^{t}v(s)ds - (\alpha+\gamma)v(t).$$

We are going to proceed with the same steps like in the previous section.

To get variance swap we have to find a solution for the last equation, namely, nonhomogeneous integro-differential equation with delay.

After taking the first derivative of this equation we obtain

$$v''(t) = -(\xi+\lambda\beta)\gamma e^{-(\xi+\lambda\beta)t}V_0$$

$$+ \frac{\xi+\lambda\beta}{\xi+\lambda\beta+\beta_1}L_0[(\beta_1-\lambda_1\eta)e^{(\beta_1-\lambda_1\eta)t} + (\xi+\lambda\beta)e^{-(\xi+\lambda\beta)t}]$$

$$+ \frac{\alpha}{\tau}[v(t) - v(t-\tau)] - (\alpha+\gamma)v'(t).$$

To solve this equation we rewrite it in vector form:

$$\vec{v}'(t) = Av(t) + B\vec{v}(t-\tau) + \vec{f}(t),$$

where

$$\vec{v}(t) := \begin{pmatrix} v_1(t) \\ v_2(t) \end{pmatrix} := \begin{pmatrix} v(t) \\ v'(t) \end{pmatrix}, A := \begin{pmatrix} 0 & 0 \\ \frac{\alpha}{\tau} & -(\alpha+\gamma) \end{pmatrix}, B := \begin{pmatrix} 0 & 0 \\ 0 & -\frac{\alpha}{\tau} \end{pmatrix},$$

$$\vec{f}(t) := \begin{pmatrix} 0 \\ -(\xi+\lambda\beta)\gamma e^{-(\xi+\lambda\beta)t}V_0 + \frac{\xi+\lambda\beta}{\xi+\lambda\beta+\beta_1}L_0[(\beta_1-\lambda_1\eta)e^{(\beta_1-\lambda_1\eta)t}+(\xi+\lambda\beta)e^{-(\xi+\lambda\beta)t}] \end{pmatrix}.$$

This is a nonhomogeneous first-order differential equation with delay.

Complementary homogeneous equation has the following look:

$$\vec{v}_h'(t) = Av(t) + B\vec{v}_h(t-\tau),$$

and has the following approximated solution (see Hale (1977) and previous section)

$$v_h(t) \approx X + Ce^{-\gamma t},$$

where

$$X := V_0 + \alpha\tau(\mu-r)^2/\gamma, \quad C := \sigma_0^2 - V_0 - \alpha\tau(\mu-r)^2/\gamma. \qquad (8.23)$$

Therefore, the nonhomogeneous eqaution has the following approximated solution (see Hale (1977))

$$v(t) \approx X + Ce^{-\gamma t} - (\xi+\lambda\beta)\gamma V_0\int_0^t (X + Ce^{\gamma(t-s)})e^{-(\xi+\lambda\beta)s}ds$$

$$+ \frac{\xi+\lambda\beta}{\xi+\lambda\beta+\beta_1}L_0\int_0^t(X + Ce^{(t-s)\gamma})((\beta_1-\lambda_1\eta)e^{(\beta_1-\lambda_1\eta)s}$$

$$+ (\xi+\lambda\beta)e^{-(\xi+\lambda\beta)s})ds. \qquad (8.24)$$

The calculations for the first two terms in (8.24) are the same like in the previous section, as for the last term in (8.24) we have

$$\frac{\xi+\lambda\beta}{\xi+\lambda\beta+\beta_1}L_0\int_0^t (X+Ce^{(t-s)\gamma})((\beta_1-\lambda_1\eta)e^{(\beta_1-\lambda_1\eta)s}+(\xi+\lambda\beta)e^{-(\xi+\lambda\beta)s})ds$$

$$=\frac{(\xi+\lambda\beta)L_0}{\xi+\lambda\beta+\beta_1}\left[X(e^{(\beta_1-\lambda_1\eta)t}-1)-X(e^{-(\xi+\lambda\beta)t}-1)\right.$$

$$\left.+C\left(\frac{\beta_1-\lambda_1\eta}{\beta_1-\lambda_1\eta-\gamma}(e^{(\beta_1-\lambda_1\eta)t}-e^{\gamma t})-\frac{\xi+\lambda\beta}{\xi+\lambda\beta+\gamma}(e^{-(\xi+\lambda\beta)t}-e^{\gamma t})\right)\right].$$

In this way, the function $v(t)$ in (8.24) takes the following look:

$$v(t)\approx X+Ce^{-\gamma t}$$

$$-(\xi+\lambda\beta)\gamma V_0\left[\frac{X}{\xi+\lambda\beta}(1-e^{-(\xi+\lambda\beta)t})+\frac{C}{\xi+\lambda\beta+\gamma}(e^{\gamma t}-e^{-(\xi+\lambda\beta)t})\right]$$

$$+L_0\frac{\xi+\lambda\beta}{\xi+\lambda\beta+\beta_1}\left[X(e^{(\beta_1-\lambda_1\eta)t}-e^{-(\xi+\lambda\beta)t})\right.$$

$$\left.+\frac{C(\beta_1-\lambda_1\eta)}{(\beta_1-\lambda_1\eta-\gamma)}(e^{(\beta_1-\lambda_1\eta)t}-e^{\gamma t})+\frac{C(\xi+\lambda\beta)}{(\xi+\lambda\beta+\gamma)}(e^{\gamma t}-e^{-(\xi+\lambda\beta)t})\right].$$

$$(8.25)$$

We note, that

$$v(t):=E_{P^*}[\sigma^2(t,S_t)],$$

and to find swap we have to calculate

$$E_{P^*}[Var(S)]:=\frac{1}{T}\int_0^T E_{P^*}[\sigma^2(t,S_t)]dt=\frac{1}{T}\int_0^T v(t)dt. \qquad (8.26)$$

After substitution of $v(t)$ into the above expression we obtain the price $\mathcal{P}^*$ of variance swap for three-factor stochastic volatility with delay and with Pilipovic mean-reversion:

$$\mathcal{P}^*=e^{-rT}\left\{\left[X-K+C\frac{1-e^{-\gamma T}}{T\gamma}\right]\right.$$

$$-\frac{(\xi+\lambda\beta)\gamma V_0}{T}\left[\frac{X}{(\xi+\lambda\beta)}\left(\frac{e^{-(\xi+\lambda\beta)T}-1}{(\xi+\lambda\beta)}+T\right)+\frac{C(e^{-(\xi+\lambda\beta)T}-1)}{(\xi+\lambda\beta)(\xi+\lambda\beta+\gamma)}+\frac{C(e^{\gamma T}-1)}{\gamma(\gamma+\xi+\lambda\beta)}\right]$$

$$+\frac{(\xi+\lambda\beta)L_0}{(\xi+\lambda\beta+\beta_1)T}\left[\frac{X(e^{(\beta_1-\lambda_1\eta)T}-1-(\beta_1-\lambda_1\eta)T)}{\beta_1-\lambda_1\eta}+\frac{X(e^{-(\xi+\lambda\beta)T}-1+(\xi+\lambda\beta)T)}{(\xi+\lambda\beta)}\right.$$

$$\left.\left.+C\left(\frac{\beta_1-\lambda_1\eta}{\beta_1-\lambda_1\eta-\gamma}\left(\frac{e^{(\beta_1-\lambda_1\eta)T}-1}{\beta_1-\lambda_1\eta}\right)-\frac{e^{\gamma T}-1}{\gamma}\right)+\frac{\xi+\lambda\beta}{\xi+\lambda\beta+\gamma}\left(\frac{e^{-(\xi+\lambda\beta)T}-1}{(\xi+\lambda\beta)}+\frac{e^{\gamma T}-1}{\gamma}\right)\right)\right]\right\},$$

$$(8.27)$$

where constants X and C are defined in (8.23).

8.5   Numerical Example 1: *S&P*60 Canada Index

In this Section, we apply the analytical solutions from previous Section to price the variance swap of the *S&P*60 Canada index for five years (January 1998–February 2002) (see Theoret, Zabre and Rostan (2002)).

In the end of February 2002, we wanted to price the fixed leg of a variance swap based on the *S&P*60 Canada index. The statistics on log returns *S&P*60 Canada Index for 5 years (January 1997–February 2002) is presented in Table 8.1:

Table 8.1

Statistics on Log Returns *S&P*60 Canada Index	
Series:	LOG RETURNS *S&P*60 CANADA INDEX
Sample:	1 1300
Observations:	1300
Mean	0.000235
Median	0.000593
Maximum	0.051983
Minimum	−0.101108
Std. Dev.	0.013567
Skewness	−0.665741
Kurtosis	7.787327

From the histogram of the *S&P*60 Canada index log returns on a 5-year historical period (1,300 observations from January 1998 to February 2002) it may be seen leptokurtosis in the histogram. If we take a look at the graph of the *S&P*60 Canada index log returns on a 5-year historical period we may see volatility clustering in the returns series. These facts indicate about the conditional heteroscedasticity.

A GARCH(1,1) regression is applied to the series and the results is obtained as in the next Table 8.2:

Table 8.2

Estimation of the GARCH(1,1) process				
Dependent Variable: Log returns of S&P60 Canada Index Prices				
Method: ML-ARCH				
Included Observations: 1300				
Convergence achieved after 28 observations				
−	**Coefficient:**	**Std. error:**	***z*-statistic:**	**Prob.**
C	0.000617	0.000338	1.824378	0.0681
Variance Equation				
C	2.58E-06	3.91E-07	6.597337	0
ARCH(1)	0.060445	0.007336	8.238968	0
GARCH(1)	0.927264	0.006554	141.4812	0
R-squared	−0.000791	Mean dependent var	−	0.000235
Adjusted R-squared	−0.003108	S.D. dependent var	−	0.013567
S.E. of regression	0.013588	Akaike info criterion	−	−5.928474
Sum squared resid	0.239283	Schwartz criterion	−	−5.912566
Log likelihood	3857.508	Durbin-Watson stat	−	1.886028

This table allows to generate different input variables to the volatility swap model.

We use the following relationship

$$\theta = \frac{V}{dt},$$

$$k = \frac{1 - \alpha - \beta}{dt},$$

to calculate the following discrete GARCH(1,1) parameters:
ARCH(1,1) coefficient $\alpha = 0.060445$;
GARCH(1,1) coefficient $\beta = 0.927264$;
GARCH(1,1) coefficient $\gamma = 0.012391$;
the Pearson kurtosis (fourth moment of the drift-adjusted stock return) $\xi = 7.787327$;
long volatility $\theta = 0.05289724$;
$k = 3.09733$;
a short volatility σ_0 equals to 0.01;
$\mu = 0.000235$,
$r = 0.02$ and $\tau = 1(day)$.
Parameter V may be found from the expression $V = \frac{C}{1-\alpha-\beta}$, where $C = 2.58 \times 10^{-6}$ is defined in Table 8.2. Thus, $V = 0.00020991$;
$dt = 1/252 = 0.003968254$.
Now, applying the expression (8.7) for our data, we find the following value:

$$X = V + \frac{\alpha\tau(\mu - r)^2}{\gamma} = 0.0002$$
$$C = \sigma_0^2 - V - \alpha\tau(\mu - r)^2/\gamma = 0.007.$$

For series of maturities up to 100 years we obtain the following plots for variances and prices of variance swaps for four multi-factor stochastic volatilites with delay. Figures 8.1–8.10 depicts the dependence of the different variance swaps with delay on maturity.

Figure 8.1 depicts the dependence of Variance for One-Factor SV with Delay on Maturity (years) for $S\&P60$ Canada Index (1998–2002) (formula (7.26), Chapter 7).

Figure 8.2 depicts the dependence of The Price of Variance Swap for One-Factor SV with Delay on Maturity (years) for $S\&P60$ Canada Index(1998–2002) (formula P^*, see Chapter 7).

Figure 8.3 depicts the dependence of Variance for Multi-Factor SV with Delay and with GBM Mean-Reversion on Maturity (years) for $S\&P60$ Canada Index (1998–2002) (formula (8.8)).

Figure 8.4 depicts the dependence of The Price of Variance Swap for Multi-Factor SV with Delay and with GBM Mean-Reversion on Maturity (years) for $S\&P60$ Canada Index(1998–2002) (formula (8.10)).

Figure 8.5 depicts the dependence of Variance for Multi-Factor SV with Delay and with OU Mean-Reversion on Maturity (years) for $S\&P60$ Canada Index (1998–2002) (formula (8.14)).

Figure 8.6 depicts the dependence of The Price of Variance Swap for Multi-Factor SV with Delay and with OU Mean-Reversion on Maturity (years) for $S\&P60$ Canada Index (1998–2002) (formula (8.16)).

Figure 8.7 depicts the dependence of Variance Multi-Factor SV with Delay and with One-factor Pilipovic Mean-Reversion on Maturity (years) for $S\&P60$ Canada Index (1998–2002) (formula (8.19)).

Figure 8.8 depicts the dependence of The Price of Variance Swap for Multi-Factor SV with Delay and with One-Factor Pilipovic Mean-Reversion on Maturity (years) for $S\&P60$ Canada Index (1998–2002) (formula 8.20).

Figure 8.9 depicts the dependence of Variance Multi-Factor SV with Delay and with Two-factor Pilipovic Mean-Reversion on Maturity (years) for $S\&P60$ Canada Index (1998–2002) (formula (8.25)).

Figure 8.10 depicts the dependence of The Price of Variance Swap for Multi-Factor SV with Delay and with Two-Factor Pilipovic Mean-Reversion on Maturity (years) for $S\&P60$ Canada Index (1998–2002) (formula (8.27)).

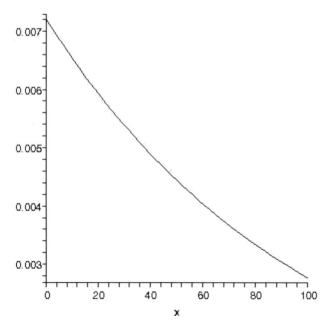

Fig. 8.1 Variance of One-Factor SV with Delay: $S\&P60$ Canada Index with $V_0 = 0.0002$, $\xi = 0.02$, $\gamma = 0.01$, $X = 0.0002$, $C = 0.007$ and delay $\tau = 1$.

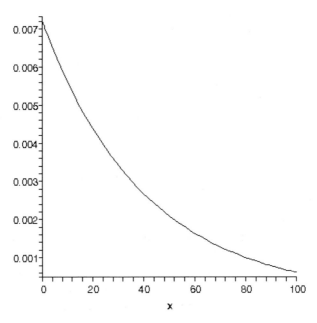

Fig. 8.2   The Price of Variance Swap for One-Factor SV with Delay: $S\&P60$ Canada Index with $V_0 = 0.0002$, $\xi = 0.02$, $\gamma = 0.01$, $X = 0.0002$, $C = 0.007$, $K = 0.00015$, $r = 0.02$ and delay $\tau = 1$.

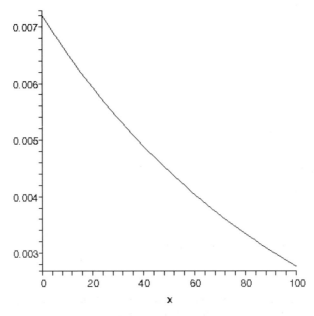

Fig. 8.3   Variance Multi-Factor SV with Delay and with GBM Mean-Reversion $S\&P60$ Canada Index: $V_0 = 0.0002$, $\xi = 0.02$, $\gamma = 0.01$, $X = 0.0002$, $C = 0.007$ and delay $\tau = 1$.

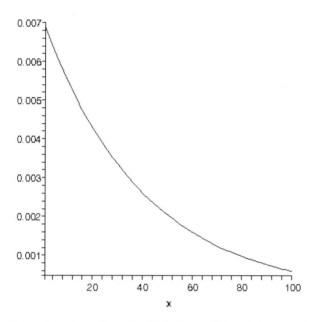

Fig. 8.4 The Price of Variance Swap for Multi-Factor SV with Delay and with GBM Mean-Reversion: $S\&P60$ Canada Index with $V_0 = 0.0002$, $\xi = 0.02$, $\gamma = 0.01$, $X = 0.0002$, $C = 0.007$, $K = 0.00015$, $\lambda = 0$, $r = 0.02$ and delay $\tau = 1$.

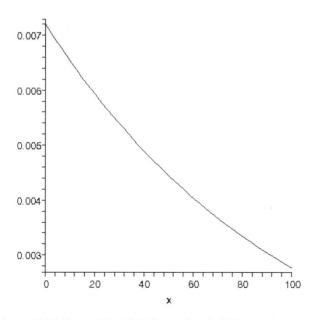

Fig. 8.5 Variance Multi-Factor SV with Delay and with OU Mean-Reversion: $S\&P60$ Canada Index with $V_0 = 0.0002$, $\xi = 0.02$, $\gamma = 0.01$, $X = 0.0002$, $C = 0.007$, $L = 0.0001$, $\lambda = 0$ and delay $\tau = 1$.

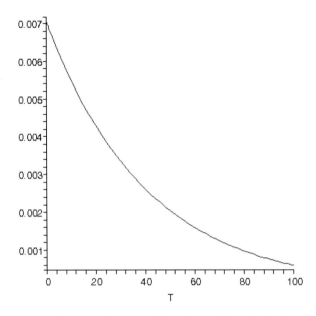

Fig. 8.6 The Price of Variance Swap for Multi-Factor SV with Delay and with OU Mean-Reversion: $S\&P60$ Canada Index with $V_0 = 0.0002$, $\xi = 0.02$, $\gamma = 0.01$, $X = 0.0002$, $C = 0.007$, $L = 0.0001$, $K = 0.00015$, $\lambda = 0$, $r = 0.02$ and delay $\tau = 1$.

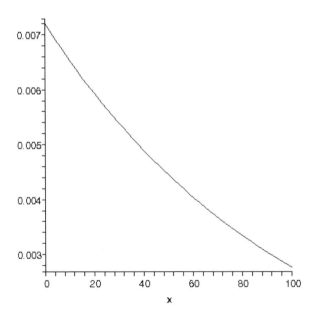

Fig. 8.7 Variance Multi-Factor SV with Delay and with One-factor Pilipovic Mean-Reversion: $S\&P60$ Canada Index with $V_0 = 0.0002$, $\xi = 0.02$, $\gamma = 0.01$, $X = 0.0002$, $C = 0.007$, $L = 0.0001$, $\lambda = 0$ and delay $\tau = 1$.

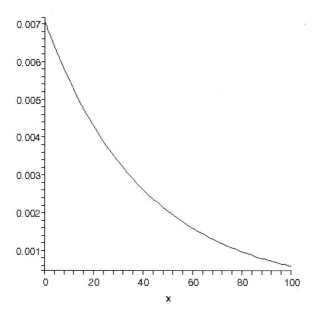

Fig. 8.8 The Price of Variance Swap for Multi-Factor SV with Delay and with One-Factor Pilipovic Mean-Reversion: $S\&P60$ Canada Index with $V_0 = 0.0002$, $\xi = 0.02$, $\gamma = 0.01$, $X = 0.0002$, $C = 0.007$, $L = 0.0001$, $K = 0.00015$, $\lambda = 0$, $r = 0.02$ and delay $\tau = 1$.

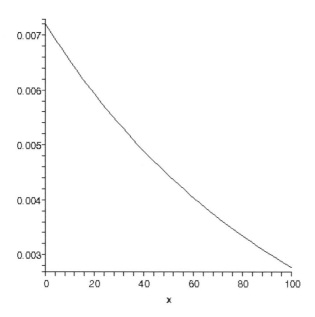

Fig. 8.9 Variance for Multi-Factor SV with Delay and with Two-factor Pilipovic Mean-Reversion: $S\&P60$ Canada Index with $V_0 = 0.0002$, $\xi = 0.024$, $\gamma = 0.01$, $X = 0.0002$, $C = 0.007$, $L_0 = 0.0001$, $\lambda = \lambda_1 = 0$, $\beta_1 = 0.02$ and delay $\tau = 1$.

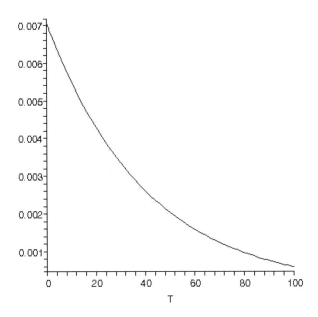

Fig. 8.10 The Price of Variance Swap for Multi-Factor SV with Delay and with Two-Factor Pilipovic Mean-Reversion: $S\&P60$ Canada Index with $V_0 = 0.0002$, $\xi = 0.02$, $\gamma = 0.01$, $X = 0.0002$, $C = 0.007$, $K = 0.00015$, $\lambda = \lambda_1 = 0$, $\beta_1 = 0.02$, $r = 0.02$ and delay $\tau = 1$.

8.6   Summary

– In the Chapter we studied multi-factor, i.e., two- and three-factors, stochastic volatility models with delay to model and price variance swaps.
– We found some analytical close forms for expectation and variance of the realized continuously sampled variances for multi-factor stochastic volatilities with delay.
– As an application of our analytical solutions, we provided a numerical examples using $S\&P60$ Canada Index (1998–2002) to price variance swaps for four different multi-factor stochastic volatility models with delay.

Bibliography

Avellaneda, M., Levy, A. and Paras, A. (1995). Pricing and hedging derivative securities in markets with uncertain volatility. *Applied Mathematical Finance*, 2, 73-88.
Bates, D. (1996). Jumps and stochastic volatility: the exchange rate processes implicit in Deutschemark options. *Review Finance Studies*, 9, 69-107.
Black, F. and Scholes, M. (1973). The pricing of options and corporate liabilities *Journal of Political Economy*, 81, 637-54.
Bollerslev, T. (1986). Generalized autoregressive conditional heteroscedasticity. *Journal of Economics*, 31, 307-27.

Brockhaus, O. and Long, D. (2000). Volatility swaps made simple. *Risk Magazine*, January, 92-96.

Buff, R. (2002). *Uncertain Volatility Model. Theory and Applications*. New York: Springer.

Carr, P. and Madan, D. (1998). Towards a theory of volatility trading. In Jarrow, R. (ed.), *Volatility*, Risk Book Publications.

Chernov, R., Gallant, E., Ghysels, E. and Tauchen, G. (2003). Alternative models for stock price dynamics. *Journal of Econometrics*, 116, 225-257.

Demeterfi, K., Derman, E., Kamal, M. and Zou, J. (1999). A guide to volatility and variance swaps. *The Journal of Derivatives*, Summer, 9-32.

Duan, J. (1995). The GARCH option pricing model, *Mathematical Finance*, 5, 13-32.

Elliott, R. and Swishchuk, A. (2004). Pricing options and variance swaps in Brownian and fractional Brownian markets. Working Paper, University of Calgary, Calgary, AB, Canada, 20 pages.

Fouque, J.-P., Papanicolaou, G. and Sircar, K. R. (2000). *Derivatives in Financial Markets with Stochastic Volatilities*. Berlin-Heidelberg: Springer-Verlag.

Fouque, J.-P. and Han, C.-H. (2003). A control variate method to evaluate option prices under multi-factor stochastic volatility models. Working Paper.

Eydeland, L. and Geman, H. (1998). Pricing power derivatives. *Risk Magazine*, 11, 10, October.

Geman, H., El Karoui, N. and Lacoste. V. (2000). On the role of state variables in interest rate models. *Applied Stochastic Models in Business and Industry*, 16

Gibson, R. and Schwartz, E. (1990). Stochastic convenience yield and the pricing of oil contigent claims. *The Journal of Finance*, 45, 959-976.

Griego, R. and Swishchuk, A. (2000). Black-Scholes formula for a market in a Markov environment. *Theory of Probability and Mathematical Statistics*, 62, 9-18.

Hale, J. (1977). *Theory of Functional Differential Equations*. Berlin-Heidelberg: Springer-Verlag.

Heston, S. (1993). A closed-form solution for options with stochastic volatility with applications to bond and currency options. *Review of Financial Studies*, 6, 327-343

Hilliard, J. and Reis, J. (1998). Valuation of commodity futures and options under stochastic convenience yields, interest rates, and jump diffusion in the spot. *Journal of Financial and Quantitative Analysis*, 33, 61-86.

Hull, J. (2000). *Options, Futures and Other Derivatives* (4th edition). New Jersey: Prentice Hall.

Hull, J., and White, A. (1987). The pricing of options on assets with stochastic volatilities. *Journal of Finance*, 42, 281-300.

Javaheri, A., Wilmott, P. and Haug, E. (2002). GARCH and volatility swaps. *Wilmott Technical Article*, January, 17 pages.

Kallsen, Taqqu (1995). Option Pricing in ARCH-type Models: with Detailed Proofs. Technical Report No. 10, University of Freiburg.

Kallianpur, G. and Karandikar, R. (2000). *Introduction to Option Pricing Theory*. Birkhäuser.

Kazmerchuk, Y., Swishchuk, A., Wu, J. (2002). The pricing of options for security markets with delayed response (submitted to *Mathematical Finance Journal*).

Kazmerchuk, Y., Swishchuk, A. and Wu, J. (2005). A continuous-time GARCH model for stochastic volatility with delay. *Canadian Applied Mathematics Quarterly*, 13, 2, Summer.

Lari-Lavassani, A., Sadeghi, A. and Ware, A. (2001). Modeling and implementing mean reverting price processes in energy markets, 27 pages. Electronic Publications of the International Energy Credit Association. Available at www.ieca.net.

Merton, R. (1973). Theory of rational option pricing. *Bell Journal of Economic Management Science*, 4, 141-183.

Mohammed, S.E.A, Arriojas, M. and Pap, Y. (2001). A delayed Black and Scholes formula. Preprint, Southern Illinois University.

Mohammed, S.E.A. (1998). Stochastic differential systems with memory: Theory, examples and applications. *Stochastic Analysis and Related Topics*, 6, 1-77.

Molina, G., Han, C.-H. and Fouque, J.-P. (2003). *MCMC Estimation of Multiscale Stochastic Volatility Models*, Preprint.

Pilipovic, D. (1997). *Valuing and Managing Energy Derivatives*. New York: McGraw-Hill.

Schoutens, W., Simons, E. and Tistaert, J. (2003). A perfect calibration! Now what? Working Paper, November 20, 30 pages.

Schwartz, E. (1997). The stochastic behaviour of commodity Prices: Implications for valuation and hedging. *The Journal of Finance*, 52, 3, July, 923-973.

Swishchuk, A. (1995). Hedging of options under mean-square criterion and with semi-Markov volatility. *Ukrainian Mathematics Journal* 47, 7, 1119-1127.

Swishchuk, A. (2000). *Random Evolutions and Their Applications. New Trends*. Dordrecht, The Netherlands: Kluwer Academic Publishers.

Swishchuk, A. (2004). Modeling of variance and volatility swaps for financial markets with stochastic volatilities. *Wilmott Magazine*, September Issue, Technical article No. 2, 64-72.

Swishchuk, A. (2005). Modeling and pricing of variance swaps for stochastic volatilities with delay. *Wilmott Magazine*, September, Issue 19, 63-73.

Theoret, R., Zabre, L. and Rostan, P. (2002). Pricing volatility swaps: Empirical testing with Canadian data. *Working Paper*, Centre de Recherche en Gestion, Document 17-2002, July 2002.

Vasiček, O. (1977). An equilibrium characterization of the term structure. *Journal of Financial Economics*, 5, 177-188.

Wilmott, P. (2000). *Quantitative Finance (Volumes I and II)*. Chichester, West Sussex, England: Wiley.

Wilmott, P. (1999). *Derivatives: The Theory and Practice of Financial Engineering*. Chichester, West Sussex, England: Wiley.

Wilmott, P., Howison, S. and Dewynne, J. (1995). *Option Pricing: Mathematical Models and Computations*. Oxford: Oxford Financial Press.

Chapter 9

Pricing Variance Swaps for Stochastic Volatilities with Delay and Jumps

9.1 Introduction

The valuation of the variance swaps for stochastic volatility with delay and jumps is discussed in this Chapter. We provide some analytical closed forms for the expectation of the realized variance for the stochastic volatility with delay and jumps. The jump part in our model is finally represented by a general version of compound Poisson processes. Besides, we also present a lower bound for delay as a measure of risk. As applications of our analytical solutions, a numerical example using $S\&P60$ Canada Index (1998–2002) is then provided to price variance swaps with delay and jumps. Finally, we find that this model not only keeps some good features of the previous model without jumps but is also easy and quick to implement.

Variance swaps are forward contracts on future realized variance, the square of the realized volatility, which provide an easy way for investors to gain exposure to the future realized variance of the asset returns instead of direct exposure to the underlying assets. The market for such derivatives develops quickly after the collapse of LCTM in 1998 when the volatilities increased to a unprecedented high level and many investors are now interested in these derivatives to hedge volatility. Recently, several papers address the valuation of variance swaps or other volatility derivatives (see Carr and Madan (1998), Demeterifi, Derman, Kamal, and Zou (1999), Heston and Nandi (2000), Brockhaus and Long (2000) and Javaheri, Wilmott and Haug (2002)).

It is known that the assumption of constant volatility in Black-Scholes model (Black and Scholes (1973)) is incompatible with some empirical study in the real market. Several stochastic volatility models (see Hull and White (1987), Heston (1993)) are developed and able to fit some important characteristics in the market like skews and smirks. Kazmerchuk, Swishchuk and Wu (2005, 2006) proposed a stochastic volatility model assuming that the volatility $\sigma(t, S_t)$ depends on $S_t = S(t + \theta)$ for $\theta \in [-\tau, 0]$, namely, stochastic volatility with delay. In Kazmerchuk, Swishchuk and Wu (2006), the Black-Scholes formula for secrity markets with delayed response was found and in Kazmerchuk, Swishchuk and Wu (2005), the continuous-time GARCH model for stochastic volatility with delay was proposed and studied. In Swishchuk (2005), the stochastic volatility model with delay to price variance swaps was discussed and some analytical close forms for expectation of the realized continuously sampled variance were found both in stationary regime and in general case. The key features of this model are the following: i) continuous-time analogue of discrete-time GARCH model; ii) mean-reversion; iii) contains the same source of randomness as stock price; iv) market is complete; v) incorporates the expectation of log-return. However, there are two ways to develop this model and make it more realistic. One way is to consider multi-factor stochastic volatility with delay which is developed in Swishchuk (2006). The other way is to consider stochastic volatility with delay and jumps which is the major work of this paper. During the last decade, financial models based on jumps processes have acquired increasing popularity in risk management and option pricing applications. A good reference is Cont and Tankov (2003), which provides a self-contained overview of the theoretical, numerical, and empirical aspects of using jump processes in financial modeling. Stochastic volatility models with jumps are also included in this book. Some attempts have been made to incorporate jumps in stochastic volatility to price variance and volatility swaps (see Howison, Rafailidis, Rasmussen (2004)).

In this Chapter, we incorporate a jump part in the stochastic volatility model with delay proposed by Swishchuk (2005) to price variance swaps. We find some analytical closed forms for the expectation of the realized continuously sampled variance for stochastic volatility with delay and jumps. The jump part in our model is finally represented by a general version of compound Poisson processes and the expectation and the covariance of the jump sizes are assumed to be deterministic functions. We note that after adding jumps, the model still keeps those good features of the previous model such as continuous-time analogue of GARCH model, mean-reversion and so on. But it is more realistic and still quick to implement. Besides, we also present a lower bound for delay as a measure of risk. As applications of our analytical solutions, a numerical example using $S\&P60$ Canada Index (1998–2002) is also provided to price variance swaps with delay and jumps.

The rest of the paper is organized as follows: In Section 9.2, we introduce some basic concepts of variance swaps and in Section 9.3, we recall the pricing model of variance swaps for stochastic volatility with delay presented by Swishchuk (2005). The pricing model and analytical formulae for stochastic volatility with delay and jumps are discussed in Section 9.4 and a lower bound for delay as a measure of risk is presented in Section 9.5. Finally, we give a numerical example in Section 9.6 and conclude in Section 9.7.

9.2   Stochastic Volatility with Delay

In this Section, we recall the model and approach of pricing variance swaps for stochastic volatility with delay presented in the paper of Swishchuk (2005).

In our model, we assume that the price of the underlying asset $S(t)$ follows the following stochastic delay differential equation (SDDE):

$$dS(t) = \mu S(t)dt + \sigma(t, S(t - \tau))dW(t), \quad t > 0, \tag{9.1}$$

where $\mu \in \mathbb{R}$ is the mean rate of return, the volatility term $\sigma > 0$ is a continuous and bounded function and $W(t)$ is a Brownian motion on a probability space $(\Omega, \mathcal{F}, \mathbb{P})$ with a filtration $\mathcal{F}(t)$. We also let $r > 0$ be the risk-free rate of return of the market.

Throughout the paper, we denote $S_t = S(t - \tau), \quad t > 0$ and the initial data of $S(t)$ is defined by $S(t) = \varphi(t)$, where $\varphi(t)$ is a deterministic function with $t \in [-\tau, 0], \quad \tau > 0$.

Now we can rewrite the equation of $S(t)$ as the following expression:

$$dS(t) = \mu S(t)dt + \sigma(t, S_t)dW(t). \tag{9.2}$$

The asset volatility is defined as the solution of the following equation:

$$\frac{d\sigma^2(t, S_t)}{dt} = \gamma V + \frac{\alpha}{\tau} \left[\int_{t-\tau}^{t} \sigma(s, S_s)dW(s) \right]^2 - (\alpha + \gamma)\sigma^2(t, S_t) \tag{9.3}$$

where the parameters α, γ, τ, V are positive constants and $0 < \alpha + \gamma < 1$. The Brownian motion $W(t)$ is the same as the one in (9.2).

From Swishchuk (2005), the key features of the *stochastic volatility model with delay* are the following:

i) *continuous-time analogue of discrete-time GARCH model;*

ii) *mean-reversion;*

iii) *does not contain another Brownian motion;*

iv) *market is complete;*

v) *incorporates the expectation of log-return.*

To price the variance swaps, we need to calculate $\mathbb{E}^*[v] = \mathbb{E}^*[\sigma^2]$, that is, the expectation of the variance under the risk-neutral probability measure. We should find such probability measure that makes the discounted asset price $D(t) = e^{-rt}S(t)$ be a local martingale and then the risk-neutral pricing formula applies.

Let $\theta(t) = \frac{\mu - r}{\sigma(t, S_t)}$ be the market price of risk, which is adapted to the filtration $\mathcal{F}(t)$, then by Girsanov's theorem for single Brownian motion, we obtain that:

1) There is a probability measure $\mathbb{P}^*$ equivalent to $\mathbb{P}$ such that

$$\frac{d\mathbb{P}^*}{d\mathbb{P}} = \exp \left\{ - \int_0^T \theta(s)dW(s) - \frac{1}{2} \int_0^T \theta^2(s)ds \right\} \tag{9.4}$$

is its Radon-Nikodym density.

2) The discounted asset price $D(t)$ is a positive local martingale with respect to $\mathbb{P}^*$, and

$$W^*(t) = \int_0^t \theta(s)ds + W(t) \tag{9.5}$$

is a standard Brownian motion with respect to $\mathbb{P}^*$.

Therefore, in the risk-neutral world, the underlying asset price $S(t)$ follows the process

$$dS(t) = rS(t)dt + \sigma(t, S_t)dW^*(t) \tag{9.6}$$

and the asset volatility is defined then as follows:

$$\frac{d\sigma^2(t, S_t)}{dt} = \gamma V + \frac{\alpha}{\tau} \left[\int_{t-\tau}^{t} \sigma(s, S_s)dW^*(s) - \int_{t-\tau}^{t} \theta(s)\sigma(s, S_s)ds \right]^2$$
$$- (\alpha + \gamma)\sigma^2(t, S_t)$$
$$= \gamma V + \frac{\alpha}{\tau} \left[\int_{t-\tau}^{t} \sigma(s, S_s)dW^*(s) - (\mu - r)\tau \right]^2$$
$$- (\alpha + \gamma)\sigma^2(t, S_t). \tag{9.7}$$

where $W^*(t)$ is defined in (9.5).

Let us take the expectations under risk-neutral measure $\mathbb{P}^*$ on the both sides of the equation above. Denoting $v(t) = \mathbb{E}^*[\sigma^2(t, S_t)]$, we obtain the following deterministic delay differential equation:

$$\frac{dv(t)}{dt} = \gamma V + \alpha\tau(\mu - r)^2 + \frac{\alpha}{\tau} \int_{t-\tau}^{t} v(s)ds - (\alpha + \gamma)v(t). \tag{9.8}$$

Notice that (9.8) has a stationary solution

$$v(t) \equiv X = V + \alpha\tau(\mu - r)^2/\gamma. \tag{9.9}$$

Hence, the expectation of the realized variance, or say the fair delivery price K_{var} *of a variance swap for stochastic volatility with delay in stationary regime under risk-neutral measure* $\mathbb{P}^*$ equals to

$$K_{var} = \mathbb{E}^*[v] = \frac{1}{T} \int_{0}^{T} v(t)dt$$
$$= V + \alpha\tau(\mu - r)^2/\gamma \tag{9.10}$$

and the price P of a variance swap at time t given delivery price K in this case should be:

$$P = e^{-r(T-t)}[V - K + \alpha\tau(\mu - r)^2/\gamma].$$

In general case, there is no way to write a solution in explicit form for arbitrarily given initial data. But we can write an approximate solution for $v(t)$ when t has large values:

$$v(t) \approx X + Ce^{-\gamma t} = V + \alpha\tau(\mu - r)^2/\gamma + Ce^{-\gamma t} \tag{9.11}$$

where

$$C = v(0) - X = \sigma_0^2 - V - \alpha\tau(\mu - r)^2/\gamma. \tag{9.12}$$

Hence, the expectation of the realized variance, or say the fair delivery price K_{var} of *variance swap for stochastic volatility with delay in general case under risk-neutral measure* $\mathbb{P}^*$ equals to

$$K_{var} = \mathbb{E}^*[v] = \frac{1}{T} \int_0^T v(t)dt$$

$$\approx \frac{1}{T} \int_0^T [V + \alpha\tau(\mu - r)^2/\gamma + (\sigma_0^2 - V - \alpha\tau(\mu - r)^2/\gamma)e^{-\gamma t}]dt$$

$$= V + \alpha\tau(\mu - r)^2/\gamma + (\sigma_0^2 - V - \alpha\tau(\mu - r)^2/\gamma)\frac{1 - e^{-\gamma T}}{T\gamma} \qquad (9.13)$$

and the price P of a variance swap at time t given delivery price K in this case should be:

$$P \approx e^{-r(T-t)} \left[V - K + \alpha\tau(\mu - r)^2/\gamma + (\sigma_0^2 - V - \alpha\tau(\mu - r)^2/\gamma)\frac{1 - e^{-\gamma T}}{T\gamma} \right].$$

9.3 Pricing Model of Variance Swaps for Stochastic Volatility with Delay and Jumps

In the following Section, we will derive some analytical closed formulae for the expectation of the realized variance for stochastic volatility with delay and jumps. First of all, we need to define the jumps and add them to the stochastic volatility model with delay. In equation (9.3), we find that the volatility is driven by a Brownian motion $W(t)$, but it is more realistic to consider it is driven by a process which consists a jump part. We represent jumps in stochastic volatility by general compound Poisson processes, and try to write the stochastic volatility in the following form:

$$\frac{d\sigma^2(t, S_t)}{dt} = \gamma V + \frac{\alpha}{\tau} \left[\int_{t-\tau}^t \sigma(s, S_s)dW(s) + \int_{t-\tau}^t y_s dN(s) \right]^2 - (\alpha + \gamma)\sigma^2(t, S_t)$$

$$(9.14)$$

where $W(t)$ is a Brownian motion, $N(t)$ is a Poisson process with intensity λ and y_t is the jump size at time t. We assume that $\mathbb{E}[y_t] = A(t)$, $\mathbb{E}[y_s y_t] = C(s, t), s < t$, $\mathbb{E}[y_t^2] = B(t) = C(t, t)$ and $A(t), B(t), C(s, t)$ are all deterministic functions. Our purpose is to valuate variance swaps when the stochastic volatility satisfies this general equation. In order to get and check the results, we first consider two simple cases which is easier to model and implement but fundamental and still capture some characteristics of the market. In Section 4.1, we discuss the case that the jump size y_t always equals to constant one, that is, the jump part is represented by $\int_{t-\tau}^t dN(s)$, just simple Poisson processes. In Section 4.2, the jump part is still compound Poisson processes denoted as $\int_{t-\tau}^t y_s dN(s)$ but the jump size y_t is assumed to be identically independent distributed random variable with mean value ξ and variance η. The general case is discussed in Section 4.3. We can compare our results with the model in Swishchuk (2005) and will see that it is a special case of our model after adding jumps in stochastic volatility. Finally, in Section 4.4 we will show that the model for stochastic volatility with delay and jumps keeps those good features of the model in Swishchuk (2005).

9.3.1 *Simple Poisson Process Case*

We assume that the price of the underlying asset $S(t)$ follows the the following stochastic delay differential equation (SDDE):

$$dS(t) = \mu S(t)dt + \sigma(t, S_t)dW(t), \quad t > 0, \tag{9.15}$$

and the asset volatility is defined as the solution of the following equation:

$$\frac{d\sigma^2(t, S_t)}{dt} = \gamma V + \frac{\alpha}{\tau} \left[\int_{t-\tau}^{t} \sigma(s, S_s)dW(s) + \int_{t-\tau}^{t} dN(s) \right]^2 - (\alpha + \gamma)\sigma^2(t, S_t) \tag{9.16}$$

where $W(t)$ is a Brownian motion and $N(t)$ is a Poisson process with intensity λ.

Recall that our purpose is to calculate $\mathbb{E}^*[v] = \mathbb{E}^*[\sigma^2(t, S_t)]$, the expectation of the variance under the risk-neutral measure. Since we assume that there is no Poisson process in the asset price, the change of measure is no different from the model we discussed in Section 9.2 to make it risk-neutral.

Let $\theta(t) = \frac{\mu - r}{\sigma(t, S_t)}$ be the market price of risk, which is adapted to the filtration $\mathcal{F}(t)$, and then by Girsanov's theorem for single Brownian motion, we obtain the risk-neutral probability measure which makes the discounted asset price be a local martingale and $W^*(t) = \int_0^t \theta(s)ds + W(t)$ is a Brownian motion under this probability measure.

Note that the change of measure does not change the Poisson intensity λ since it is independent to the Brownian motion.

In the risk-neutral world, the volatility can be defined as follows:

$$\frac{d\sigma^2(t, S_t)}{dt} = \gamma V + \frac{\alpha}{\tau} \left[\int_{t-\tau}^{t} \sigma(s, S_s)dW^*(s) + \int_{t-\tau}^{t} dN(s) - (\mu - r)\tau \right]^2$$

$$- (\alpha + \gamma)\sigma^2(t, S_t) \tag{9.17}$$

which can be expanded as follows:

$$\frac{d\sigma^2(t, S_t)}{dt} = \gamma V + \frac{\alpha}{\tau} \left[\left(\int_{t-\tau}^{t} \sigma(s, S_s)dW^*(s) \right)^2 + \left(N(t) - N(t-\tau) \right)^2 + (\mu - r)^2\tau^2 \right.$$

$$+ 2 \left(\int_{t-\tau}^{t} \sigma(s, S_s)dW^*(s) \right) \left(N(t) - N(t-\tau) \right)$$

$$\left. - 2 \left(\int_{t-\tau}^{t} \sigma(s, S_s)dW^*(s) \right) (\mu - r)\tau - 2 \left(N(t) - N(t-\tau) \right)(\mu - r)\tau \right]$$

$$- (\alpha + \gamma)\sigma^2(t, S_t). \tag{9.18}$$

Now let us take the expectation under risk-neutral probability $\mathbb{P}^*$ on both sides of the equation. Note that the Brownian motion and the Poisson process are independent. Let $v(t) = \mathbb{E}^*[\sigma^2(t, S_t)]$, we obtain the following deterministic delay differential equation:

$$\frac{dv(t)}{dt} = \gamma V + \frac{\alpha}{\tau}\left[\mathbb{E}^*\left(\int_{t-\tau}^t \sigma(s, S_s)dW^*(s)\right)^2 + \mathbb{E}^*\left((N(t) - N(t-\tau))\right)^2\right.$$

$$\left. + (\mu - r)^2\tau^2 - 2\mathbb{E}^*\left(N(t) - N(t-\tau)\right)(\mu - r)\tau\right] - (\alpha + \gamma)v(t)$$

$$= \gamma V + \frac{\alpha}{\tau}\left[\int_{t-\tau}^t v(s)ds + Var^*\left(N(t) - N(t-\tau)\right) + \left(\mathbb{E}^*(N(t) - N(t-\tau))\right)^2\right.$$

$$\left. + (\mu - r)^2\tau^2 - 2\mathbb{E}^*\left(N(t) - N(t-\tau)\right)(\mu - r)\tau\right] - (\alpha + \gamma)v(t)$$

$$= \gamma V + \alpha\lambda + \alpha\lambda^2\tau - 2\alpha\lambda\tau(\mu - r) + \alpha\tau(\mu - r)^2 + \frac{\alpha}{\tau}\int_{t-\tau}^t v(s)ds$$

$$- (\alpha + \gamma)v(t). \tag{9.19}$$

From this equation, if the intensity of the Poisson process $\lambda = 0$, then it is the same as the equation (9.8), the case of stochastic volatility with delay and without jumps.

Notice that the equation (9.19) has a stationary solution

$$v(t) \equiv X = V + [\alpha\lambda + \alpha\lambda^2\tau - 2\alpha\lambda\tau(\mu - r) + \alpha\tau(\mu - r)^2]/\gamma$$

$$= V + \frac{\alpha}{\gamma}[\lambda + \tau(\lambda - \mu + r)^2]. \tag{9.20}$$

Hence, the expectation of the realized variance, or say the fair delivery price K_{var} of *variance swap for stochastic volatility with delay and Poisson jump in stationary regime under risk-neutral measure* $\mathbb{P}^*$ equals to

$$K_{var} = \mathbb{E}^*[v] = \frac{1}{T}\int_0^T v(t)dt$$

$$= V + \frac{\alpha}{\gamma}[\lambda + \tau(\lambda - \mu + r)^2] \tag{9.21}$$

and the price P of a variance swap at time t given delivery price K in this case should be:

$$P = e^{-r(T-t)}\left\{V - K + \frac{\alpha}{\gamma}\left[\lambda + \tau(\lambda - \mu + r)^2\right]\right\}.$$

There is no way to write a solution in explicit form for arbitrarily given initial data. But we can understand an approximate behavior of solutions of (9.19) by looking at its eigenvalues. Let us substitute $v(t) = X + Ce^{\rho t}$ into (9.19), where X is defined in (9.20). Then, the characteristic equation for ρ is:

$$\rho = \frac{\alpha}{\rho\tau}(1 - e^{-\rho\tau}) - (\alpha + \gamma) \qquad (9.22)$$

which is equivalent to (when $\rho \neq 0$):

$$\rho^2 = \frac{\alpha}{\tau} - \frac{\alpha}{\tau}e^{-\rho\tau} - (\alpha + \gamma)\rho.$$

The only solution to this equation is $\rho \approx -\gamma$, assuming that γ is sufficiently small. Then the behavior of any solution is stable near X, and

$$v(t) \approx X + Ce^{-\gamma t} \qquad (9.23)$$

for large values of t.

In this way, we have:

$$\begin{aligned} v(t) &\approx X + Ce^{-\gamma t} \\ &= V + \frac{\alpha}{\gamma}\left[\lambda + \tau(\lambda - \mu + r)^2\right] + Ce^{-\gamma t}. \end{aligned} \qquad (9.24)$$

Since $v(0) = \sigma(0, S(0 - \tau)) = \sigma(0, \varphi(-\tau)) = \sigma_0$, we can find the value of C:

$$C = \sigma_0^2 - V - \frac{\alpha}{\gamma}\left[\lambda + \tau(\lambda - \mu + r)^2\right]. \qquad (9.25)$$

Hence, the expectation of the realized variance, or say the fair delivery price K_{var} of *variance swap for stochastic volatility with delay and Poisson jump in general case under risk-neutral measure* $\mathbb{P}^*$ equals to

$$\begin{aligned} K_{var} &= \mathbb{E}^*[v] = \frac{1}{T}\int_0^T v(t)dt \\ &\approx V + \frac{\alpha}{\gamma}\left[\lambda + \tau(\lambda - \mu + r)^2\right] + C\frac{1 - e^{-\gamma T}}{\gamma T} \end{aligned} \qquad (9.26)$$

where C is give by (9.25).

Of course, the equation (9.26) can also be written as:

$$K_{var} \approx X + (\sigma_0^2 - X)\frac{1 - e^{-\gamma T}}{\gamma T} \qquad (9.27)$$

and the price P of a variance swap at time t given delivery price K in this case should be:

$$P \approx e^{-r(T-t)}\left[X - K + (\sigma_0^2 - X)\frac{1 - e^{-\gamma T}}{\gamma T}\right]$$

where X is given by (9.20).

9.3.2 Compound Poisson Process Case

In the Section we will consider the jumps represented by a compound Poisson precess, since it allows the jumps size to be a random number but not always one in Poisson process, the model is more realistic. Our approach in the last section can be easily used in compound Poisson process case.

In the risk-neutral world, the volatility can be defined as follows:

$$\frac{d\sigma^2(t, S_t)}{dt} = \gamma V + \frac{\alpha}{\tau} \left[\int_{t-\tau}^{t} \sigma(s, S_s) dW^*(s) + \int_{t-\tau}^{t} y_s dN(s) - (\mu - r)\tau \right]^2$$

$$- (\alpha + \gamma)\sigma^2(t, S_t) \tag{9.28}$$

where $W^*(t)$ is a Brownian motion, $N(t)$ is a Poisson process with intensity λ and y_t is the jump size at time t which is identically independent distributed random variable. We assume that the mean of y_t is ξ and the variance of y_t is η. Note that the Poisson intensity λ and the jump size y_t does not change in risk-neutral world, since they are independent to the Bownian motion.

Our first step is still take the expectation under risk-neutral probability $\mathbb{P}^*$ on both sides of the equation. Note that the Brownian motion and the compound Poisson process are independent. Let $v(t) = \mathbb{E}^*[\sigma^2(t, S_t)]$, we obtain the following equation:

$$\frac{dv(t)}{dt} = \gamma V + \frac{\alpha}{\tau} \left[\int_{t-\tau}^{t} v(s)ds + Var^*\left(\sum_{t-\tau \leq s \leq t} y_s \right) + \left(\mathbb{E}^*\left(\sum_{t-\tau \leq s \leq t} y_s \right) \right)^2 \right.$$

$$\left. + (\mu - r)^2\tau^2 - 2\mathbb{E}^*\left(\sum_{t-\tau \leq s \leq t} y_s \right)(\mu - r)\tau \right] - (\alpha + \gamma)v(t)$$

$$= \gamma V + \alpha\lambda(\xi^2 + \eta) + \alpha\lambda^2\tau\xi^2 - 2\alpha\lambda\tau\xi(\mu - r) + \alpha\tau(\mu - r)^2 + \frac{\alpha}{\tau} \int_{t-\tau}^{t} v(s)ds$$

$$- (\alpha + \gamma)v(t). \tag{9.29}$$

From this equation, if $\xi = 1$ and $\eta = 0$, the compound Poisson process is just a Poisson process, then the equation (9.26) becomes to:

$$\frac{dv(t)}{dt} = \gamma V + \alpha\lambda + \alpha\lambda^2\tau - 2\alpha\lambda\tau(\mu - r) + \alpha\tau(\mu - r)^2 + \frac{\alpha}{\tau} \int_{t-\tau}^{t} v(s)ds$$

$$- (\alpha + \gamma)v(t)$$

which is the same as equation (9.8).

The equation (9.26) has a stationary solution

$$v(t) \equiv X = V + \left[\alpha\lambda(\xi^2 + \eta) + \alpha\lambda^2\tau\xi^2 - 2\alpha\lambda\tau\xi(\mu - r) + \alpha\tau(\mu - r)^2 \right]/\gamma$$

$$= V + \frac{\alpha}{\gamma} \left[\lambda(\xi^2 + \eta) + \tau(\lambda\xi - \mu + r)^2 \right]. \tag{9.30}$$

Hence, the expectation of the realized variance, or say the fair delivery price K_{var} of *variance swap for stochastic volatility with delay and compound Poisson jump in stationary regime under risk-neutral measure* $\mathbb{P}^*$ equals to

$$K_{var} = \mathbb{E}^*[v] = \frac{1}{T} \int_0^T v(t)dt$$

$$= V + \frac{\alpha}{\gamma} \left[\lambda(\xi^2 + \eta) + \tau(\lambda\xi - \mu + r)^2 \right] \tag{9.31}$$

and the price P of a variance swap at time t given delivery price K in this case should be:

$$P = e^{-r(T-t)} \left\{ V - K + \frac{\alpha}{\gamma} \left[\lambda(\xi^2 + \eta) + \tau(\lambda\xi - \mu + r)^2 \right] \right\}.$$

In general case, we substitute $v(t) = X + Ce^{\rho t}$ in (9.26) where X is defined in (9.30). Then the characteristic equation for ρ is:

$$\rho = \frac{\alpha}{\rho\tau}(1 - e^{-\rho\tau}) - (\alpha + \gamma) \tag{9.32}$$

which is also the same as it in the last section.

Therefore, the only solution to this equation is $\rho \approx -\gamma$, and by the same method, we have

$$v(t) \approx X + Ce^{-\gamma t}$$

$$= V + \frac{\alpha}{\gamma} \left[\lambda(\xi^2 + \eta) + \tau(\lambda\xi - \mu + r)^2 \right] + Ce^{-\gamma t} \tag{9.33}$$

and

$$C = \sigma_0^2 - V - \frac{\alpha}{\gamma} \left[\lambda(\xi^2 + \eta) + \tau(\lambda\xi - \mu + r)^2 \right]. \tag{9.34}$$

Hence, the expectation of the realized variance, or say the fair delivery price K_{var} of *variance swap for stochastic volatility with delay and compound Poisson jump in general case under risk-neutral measure* $\mathbb{P}^*$ equals to

$$K_{var} = \mathbb{E}^*[v] = \frac{1}{T} \int_0^T v(t)dt$$

$$\approx V + \frac{\alpha}{\gamma} \left[\lambda(\xi^2 + \eta) + \tau(\lambda\xi - \mu + r)^2 \right] + C\frac{1 - e^{-\gamma T}}{\gamma T} \tag{9.35}$$

where C is given by (9.34).

Of course, the equation (9.35) can also be written as:

$$K_{var} \approx X + (\sigma_0^2 - X)\frac{1 - e^{-\gamma T}}{\gamma T} \tag{9.36}$$

and the price P of a variance swap at time t given delivery price K in this case should be:

$$P \approx e^{-r(T-t)} \left[X - K + (\sigma_0^2 - X)\frac{1 - e^{-\gamma T}}{\gamma T} \right]$$

where X is given by (9.30).

It is interesting to see that when $\tau = 0$, which means there is no delay in the model, we have

$$\mathbb{E}^*[v] \approx \frac{1 - e^{-\gamma T}}{\gamma T}\left(\sigma_0^2 - V - \frac{\alpha\lambda(\xi^2 + \eta)}{\gamma}\right) + V + \frac{\alpha\lambda(\xi^2 + \eta)}{\gamma}. \tag{9.37}$$

9.3.3 *More General Case*

In the previous Section, we assume that the mean value and variance of the jump size y_t in the compound Poisson process are constants. Now we consider a more general case in which they are deterministic functions. The approach used in this Section is different from the previous ones, which is a more general method and can be applied to derive the same formulae in the previous simple cases.

In the risk-neutral world, the volatility still satisfies the following equation:

$$\frac{d\sigma^2(t, S_t)}{dt} = \gamma V + \frac{\alpha}{\tau}\left[\int_{t-\tau}^t \sigma(s, S_s)dW^*(s) + \int_{t-\tau}^t y_s dN(s) - (\mu - r)\tau\right]^2$$
$$- (\alpha + \gamma)\sigma^2(t, S_t) \tag{9.38}$$

where $W^*(t)$ is a Brownian motion, $N(t)$ is a Poisson process with intensity λ and y_t is the jump size at time t. We assume that $\mathbb{E}[y_t] = A(t)$, $\mathbb{E}[y_s y_t] = C(s, t), s < t$ and $\mathbb{E}[y_t^2] = B(t) = C(t, t)$, where $A(t), B(t), C(s, t)$ are all deterministic functions. Note that the change of measure does not change the Poisson intensity λ and the distribution of jump size y_t, since they are independent to the Brownian motion.

Let $v(t) = \mathbb{E}^*[\sigma^2(t, S_t)]$ and take the expectation under risk-neutral probability $\mathbb{P}^*$ on both sides of the equation (9.38). Note that the Brownian motion and the Poisson process are independent, we obtain the following equation:

$$\frac{dv(t)}{dt} = \gamma V + \frac{\alpha}{\tau}\left[\int_{t-\tau}^t v(s)ds + \mathbb{E}^*\left(\int_{t-\tau}^t y_s dN(s)\right)^2 + (\mu - r)^2\tau^2\right.$$
$$\left. - 2\mathbb{E}^*\left(\int_{t-\tau}^t y_s dN(s)\right)(\mu - r)\tau\right] - (\alpha + \gamma)v(t). \tag{9.39}$$

In order to compute the two expectations in this equation, we first introduce two lemmas as follows (see Lamberton and Lapeyre (1996)):

Lemma 1. *Define* $I(t) = \int_0^t y_s d(N(s) - \lambda s)$, *then* $I(t)$ *is a martingale and* $\mathbb{E}I(t) = 0$.

Lemma 2. *Define* $I(t) = \int_0^t y_s d(N(s) - \lambda s)$, *then* $\mathbb{E}I^2(t) = \lambda\mathbb{E}\int_0^t y_s^2 ds$.

Therefore,

$$\mathbb{E}^*\left(\int_{t-\tau}^t y_s dN(s)\right) = \mathbb{E}^*\left(\int_{t-\tau}^t y_s d(N(s)-\lambda s)\right) + \mathbb{E}^*\left(\int_{t-\tau}^t y_s d\lambda s\right)$$

$$= \lambda\mathbb{E}^*\left(\int_{t-\tau}^t y_s ds\right)$$

$$= \lambda\int_{t-\tau}^t A(s)ds \tag{9.40}$$

and

$$\mathbb{E}^*\left(\int_{t-\tau}^t y_s dN(s)\right)^2 = \mathbb{E}^*\left(\int_{t-\tau}^t y_s d(N(s)-\lambda s)\right)^2 + \mathbb{E}^*\left(\int_{t-\tau}^t y_s d\lambda s\right)^2$$

$$= \lambda\int_{t-\tau}^t \mathbb{E}^* y_s^2 ds + \lambda^2\mathbb{E}^*\left(\int_{t-\tau}^t y_s ds\right)^2$$

$$= \lambda\int_{t-\tau}^t B(s)ds + \lambda^2\mathbb{E}^*\left(\int_{t-\tau}^t y_s ds\right)^2. \tag{9.41}$$

To compute $\mathbb{E}^*(\int_{t-\tau}^t y_s ds)^2$, we take the derivative of $(\int_{t-\tau}^t y_s ds)^2$ and then integral it:

$$d\left(\int_{t-\tau}^t y_s ds\right)^2 = 2\int_{t-\tau}^t y_s ds(y_t - y_{t-\tau})dt$$

$$= 2\int_{t-\tau}^t (y_s y_t - y_s y_{t-\tau})dsdt \tag{9.42}$$

$$\left(\int_{t-\tau}^t y_s ds\right)^2 = 2\int_0^t\int_{u-\tau}^u (y_s y_u - y_s y_{u-\tau})dsdu + \left(\int_{-\tau}^0 y_s ds\right)^2$$

$$= 2\int_0^t\int_{u-\tau}^u (y_s y_u - y_s y_{u-\tau})dsdu + \int_{-\tau}^0\int_{-\tau}^0 y_s y_u dsdu. \tag{9.43}$$

Now take the expectation under risk-neutral probability and we have

$$\mathbb{E}^*\left(\int_{t-\tau}^t y_s ds\right)^2 = 2\int_0^t\int_{u-\tau}^u (C(s,u)-C(s,u-\tau))dsdu + \int_{-\tau}^0\int_{-\tau}^0 C(s,u)dsdu$$

$$= K(t,\tau) + G \tag{9.44}$$

where $K(t,\tau) = 2\int_0^t\int_{u-\tau}^u (C(s,u)-C(s,u-\tau))dsdu$ and $G = \int_{-\tau}^0\int_{-\tau}^0 C(s,u)dsdu$.

Take into account equation (9.40), (9.41) and (9.44), equation (9.39) becomes to

$$\frac{dv(t)}{dt} = \gamma V + \frac{\alpha}{\tau}\left[\int_{t-\tau}^t v(s)ds + \lambda\int_{t-\tau}^t B(s)ds + \lambda^2(K(t,\tau)+G)\right.$$

$$\left. + (\mu-r)^2\tau^2 - 2\lambda\tau(\mu-r)\int_{t-\tau}^t A(s)ds\right] - (\alpha+\gamma)v(t). \tag{9.45}$$

We can check that the equation (9.26) in Section 9.3.2 is a special case of equation (11.21) with $A(t) = \mathbb{E}[y_t] = \xi$, $B(t) = \mathbb{E}[y_t^2] = Var[y_t] + (\mathbb{E}[y_t])^2 = \eta + \xi^2$ and $C(s,t) = \mathbb{E}[y_s y_t] = \mathbb{E}[y_s]\mathbb{E}[y_t] = \xi^2$.

To get the expectation of the realized variance in the risk-neutral world $\mathbb{E}^*[v]$, we have to find a solution to equation (9.45), a nonhomogeneous integro-differential equation with delay.

After taking the first derivative of this equation, we obtain

$$v''(t) = \frac{\alpha}{\tau}\Big[v(t) - v(t-\tau)\Big] - (\alpha + \gamma)v'(t) + h(t,\tau) \tag{9.46}$$

where $h(t,\tau) = \frac{\alpha}{\tau}\Big[\lambda\big(B(t) - B(t-\tau)\big) + \lambda^2 K'(t,\tau) - 2\lambda\tau(\mu - r)\big(A(t) - A(t-\tau)\big)\Big]$. This is a second order delay differential equation with constant coefficients, so Laplace transform can be applied to find its solution with initial condition $v(t) = \sigma(t, S_t), t \in [-\tau, 0]$, which is already known (see Hale and Lunel (1993) or Bellman and Cooke (1963)).

Let us denote the Laplace transform of a function $f(t)$ as:

$$\mathcal{L}\{f(t)\} = \int_0^\infty f(t)e^{-st}dt,$$

and do the Laplace transform for equation (9.46):

$$\mathcal{L}\{v''(t)\} = \frac{\alpha}{\tau}\Big[\mathcal{L}\{v(t)\} - \mathcal{L}\{v(t-\tau)\}\Big] - (\alpha + \gamma)\mathcal{L}\{v'(t)\} + \mathcal{L}\{h(t,\tau)\}. \tag{9.47}$$

By change of variable and the property of Laplace transform, the equation (9.47) yields:

$$\Big[s^2 + (\alpha + \gamma)s - \frac{\alpha}{\tau}(1 - e^{-s\tau})\Big]\mathcal{L}\{v(t)\} = v'(0) + (s + \alpha + \gamma)v(0)$$

$$- \frac{\alpha}{\tau}e^{-s\tau}\int_{-\tau}^0 v(t)e^{-st}dt + \mathcal{L}\{h(t,\tau)\}.$$

The characteristic function of equation (9.46) is:

$$C(s) = s^2 + (\alpha + \gamma)s - \frac{\alpha}{\tau}(1 - e^{-s\tau}) \approx s^2 + \gamma s$$

Therefore,

$$\mathcal{L}\{v(t)\} = C^{-1}(s)\Big[v'(0) + (s + \alpha + \gamma)v(0) - \frac{\alpha}{\tau}e^{-s\tau}\int_{-\tau}^0 v(t)e^{-st}dt + \mathcal{L}\{h(t,\tau)\}\Big]. \tag{9.48}$$

Applying the inverse transform (see Bellman and Cooke (1963)), we have:

$$v(t) \approx \frac{1 - e^{-\gamma t}}{\gamma}v'(0) + \Big[\frac{\alpha}{\gamma}(1 - e^{-\gamma t}) + 1\Big]v(0) - \frac{\alpha}{\gamma\tau}\int_{-\tau}^0 v(s)[1 - e^{-\gamma(t-s-\tau)}]ds$$

$$+ \frac{1}{\gamma}\int_0^t h(s,\tau)[1 - e^{-\gamma(t-s)}]ds + C. \tag{9.49}$$

By the initial condition,

$$C = \frac{\alpha}{\gamma\tau} \int_{-\tau}^{0} v(s)[1 - e^{\gamma(s+\tau)}]ds.$$

Hence, the expectation of the realized variance, or say the fair delivery price K_{var} of *variance swap for stochastic volatility with delay and compound Poisson jump under risk-neutral measure* $\mathbb{P}^*$ can be obtained by:

$$K_{var} = \mathbb{E}^*[v] = \frac{1}{T}\int_0^T v(t)dt \qquad (9.50)$$

and the price P of a variance swap at time t given delivery price K in this case should be:

$$P = e^{-r(T-t)}\{\mathbb{E}^*[v] - K\}.$$

Of course, the equation (9.46) can also be solved numerically.

Remark. From expression (9.43), we note that only a double integral need to be solved to get the expectation of the realized variance. If the functions $A(t), B(t)$ and $C(s,t)$ are in simple forms, say $\sin(t)$ or $\cos(t)$, the integral can even be solved explicitly. If we only consider the simple case in Section 4.1 or 4.2, the final analytical formulae only need algebraic computation and no numerical approximation is needed. Thus, we can say that the stochastic volatility model with delay and jumps is easy to implement and time saving.

9.4 Delay as a Measure of Risk

The compound Poisson case in Section 4.2 is a special case of the model we discussed in Section 4.3, but it is more basic and also shows some important characteristics. Therefore, in this Section, we focus on this case.

By Section 4.2, we obtain that:

$$\mathbb{E}^*[v] \approx V + \frac{\alpha}{\gamma}\left[\lambda(\xi^2 + \eta) + \tau(\lambda\xi - \mu + r)^2\right]$$

$$+ \left\{\sigma_0^2 - V - \frac{\alpha}{\gamma}\left[\lambda(\xi^2 + \eta) + \tau(\lambda\xi - \mu + r)^2\right]\right\}\frac{1 - e^{-\gamma T}}{\gamma T}. \qquad (9.51)$$

This expression contains all the information about our model, since it contains all the initial parameters. We note that $\sigma_0^2 = \sigma^2(0, \varphi(-\tau))$. So the sign of the second term in (9.51) depends on the relationship between σ_0^2 and $V + \frac{\alpha}{\gamma}\left[\lambda(\xi^2 + \eta) + \tau(\lambda\xi - \mu + r)^2\right]$.

If $\sigma_0^2 > V + \frac{\alpha}{\gamma}\left[\lambda(\xi^2 + \eta) + \tau(\lambda\xi - \mu + r)^2\right]$, the second term in (9.51) is positive and $\mathbb{E}^*[v]$ stays above $V + \frac{\alpha}{\gamma}\left[\lambda(\xi^2 + \eta) + \tau(\lambda\xi - \mu + r)^2\right]$, which means the risk is high.

If $\sigma_0^2 < V + \frac{\alpha}{\gamma} \left[\lambda(\xi^2 + \eta) + \tau(\lambda\xi - \mu + r)^2\right]$, the second term in (9.51) is negative and $\mathbb{E}^*[v]$ stays below $V + \frac{\alpha}{\gamma} \left[\lambda(\xi^2 + \eta) + \tau(\lambda\xi - \mu + r)^2\right]$, which means the risk is low.

Therefore,

$$\sigma_0^2 = V + \frac{\alpha}{\gamma} \left[\lambda(\xi^2 + \eta) + \tau(\lambda\xi - \mu + r)^2\right] \tag{9.52}$$

defines the measure of risk in the stochastic volatility model with delay and jumps.

To reduce the risk we need to take into account the following relationship with respect to the delay τ (which follows from (9.51)):

$$\tau > \frac{(\sigma_0^2 - V)r - \alpha\lambda(\xi^2 + \eta)}{\alpha(\lambda\xi - \mu + r)^2}. \tag{9.53}$$

9.5 Numerical Example

In this Section, we apply the analytical solutions from Section 4.2 to price the variance swap of the S&P60 Canada index for five years (January 1998–February 2002) (see Theoret, Zabre and Rostan (2002)).

At the end of February 2002, we wanted to price the fixed leg of a variance swap based on the *S&P*60 Canada index. The statistics on log returns *S&P*60 Canada Index for 5 years (January 1997–February 2002) is presented in Table 9.1:

Table 9.1

Statistics on Log Returns *S&P*60 **Canada Index**	
Series:	LOG RETURNS *S&P*60 CANADA INDEX
Sample:	1 1300
Observations:	1300
Mean	0.000235
Median	0.000593
Maximum	0.051983
Minimum	−0.101108
Std. Dev.	0.013567
Skewness	−0.665741
Kurtosis	7.787327

From the histogram of the *S&P*60 Canada index log returns on a 5-year historical period (1,300 observations from January 1998 to February 2002) it may be seen leptokurtosis in the histogram. If we take a look at the graph of the *S&P*60 Canada index log returns on a 5-year historical period we may see volatility clustering in the returns series. These facts indicate the conditional heteroscedasticity.

There are several parameters which need to be estimated from the data, the jump intensity λ, the mean value ξ and the variance η of the jump size. We use the following method to detect the jumps. If the difference between the spot log return and the mean of our log return series is larger than triple of the standard deviation, then we say that a jump occurs at that time point (see Clewlow and Strickland (2000)). We count the number of the jumps, denoted as N and the estimation of jump intensity $\bar{\lambda} = N/1300$. The jump size is defined as the difference between the log returns at a jump time point and the previous time point and the sample mean and variance of these data are unbias estimation of ξ and η.

From the data in Table 9.1, we get the following estimation:

$N = 15$; $\lambda = N/1300 = 0.0115$; $\xi = -0.003$ and $\eta = 0.0035$.

A GARCH(1,1) regression is applied to the series and the results is obtained as in Table 9.2:

Table 9.2

Estimation of the GARCH(1,1) process				
Dependent Variable: Log returns of S&P60 Canada Index Prices				
Method: ML-ARCH				
Included Observations: 1300				
Convergence achieved after 28 observations				
–	**Coefficient:**	**Std. error:**	**z-statistic:**	**Prob.**
C	0.000617	0.000338	1.824378	0.0681
Variance Equation				
C	2.58E-06	3.91E-07	6.597337	0
ARCH(1)	0.060445	0.007336	8.238968	0
GARCH(1)	0.927264	0.006554	141.4812	0
R-squared	−0.000791	Mean dependent var	–	0.000235
Adjusted R-squared	−0.003108	S.D. dependent var	–	0.013567
S.E. of regression	0.013588	Akaike info criterion	–	−5.928474
Sum squared resid	0.239283	Schwartz criterion	–	−5.912566
Log likelihood	3857.508	Durbin-Watson stat	–	1.886028

This table allows to generate different input variables to the volatility swap model.

We use the following relationship

$$\theta = \frac{V}{dt},$$

$$k = \frac{1 - \alpha - \beta}{dt},$$

to calculate the following discrete GARCH(1,1) parameters:

ARCH(1,1) coefficient $\alpha = 0.060445$;

GARCH(1,1) coefficient $\beta = 0.927264$;

GARCH(1,1) coefficient $\gamma = 0.012391$;

the Pearson kurtosis (fourth moment of the drift-adjusted stock return) $\varepsilon = 7.787327$;

long volatility $\theta = 0.05289724$;

$k = 3.09733$;

a short volatility σ_0 equals to 0.01;

$\mu = 0.000235$,

$r = 0.02$ and $\tau = 1(day)$.

Parameter V may be found from the expression $V = \frac{C}{1-\alpha-\beta}$, where $C = 2.58 \times 10^{-6}$ is defined in Table 2. Thus, $V = 0.00020991$;

$dt = 1/252 = 0.003968254$.

Now, applying the analytical solutions (9.35) for a variance swap maturity T of 1 year, we find the following value:

$$V + \frac{\alpha}{\gamma}[\lambda(\xi^2 + \eta) + \tau(\lambda\xi - \mu + r)^2]$$

$$= 0.0002 + 0.0604/0.0124 \times [0.0115 \times [(-0.003)^2 + 0.0035]$$

$$+ (0.0115 \times (-0.003) - 0.0002 + 0.0124)^2]$$

$$= 0.0023.$$

$$\mathbb{E}^*[v] \approx V + \frac{\alpha}{\gamma}\left[\lambda(\xi^2 + \eta) + \tau(\lambda\xi - \mu + r)^2\right]$$

$$+ \left\{\sigma_0^2 - V - \frac{\alpha}{\gamma}\left[\lambda(\xi^2 + \eta) + \tau(\lambda\xi - \mu + r)^2\right]\right\}\frac{1 - e^{-\gamma T}}{\gamma T}$$

$$= 0.0023 + (0.0001 - 0.0023) \times \frac{1 - e^{-0.0124}}{0.0124}$$

$$= 0.0001136.$$

Repeating this approach for a series of delays up to 5 days and series of maturities up to 100 years, we compare our results with the no jump model and obtain the following plots: (see Figures 9.1 and 9.2) of $S\&P60$ Canada Index Variance Swap. Figure 9.3 depicts the dependence of delivery price of variance swaps on jump intensity. Figures 9.4, 9.5 and 9.6 depict the dependence of delivery price of variance swaps on two variables of maturity, delay and jump intensity.

Figure 9.1 depicts the Dependence of Delivery Price on Maturity for $S\&P60$ Canada Index (1998–2002).

Figure 9.2 depicts the Dependence of Delivery Price on Delay for $S\&P60$ Canada Index (1998–2002).

Figure 9.3 depicts the Dependence of Delivery Price on Jump Intensity for $S\&P60$ Canada Index (1998–2002).

Figure 9.4 depicts the Dependence of Delivery Price on Delay and Jump Intensity for $S\&P60$ Canada Index (1998–2002).

Figure 9.5 depicts the Dependence of Delivery Price on Delay and Maturity for $S\&P60$ Canada Index (1998–2002).

Figure 9.6 depicts the Dependence of Delivery Price on Jump Intensity and Maturity for $S\&P60$ Canada Index (1998–2002).

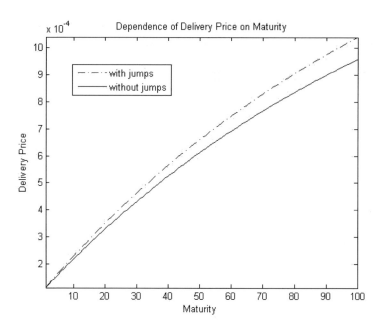

Fig. 9.1   Dependence of Delivery Price on Maturity (*S&P*60 Canada Index).

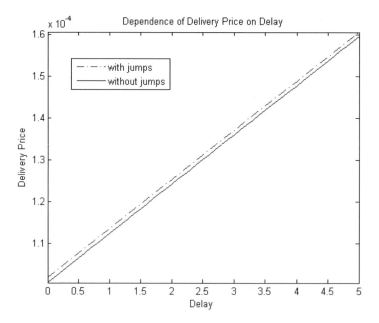

Fig. 9.2   Dependence of Delivery Price on Delay (*S&P*60 Canada Index).

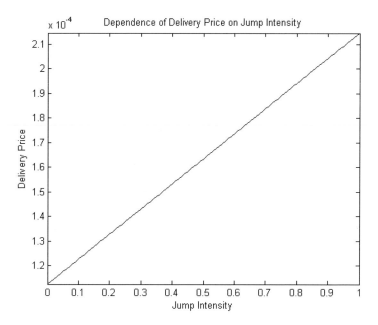

Fig. 9.3 Dependence of Delivery Price on Jump Intensity (*S&P*60 Canada Index).

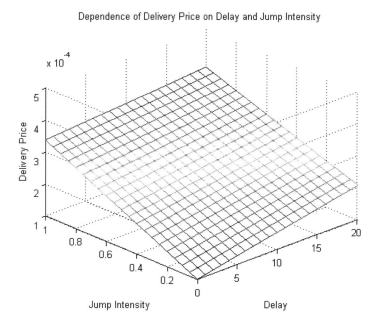

Fig. 9.4 Dependence of Delivery Price on Delay and Jump Intensity (*S&P*60 Canada Index).

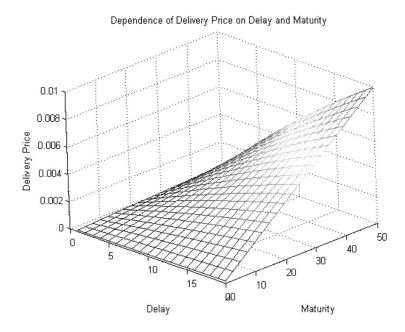

Fig. 9.5   Dependence of Delivery Price on Delay and Maturity (*S&P*60 Canada Index).

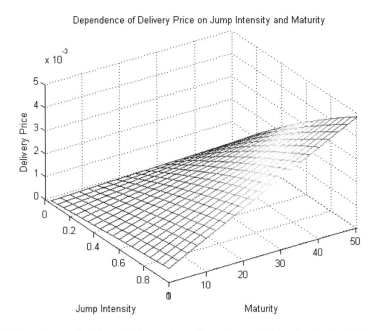

Fig. 9.6   Dependence of Delivery Price on Jump Intensity and Maturity (*S&P*60 Canada Index).

9.6 Summary

- In this Chapter we studied stochastic volatility model with delay and jumps to price variance swaps.
- We applied a general approach to derive the analytical close forms for expectation of the realized continuously sampled variance for stochastic volatility with delay and jumps. The jump part in our model is represented by a general version of compound Poisson processes.
- The key features of the model are the following: i) continuous-time analogue of discrete-time GARCH model; ii) mean-reversion; iii) contains the same source of randomness as stock price; iv) market is complete; v) incorporates the expectation of log-return; vi) incorporates the jumps in volatility.
- The model is also easy to implement and time saving.
- We also presented a lower bound for delay as a measure of risk.
- From the numerical example in Section 9.5, we found that after adding jumps in volatility, the expectation of the realized variance is higher than the one without jumps for variance swaps. It is easy to explain it since the existence of jumps means the market is more risky which asks for higher cost of variance swaps.

Bibliography

Avellaneda, M., Levy, A. and Paras, A. (1995). Pricing and hedging derivative securities in markets with uncertain volatility. *Applied Mathematical Finance*, 2, 73-88.

Bellman, R. and Cooke, K. (1963). *Differential-Difference Equations*. New York: Academic Press Inc.

Black, F. and Scholes, M. (1973). The pricing of options and corporate liabilities. *Journal of Political Economy*, 81, 637-54.

Bollerslev, T. (1986). Generalized autoregressive conditional heteroscedasticity. *Journal of Economics*, 31, 307-27.

Brockhaus, O. and Long, D. (2000). Volatility swaps made simple. *Risk Magazine*, January, 92-96.

Buff, R. (2002). *Uncertain Volatility Model. Theory and Applications*. New York: Springer.

Carr, P. and Madan, D. (1998). Towards a theory of volatility trading. In Jarrow, R. (ed.), *Volatility*, Risk Book Publications.

Clewlow, L. and Strickland, C. (2000). *Energy Derivatives*. Lacima Group.

Cont, R. and Tankov, P. (2003). *Financial Modelling with Jump Processes*. New Jersey: Chapman & Hall/CRC Press.

Demeterfi, K., Derman, E., Kamal, M. and Zou, J. (1999). A guide to volatility and variance swaps. *The Journal of Derivatives*, Summer, 9-32.

Duan, J. (1995). The GARCH option pricing model, *Mathematical Finance*, 5, 13-32.

Elliott, R. and Swishchuk, A. (2007). Pricing Options and Variance Swaps in Brownian and Fractional Brownian Markets. *IMA Journal of Management Mathematics* (Special Issue on Financial Mathematics).

Elliott, R. and Swishchuk, A. (2007). Pricing options and variance swaps in Markov-modulated Brownian markets. In Elliott, R. and Mamon, R. (eds.), *Hidden Markov Models in Finance*. New York: Springer.

Fouque, J.-P., Papanicolaou, G. and Sircar, K. R. (2000). *Derivatives in Financial Markets with Stochastic Volatilities*. Berlin-Heidelberg: Springer-Verlag.

Griego, R. and Swishchuk, A. (2000). Black-Scholes formula for a market in a Markov environment. *Theory of Probability and Mathematical Statistics*, 62, 9-18.

Hale, J. and Lunel, S. (1993). *Introduction to Functional Differential Equations*. Berlin-Heidelberg: Springer-Verlag.

Heston, S. (1993). A closed-form solution for options with stochastic volatility with applications to bond and currency options. *Review of Financial Studies*, 6, 327-343.

Heston, S. and Nandi, S. (1998). Preference-free option pricing with path-dependent volatility: A closed-form approach. Discussion paper, Federal Reserve Bank of Atlanta.

Howison, S., Rafailidis, A. and Rasmussen, H. (2004). Pricing and hedging of volatility derivatives. *Applied Mathematical Finance*, 11, 4, 317-348.

Hull, J. (2000). *Options, Futures and Other Derivatives* (4th edition). New Jersey: Prentice Hall.

Hull, J., and White, A. (1987). The pricing of options on assets with stochastic volatilities. *Journal of Finance*, 42, 281-300.

Javaheri, A., Wilmott, P. and Haug, E. (2002). GARCH and volatility swaps. *Wilmott Technical Article*, January, 17 pages.

Kallsen, Taqqu (1995). Option Pricing in ARCH-type Models: With Detailed Proofs. Technical Report No. 10, University of Freiburg.

Kazmerchuk, Y., Swishchuk, A., and Wu, J. (2006). The pricing of options for security markets with delayed response. *Mathematics and Computers in Simulation*, 75, 3-4, 69-79.

Kazmerchuk, Y., Swishchuk, A. and Wu, J. (2005). A continuous-time GARCH model for stochastic volatility with delay. *Canadian Applied Mathematics Quarterly*, 13, 2.

Lamberton, D. and Lapeyre, B. (1996). *Introduction to Stochastic Calculus Applied to Finance*. New Jersey: Chapman & Hall.

Merton, R. (1973). Theory of rational option pricing. *Bell Journal of Economic Management Science*, 4, 141-183.

Mohammed, S.E.A, Arriojas, M. and Pap, Y. (2001). A delayed Black and Scholes formula. Preprint, Southern Illinois University.

Mohammed, S.E.A. (1998). Stochastic differential systems with memory: Theory, examples and applications, In *Stochastic Analysis and Related Topics*, 6, 1-77.

Schoutens, W., Simons, E. and Tistaert, J. (2003). A perfect calibration! Now what? Working paper, November 20, 30 pages.

Shreve, S.E. (2005). *Stochastic Calculus for Finance (Volumes I and II)*. New York: Springer.

Swishchuk, A. (1995). Hedging of options under mean-square criterion and with semi-Markov volatility. *Ukrainian Mathematicals Journal*, 47, 7, 1119-1127.

Swishchuk, A. (2000). *Random Evolutions and Their Applications. New Trends*. Dordrecht: The Netherlands: Kluwer Academic Publishers.

Swishchuk, A. (2004). Modeling of variance and volatility swaps for financial markets with stochastic volatilities. *Wilmott Magazine*, September Issue, Technical article No. 2, 64-72.

Swishchuk, A. (2005). Modeling and pricing of variance swaps for stochastic volatilities with delay. *Wilmott Magazine*, Issue 19, September 2005, pp. 63-73.

Swishchuk, A. (2006). Modeling and pricing of variance swaps for multi-factor stochastic volatilities with delay. *Canadian Applied Mathematics Quarterly*, 14, 4, 439-468.

Theoret, R., Zabre, L. and Rostan, P. (2002). Pricing volatility swaps: Empirical testing with Canadian data. Working Paper, Centre de Recherche en Gestion, Document 17-2002, July.

Wilmott, P. (2000). *Quantitative Finance (Volumes I and II)*. Chichester, West Sussex, England: Wiley.

Wilmott, P. (1999). *Derivatives: The Theory and Practice of Financial Engineering*. Chichester, West Sussex, England: Wiley.

Wilmott, P., Howison, S. and Dewynne, J. (1995). *Option Pricing: Mathematical Models and Computations*. Oxford: Oxford Financial Press.

Chapter 10

Variance Swap for Local Lévy-Based Stochastic Volatility with Delay

10.1 Introduction

The valuation of the variance swaps for local Lévy-based stochastic volatility with delay (LLBSVD) is discussed in this Chapter. We provide some analytical closed forms for the expectation of the realized variance for the LLBSVD. As applications of our analytical solutions, we fit our model to 10 years of S&P500 data (2000-01-01–2009-12-31) with variance gamma model and apply the obtained analytical solutions to price the variance swap.

The key risk factors considered in option pricing models, besides the diffusive price risk of the underlying asset, are stochastic volatility and jumps, both in the asset price and its volatility. Models that include some or all of these factors were developed in Merton (1973), Heston (1993), Bates (1996), Bakshi *et al.* (1997) and Duffie *et al.* (2000).

The importance of jumps in volatility has become apparent in recent studies, which try to explain the time series properties of both stock and option prices, like Eraker *et al.* (2003), or Broadie *et al.* (2007). The jumps in stock market volatility are found to be so active that this discredits many recently proposed stochastic volatility models without jumps (Bollerslev *et al.*, 2008). There is currently fairly compelling evidence for jumps in the level of financial prices. The most convincing evidence comes from recent nonparametric work using high-frequency data as in Barndorff-Nielsen and Shephard (2005), and Ait-Sahalia and Jacod (2008) among others. Also, paper Todorov and Tauchen (2008) conducts a non-parametric analysis of the market volatility dynamics using high-frequency data on the VIX index compiled by the CBOE and the S&P500 index. The results in Eraker *et al.* (2003) show that the jump-in-volatility models provide a significant better fit to the returns data. They use returns data to investigate the performance of models with jumps in volatility using the class of jump-in-volatility models proposed by Duffie *et al.* (2000). Technical issues aside, jumps are important because they represent a significant source of non-diversifiable risk as discussed at length in Bollerslev *et al.* (2008). The dynamics of return and volatility with Lévy jumps and time-changed Lévy process were considered in Junye (2009) and Li *et al.* (2007), respectively.

137

From the other side, some statistical studies of stock prices (see (Sheinkman and LeBaron, 1989) and (Akrigay, 2003)) indicate the dependence on past returns. A diffusion approximation result for processes satisfying some equations with past-dependent coefficients obtained in (Kind *et al.*, 1991), and this result they applied to a model of option pricing, in which the underlying asset price volatility depends on the past evolution to obtain a generalized (asymptotic) Black-Scholes formula. Hobson and Rogers (1998) suggested a new class of nonconstant volatility models, which can be extended to include the aforementioned level-dependent model and share many characteristics with the stochastic volatility model.

In this chapter, we incorporate a jump part into the stochastic volatility model with delay (and without jumps) proposed in Swishchuk (2005). The stock price $S(t)$ satisfies the following equation

$$dS(t) = \mu S(t)dt + \sigma(t, S_t)S(t)dW(t), \quad t > 0,$$

where $\mu \in R$ is the mean rate of return, the volatility term $\sigma > 0$ is a bounded function and $W(t)$ is a Brownian motion on a probability space $(\Omega, \mathcal{F}, P)$ with a filtration $\mathcal{F}_t$. We also let $r > 0$ be the risk-free rate of return of the market. We denote $S_t = S(t - \tau)$, $\quad t > 0$ and the initial data of $S(t)$ is defined by $S(t) = \varphi(t)$, where $\varphi(t)$ is a deterministic function with $t \in [-\tau, 0]$, $\quad \tau > 0$. The volatility $\sigma(t, S_t)$ satisfies the following equation:

$$\frac{d\sigma^2(t, S_t)}{dt} = \gamma V + \frac{\alpha}{\tau} \left[\int_{t-\tau}^{t} \sigma(u, S_u)dL(u) \right]^2$$

$$- (\alpha + \gamma)\sigma^2(t, S_t)$$

where $L(t)$ is a Lévy process independent of $W(t)$ with Lévy triplet (a, γ, ν). Here, $V > 0$ is a mean-reverting level (or long-term equilibrium of $\sigma^2(t, S_t)$), $\alpha, \gamma > 0$, and $\alpha + \gamma < 1$.

Our model of stochastic volatility exhibits jumps and also past-dependence: the behavior of a stock price right after a given time t not only depends on the situation at t, but also on the whole past (history) of the process $S(t)$ up to time t. This draws some similarities with fractional Brownian motion models (see Mandelbrot, 1997) due to a long-range dependence property. Another advantage of this model is mean-reversion. This model is also a continuous-time version of GARCH(1,1) model (see Bollerslev, 1986) with jumps:

$$\sigma_n^2 = \gamma V + \alpha \ln^2(S_{n-1}/S_{n-2}) + (1 - \alpha - \gamma)\sigma_{n-1}^2$$

or, more general,

$$\sigma_n^2 = \gamma V + \frac{\alpha}{l} \ln^2(S_{n-1}/S_{n-1-l}) + (1 - \alpha - \gamma)\sigma_{n-1}^2.$$

If we write down the last equation in differential form we can get the continuous-time GARCH with expectation of log-returns of zero:

$$\frac{d\sigma^2(t)}{dt} = \gamma V + \frac{\alpha}{\tau} \ln^2\left(\frac{S(t)}{S(t - \tau)}\right) - (\alpha + \gamma)\sigma^2(t).$$

If we incorporate non-zero expectation of log-return (using Itô Lemma for $\ln \frac{S(t)}{S(t-\tau)}$), then we arrive to our continuous-time GARCH model for stochastic volatility with delay (see Swishchuk, 2005):

$$\frac{d\sigma^2(t, S_t)}{dt} = \gamma V + \frac{\alpha}{\tau} \left[\int_{t-\tau}^{t} \sigma(s, S_s) dW(s) \right]^2 - (\alpha + \gamma)\sigma^2(t, S_t).$$

We note, that paper by Swishchuk (2005) studied the case when $\lambda = 0$ (model without jumps). The paper by Swishchuk and Li (2007) investigated the case with pure Poisson jumps in the form of $\int_{t-\tau}^{t} dN(s)$, compound Poisson jumps in the form of $\int_{t-\tau}^{t} y_s dN(s)$, and more general case with jump sizes y_t that have finite mean and variance. The chapter by Swishchuk (2009) incorporates the case of jumps into the model Swishchuk (2005) in the form of the following integral $\int_{t-\tau}^{t} \sigma(s, S_s) d\tilde{N}(s)$. As long as the stochastic volatility $\sigma(t, S_t)$ depends on t and S_t, has Lévy process as a random factor and delay we call it *local Lévy based stochastic volatility with delay* (LLBSVD).

10.2 Variance Swap for Lévy-Based Stochastic Volatility with Delay

Suppose the asset volatility is defined as the solution of the following delay equation,

$$d\sigma^2(t, S_t)/dt = \gamma V + \frac{\alpha}{\tau} \left[\int_{t-\tau}^{t} \sigma(s, S_s) dL(s) \right]^2 - (\alpha + \gamma)\sigma^2(t, S_t) \qquad (10.1)$$

where L is a Lévy process (see (Schoutens, 2003)). From (Protter, 2003) we have

$$[\sigma \cdot L, \sigma \cdot L]_t = \int_0^t \sigma_s^2 d[L, L]_s$$

hence

$$E\left(\left[\int_{t-\tau}^{t} \sigma(s, S_s) dL(s) \right]^2 \right) = E\left(\int_{t-\tau}^{t} \sigma_s^2 d[L, L]_s \right). \qquad (10.2)$$

From Cont and Tankov (2003), if L is a Lévy process with characteristic triplet (a, γ, ν) we have

$$[L, L]_t = a^2 t + \int_{[0,t]} \int_R y^2 J_L(ds, dy) \qquad (10.3)$$

where J_L is the jump measure of L. So we have

$$E\left(\left[\int_{t-\tau}^{t} \sigma(s, S_s) dL(s) \right]^2 \right) = E\left(\int_{t-\tau}^{t} a^2 \sigma_s^2 ds + \int_{[0,t]} \int_R \sigma_s^2 y^2 J_L(ds, dy) \right)$$

with $\sigma_s^2 \geq 0$ applying Fubini's Theorem we get

$$E\left(\left[\int_{t-\tau}^{t} \sigma(s, S_s) dL(s) \right]^2 \right) = E\left(\int_{t-\tau}^{t} a^2 \sigma_s^2 ds + \int_{t-\tau}^{t} \sigma_s^2 ds \int_R y^2 \nu(dy) \right)$$

$$= E\left(\int_{t-\tau}^{t} \sigma_s^2 ds \right) \left[a^2 + \int_R y^2 \nu(dy) \right].$$

Taking the expectation of (10.1) and denoting $v(t) = E[\sigma^2(t, S_t)]$ we get

$$\frac{dv(t)}{dt} = \gamma V + \alpha\tau(\mu - r)^2 + \frac{\alpha}{\tau}\left(a^2 + \int_R y^2\nu(dy)\right)\int_{t-\tau}^t v(s)ds - (\alpha + \gamma)v(t). \quad (10.4)$$

This has a stable solution of

$$v(t) \equiv X = \frac{\gamma V + \alpha\tau(\mu - r)^2}{\alpha + \gamma - \alpha\left(a^2 + \int_R y^2\nu(dy)\right)}.$$

As an approximate solution we assume $v(t) = X + Ce^{\rho t}$, substituting into (10.4), the characteristic equation for ρ is

$$\rho = \frac{\alpha\left(a^2 + \int_R y^2\nu(dy)\right)}{\rho\tau}\left(1 - e^{-\rho\tau}\right) - \gamma - \alpha.$$

Approximating $e^{-\rho\tau} \approx 1 - \rho\tau$ we get

$$\rho = \alpha\left(a^2 + \int_R y^2\nu(dy) - 1\right) - \gamma.$$

Now with $v(0) = \sigma_0^2$ we have

$$C = \sigma_0^2 - \frac{\gamma V + \alpha\tau(\mu - r)^2}{\alpha + \gamma - \alpha\left(a^2 + \int_R y^2\nu(dy)\right)}.$$

In this way, we obtained the following result.

Theorem (Variance Swap for LLBSVD). The general approximated solution for (10.4) has the following form:

$$v(t) \approx X + Ce^{\rho t}$$

$$= \frac{\gamma V + \alpha\tau(\mu - r)^2}{\alpha + \gamma - \alpha\left(a^2 + \int_R y^2\nu(dy)\right)} + \left[\sigma_0^2 - \frac{\gamma V + \alpha\tau(\mu - r)^2}{\alpha + \gamma - \alpha\left(a^2 + \int_R y^2\nu(dy)\right)}\right]$$

$$\times \exp\left[\alpha\left(a^2 + \int_R y^2\nu(dy) - 1\right) - \gamma\right]t.$$

Remark 1. As we mentioned in Introduction, the Lévy process $L(t)$ in our Lévy-based stochastic volatility model with delay is independent of the Wiener process $W(t)$ in the stock price model. The leverage effect for our model may be considered in the following way.

Let $B(t)$ be a Wiener process from the Itô-Lévy decomposition of the Lévy process $L(t)$, (for example, if we take Kou's jump-diffusion, Kou (2002) and let $B^1(t)$ be another Wiener process independent of $B(t)$. Then we can take for $W(t)$ the following process:

$$dW(t) = \rho dB(t) + \sqrt{1 - \rho^2}dB^1(t).$$

Obviously, this is a Wiener process and $dW(t)dB(t) = \rho dt$, where ρ is the leverage parameter. In this way, we can incorporate this case in our study as well (leaving it for our future work).

Remark 2. It is well known, that time-changed Lévy processes are more accurate than ordinary Lévy processes when describing the dynamics of return or volatility (see, e.g., Junye (2009) and Li *et al.* (2007)). Let us show how we can incorporate the time-changed Lévy process in our model for our Lévy-based stochastic volatility with delay.

Let (a, γ, ν) be a Lévy triplet for Lévy process $L(t)$ and let $(0, b, \mu)$ be a Lévy triplet for a subordinator $T(t)$. Then the Lévy triplet (a^Y, γ^Y, ν^Y) for the time-changed Lévy process $Y(t) := L(T(t))$ is:

$$a^Y = ba,$$

$$\gamma^Y = b\gamma + \int_0^{+\infty} \mu(ds) \int_{|x| \leq 1} x p_s^L(dx),$$

$$\nu^Y(dy) = b\nu(dy) + \int_0^{+\infty} p_s^L(dy)\mu(ds),$$

where p_s^L is a probability distribution of $L(s)$.

In this way, all our calculations may be adjusted with respect to the new triplet (a^Y, γ^Y, ν^Y) of the time-changed Lévy process $Y(t) := L(T(t))$. For example, the expression for the second moment of the delayed integral with respect to $Y(t)$ has the following expression (see equation (10.2)):

$$E\left(\left[\int_{t-\tau}^t \sigma(s, S_s)dY(s)\right]^2\right) = E\left(\int_{t-\tau}^t \sigma^2(s, S_s)ds\right)$$

$$\times \left[a^2 b^2 + \int y^2 \left(b\nu(dy) + \int_0^{+\infty} p_s^L(dy)\mu(ds)\right)\right].$$

10.3 Examples

10.3.1 *Example 1 (Variance Gamma)*

Consider a Variance Gamma process with the CGM parameterization (see Schoutens (2003)), that is, with characteristic function of the form

$$\phi_{VG}(u; C, G, M) = \left(\frac{GM}{GM + (M-G)iu + u^2}\right)^C$$

where $C > 0$, $G > 0$, and $M > 0$. With Lévy measure

$$\nu_{VG}(x) = C|x|^{-1}(\exp(Gx)1_{(X<0)} + \exp(-Mx)1_{(X>0)})dx$$

we have

$$\int_R x^2 \nu_{VG}(x) = \frac{C}{M^2} - \frac{C}{G^2}$$

so

$$X = \frac{\gamma V + \alpha\tau(\mu - r)^2}{\alpha + \gamma - \alpha\left(\frac{C}{M^2} - \frac{C}{G^2}\right)}$$

and

$$\rho = \alpha \left(\frac{C}{M^2} - \frac{C}{G^2} - 1 \right) - \gamma.$$

10.3.2   *Example 2 (Tempered Stable)*

The Tempered Stable distribution (see Schoutens (2003)) has the characteristic function

$$\phi_{\text{TS}}(u; \kappa, a, b) = \exp(ab - a(b^{1/\kappa} - 2iu)^\kappa)$$

where $a > 0$, $b \geq 0$, and $0 < \kappa < 1$. Here

$$\nu_{TS}(\text{x}) = a 2^\kappa \frac{\kappa}{\Gamma(1 - \kappa)} x^{-\kappa - 1} \exp\left(-\frac{1}{2} b^{1/\kappa} x \right) 1_{(x>0)} \mathrm{d}x$$

and hence if $b > 0$ then

$$\int_R x^2 \nu_{TS}(\text{x}) = \frac{2^{\kappa+4}\, a\, \Gamma(\kappa + 1)\, \sin(\pi\,\kappa)}{\pi\, b^{\frac{3}{\kappa}}}$$

so

$$X = \frac{\pi\, b^{\frac{3}{\kappa}} \left(\gamma\, V + \alpha\, \tau\, (\mu - r)^2 \right)}{\pi\, b^{\frac{3}{\kappa}} (\alpha + \gamma) - 2^{\kappa+4}\, \alpha\, a\, \sin(\pi\,\kappa)\, \Gamma(\kappa + 4)}$$

with

$$\rho = \alpha \left(\frac{2^{\kappa+4}\, a\, \Gamma(\kappa + 1)\, \sin(\pi\,\kappa)}{\pi\, b^{\frac{3}{\kappa}}} - 1 \right) - \gamma.$$

10.3.3   *Example 3 (Jump-Diffusion)*

Consider a process with characteristic triplet $(1, 0, \lambda\delta(1))$ we then have

$$\int_R x^2 \nu(\text{x}) = \lambda$$

so

$$X = \frac{\gamma V + \alpha\tau(\mu - r)^2}{\alpha + \gamma - \alpha(1 + \lambda)} = \frac{\gamma V + \alpha\tau(\mu - r)^2}{\gamma - \alpha\lambda}$$

with

$$\rho = \alpha\lambda - \gamma.$$

We got the result obtained in Swishchuk (2005).

10.3.4 *Example 4 (Kou's Jump-Diffusion)*

Kou's Jump-diffusion ([Kou (2002)]) has a characteristic triplet $(1, 0, \lambda f(x))$ where

$$f(x) = p\eta_1 e^{-\eta_1 x} 1_{x \geq 0} + q\eta_2 e^{\eta_2 x} 1_{<0}$$

with $\eta_1 > 1, \eta_2 > 0$, and $p, q \geq 0$, $p + q = 1$. Here we have

$$\int_R x^2 \nu(\underline{x}) = 2\lambda \left(\frac{p}{\eta_1^2} + \frac{q}{\eta_2^2} \right)$$

so

$$X = \frac{\gamma V + \alpha \tau (\mu - r)^2}{\gamma - 2\alpha\lambda \left(\frac{p}{\eta_1} + \frac{q}{\eta_2} \right)}$$

with

$$\rho = 2\alpha\lambda \left(\frac{p}{\eta_1} + \frac{q}{\eta_2} \right).$$

10.4 Parameter Estimation

As in Kazmerchuk *et al.* (2005) we consider a Maximum Likelihood for the estimation of the parameters in (10.1). The discrete time analogue of (10.1) is given by

$$\sigma_n^2 = \omega + \frac{\alpha}{l} \left(\sum_{i=1}^{l} \epsilon_{n-i} \right)^2 + \beta \sigma_{n-1}^2$$

where

$$\epsilon_n = y_n - \mu$$

and $\epsilon = \ln(S_n/S_{n-1})$ are the log-returns. Furthermore as in the GARCH model we have $\alpha + \beta + \gamma < 1$ and $l \geq 1$ is our discrete delay parameter. Here we assume ϵ_n follows a distribution with a probability density function given by $f(x; \theta)$ where θ is a vector of parameters introduced by the distribution of ϵ_n. As in Konlack *et al.* (2009) estimation of our model parameters then becomes an exercise in maximizing the likelihood function

$$L(\mu, \alpha, \beta, \omega, \theta) = \sum_{t=1}^{T} \left[\ln f(\epsilon_t \sigma_t^{-1}; \theta) - \ln \sigma_t \right]$$

for a given lag l. Following Kazmerchuk *et al.* (2005) we use the *Akaike's information criterion* to select an l. With L_{max} being the maximum likelihood we have

$$AICc = 2k - 2\ln(L_{max}) + \frac{2k(k+1)}{n - k - 1}$$

where k is the number of parameters $k = 3 + (l - 1) + v)$ with $\theta \in R^v$.

Table 10.1   S&P 500 (2000-01-01 to 2009-12-31) Statistics.

Observations	2514
Mean	-0.0001058922
Maximum	0.1095720
Minimum	-0.09469514
Std. Dev.	0.01400746
Skewness	-0.1036433
Kurtosis	7.635567

10.5   Numerical Example: $S\&P500$ (2000-01-01–2009-12-31)

Here we fit our model to 10 years of $S\&P500$ data (2000-01-01–2009-12-31) and
apply the obtained analytical solutions to price the variance swap. As an example
we'll assume a variance gamma distribution with probability density function

$$f_{VG}(x;\theta) = \frac{\gamma^{2\lambda}|x-\mu|^{\lambda-1/2}K_{\lambda-1/2}\left(\alpha|x-\mu|\right)}{\sqrt{\pi}\Gamma(\lambda)(2\alpha)^{\lambda-1/2}} \, e^{\beta(x-\mu)}$$

where $\theta = [\mu,\alpha,\gamma,\lambda]$, $\lambda > 0$, $\gamma = \sqrt{\alpha^2 - \beta^2} > 0$ and

$$K_\alpha(x) = \frac{1}{2}e^{-\frac{1}{2}\alpha\pi i}\int_{-\infty}^{+\infty} e^{-ix\sinh t - \alpha t}t$$

is the modified Bessel function of the second kind.

We can see from Table 10.1 that for this set of data the gaussian case is unstable
for most l aside from $l \in \{1,2,4,15\}$ and selecting a model using the minimum AICc
we select the variance gamma case with a discrete lag of 6.

A note should be made on the estimation of the added Variance Gamma parameters. From Table 10.1 we can see that by initializing the model parameters α,
β and ω to the gaussian case we can fit the Variance Gamma model under these
parameters such that they provide no additional increase in the likelihood. This
is an indication of over-fitting of the data, we could choose any α, β and ω and
simply fit the Variance Gamma parameters to achieve a high likelihood. However
by selecting Variance Gamma parameters to introduce certain features not present
in the normal distribution we could attain stability in the optimization and reduce
the problem of over-fitting by reducing the degrees of freedom.

To price the variance swap we can change to the CGM parameterization by

$$C = 1/v,$$

$$G = \left(\sqrt{\frac{1}{4}\theta^2 v^2 + \frac{1}{2}\sigma^2 v} - \frac{1}{2}\theta v\right)^{-1},$$

$$M = \left(\sqrt{\frac{1}{4}\theta^2 v^2 + \frac{1}{2}\sigma^2 v} + \frac{1}{2}\theta v\right)^{-1},$$

Table 10.1 Parameter Estimation.

l	Iters.	$L(\cdot)$	ω	α	β	μ	μ_{VG}	α_{VG}	β	λ	AICc	Likelihood-ratio
1	83	−7820.184	1.069882e-06	0.07432672	0.9191405	0.0003210153					−15632.35	0.003886656
2	38	−7750.7	7.064223e-06	0.1643814	0.8036003	0.0006308619					−15491.38	2.763249e-13
3	45	−7831.666	1.309638e-06	0.1028915	0.903359	0.0009952458					−15651.30	0
4	43	−7815.595	1.548323e-06	0.09192832	0.9066187	0.000955246					−15617.14	5.950452e-06
5	48	−7819.878	1.288561e-06	0.09439167	0.9146567	0.000955246					−15623.70	8.912324e-05
6	48	−7814.589	1.445509e-06	0.1019126	0.909814	0.000955246					−15611.11	3.553801e-08
7	48	−7819.471	1.534373e-06	0.09615287	0.9147871	0.0009552461					−15618.85	0
8	49	−7812.443	1.809161e-06	0.08983728	0.91699	0.000955246					−15602.78	3.829586e-08
9	48	−7809.342	2.073047e-06	0.08754855	0.9163103	0.000955246					−15594.56	2.830733e-07
10	46	−7790.932	2.652642e-06	0.1088227	0.9022083	0.000955246					−15555.72	0
11	45	−7809.248	1.979591e-06	0.09143733	0.9187724	0.000955246					−15590.33	0
12	42	−7786.518	1.93956e-06	0.1072487	0.9053027	0.000955246					−15542.84	7.124334e-13
13	40	−7812.526	1.733541e-06	0.09685873	0.9137061	0.0009552461					−15592.83	0.002946119
14	46	−7799.954	1.861192e-06	0.08488818	0.9200679	0.000955246					−15565.66	0
15	38	−7759.444	2.670334e-06	0.09193074	0.9066226	0.0005221502					−15482.61	1.482822e-20
1	145	−7826.047	1.04447e-06	0.0743266	0.9191404	0.0003210153	0.2643493	2.496919	−0.3087641	3.006923	−15638.05	0.003886656
2	36	−7780.746	7.05914e-06	0.1643814	0.8036003	0.0006308619	0.2368329	2.32664	−0.2455201	2.570153	−15545.44	2.763249e-13
3	36	−7794.82	7.05644e-06	0.1643814	0.8036003	0.0006308619	0.2514001	2.224321	−0.2508395	2.466265	−15571.57	0
4	36	−7828.326	1.54644e-06	0.09192832	0.9066187	0.000955246	0.2532552	2.112255	−0.2360894	2.358500	−15636.56	5.950452e-06
5	36	−7829.777	1.545743e-06	0.09192831	0.9066187	0.000955246	0.2545727	2.044816	−0.2304716	2.274124	−15637.45	8.912324e-05
6	37	−7832.615	1.511631e-06	0.0919282	0.9066185	0.000955246	0.2508737	2.012094	−0.2220731	2.242436	−15641.11	3.553801e-08
7	33	−7120.043	1.548323e-06	0.09192832	0.9066187	0.000955246	−0.0932754	0.3789917	0.1709914	0.332127	−14213.94	0
8	33	−7830.393	1.335184e-06	0.09192757	0.9066175	0.000955246	0.2499842	1.911338	−0.2077163	2.143222	−15632.62	3.829586e-08
9	36	−7825.23	1.545914e-06	0.09192831	0.9066187	0.000955246	0.2456492	1.853450	−0.1966411	2.099737	−15620.27	2.830733e-07
10	33	−7206.951	1.548323e-06	0.09192832	0.9066187	0.000955246	0.07534299	0.483701	0.06209213	0.2782939	−14381.69	0
11	33	−7423.575	1.443317e-06	0.09192803	0.9066182	0.000955246	0.2063468	0.4486394	−0.06854613	0.3956365	−14812.91	0
12	36	−7815.601	1.543790e-06	0.0919283	0.9066187	0.000955246	0.2414999	1.726380	−0.1794623	1.968362	−15594.93	7.124334e-13
13	36	−7818.69	1.544083e-06	0.09192896	0.906619	0.000955246	0.2436979	1.634349	−0.1794541	1.788587	−15599.08	0.002946119
14	33	−7458.664	1.548323e-06	0.09192832	0.9066187	0.000955246	0.3019051	0.5151854	−0.1201923	0.430108	−14876.99	0
15	38	−7806.454	2.635978e-06	0.09193067	0.9066225	0.0005221502	0.2126317	1.772173	−0.1806496	1.823438	−15570.54	1.482822e-20

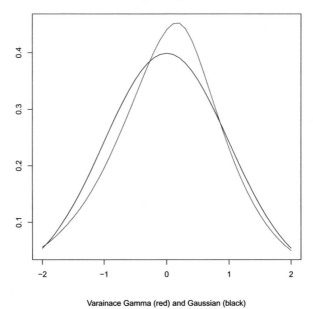

Fig. 10.1   Probability Density Functions.

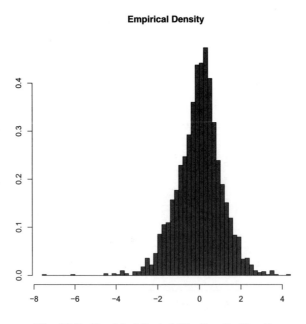

Fig. 10.2   Empirical Probability Density Function.

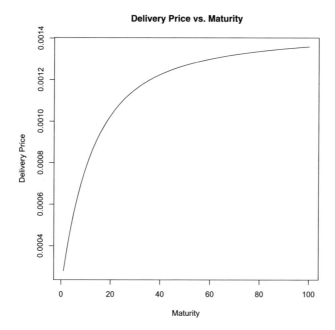

Fig. 10.3 Delivery Price vs. Maturity.

where a change of parameters gives us $c = 0.2508737$, $\sigma = 1.058981$, $\theta = -0.2490418$, and $\nu = 0.4459436$ we then get $C = 2.242436$, $G = 1.790021$, $M = 2.234167$. Figure 10.1 shows the probability density against the normal case. We can see that the variance gamma case captures the skewness observed in the empirical distribution in Figure 10.2.

Now assuming $r = 0.02$ and a maturity of $T = 1$ we get a price of 0.0002104639 under the gaussian model with a lag of 1 and a price of 0.0002048042 under the variance gamma model. Using the AICc selected model, variance gamma with lag of 6, we get a price of 0.0002879282. In Figure 10.3 we can see the price given time to maturity.

Remark 3. We note, that there are no jumps in our model for the return process $S(t)$. The need to include jumps into the volatility process was tested in the paper by Swishchuk and Li (2007), where we considered stochastic volatility with delay and particular kind of jumps, including Poisson and compound Poisson cases.

10.6 Summary

- The valuation of the variance swaps for local Lévy-based stochastic volatility with delay (LLBSVD) was discussed in this Chapter.
- We provided some analytical closed forms for the expectation of the realized variance for the LLBSVD.

– As applications of our analytical solutions, we did fit our model to 10 years of
*S&P*500 data (2000-01-01–2009-12-31) with variance gamma model and apply
the obtained analytical solutions to price the variance swap.

Bibliography

Ait-Sahalia, Y. and Jacod, J. (2008). Estimating the degree of activity of jumps in high
frequency financial data. *Annals of Statistics* (Forthcoming).

Akrigay, V. (2003). The finite moment logstable process and option pricing. *The Journal
of Finance*, 58, 2, 753-777.

Bakshi, C., Cao, C. and Chen, Z. (1997). Empirical performance of alternative option
pricing models. *Journal of Finance*, 52, 2003-2049.

Barndorff-Nielsen, O. and Shephard, N. (2005). Variation, jumps, market frictions
and high frequency data in financial econometrics, Economics Papers 2005-
W16, Economics Group, Nuffield College, University of Oxford. Available at:
http://ideas.repec.org/p/nuf/econwp/0516.html.

Bates, D. (1996). Jumps and stochastic volatility: Exchange rate processes implicit in
deutsche mark options. *Reveiw of Financial Studies*, 9, 69-107.

Bollerslev, T. (1986). Generalized autoregressive conditional heteroskedasticity. *Journal of
Econometrics*, 31, 3, 307-327.

Bollerslev, T., Law, T. and Tauchen, G. (2008). Risk, jumps, and diversification. *Journal
of Econometrics*, 144, 234-256.

Broadie, M., Chernov, M. and Johannes, M. (2007). Model specification and risk premia:
Evidence from futures options. *Journal of Finance*, 62, 1453-1490.

Cont, R. and Tankov, P. (2003). *Financial Modelling with Jump Processes*. England: Chapman & Hall/CRC Financial Mathematics Series.

Duffie, D., Pan, J. and Singleton, K. (2000). Transform analysis and asset pricing for affine
jump-diffusions. *Econometrica*, 68, 1343-1376.

Eraker, B., Johannes, M. and Polson, N. (2003). The impact of jumps in volatility and
returns. *Journal of Finance*, 58, 1269-1300.

Heston, S. (1993). A closed-form solution for options with stochastic volatility with applications to bond and currency options. *Review of Financnial Studies*, 6, 327-343.

Hobson, D. and Rogers, L. (1998). Complete models with stochastic volatility. *Mathematical Finance*, 8, 1, 27-48.

Junye, L. (2009). A bayesian estimation of time-changed infinite activity Lévy models:
Explaining return-volatility relations. Working Paper.

Kazmerchuk, Y., Swishchuk, A. and Wu, J. (2005). A continuous-time GARCH model
for stochastic volatility with delay. *Canadian Applied Mathematics Quarterly*,
13, 2, 123-149.

Kind, P., Liptser, R. and Runggaldier, W. (1991). Diffusion approximation in past-
dependent models and applications to option pricing. *Annals of Probability*, 1, 3,
379-405.

Konlack, V., Kulikova, M., Taylor, D. and Zafack-Takadong, T. (2009). Financial time
series modeling and the estimation of garch-type generalized hyperbolic models.

Kou, S. G. (2002). A jump-diffusion model for option pricing. *Management Science* 48,
1086-1101.

Li, H., Wells, M. T. and Yu, C. L. (2007). A bayesian analysis of return dynamics with
Lévy jumps. *Review of Financial Studies* (Forthcoming).

Mandelbrot, B. (1997). *Fractals and Scaling in Finance, Discontinuity, Concentration,
Risk*. New York: Springer.

Merton, P. (1973). Theory of rational option pricing. *Bell Journal of Economic Management Science*, 4, 141-183.

Protter, P. E. (2003). *Stochastic Integration and Differential Equations* (2nd edn.) New York: Springer.

Schoutens, W. (2003). *Lévy Processes in Finance. Pricing Financial Derivatives.* England: Wiley & Sons.

Sheinkman, J. and LeBaron, B. (1989). Nonlinear dynamics and stock returns. *Journal of Business*, 62, 311-337.

Swishchuk, A. (2005). Modeling and pricing of variance swaps for stochastic volatilities with delay. *Wilmott Magazine*, 19, 63-73.

Swishchuk, A. and Li, X. (2011). Pricing of variance swaps for stochastic volatility with delay and jumps. *Quantitative Finance*, Volume 2011, 27 pages.

Todorov, V. and Tauchen, G. (2008). Volatility jumps. Working Paper, Duke University, NC.

Chapter 11

Delayed Heston Model: Improvement of the Volatility Surface Fitting

11.1 Introduction

We present in this Chapter a variance drift adjusted version of the Heston model which leads to significant improvement of the market volatility surface fitting (compared to Heston). The numerical example we performed with recent market data shows a significant (44%) reduction of the average absolute calibration error[1] (calibration on 30th September 2011 for underlying EURUSD). Our model has two additional parameters compared to the Heston model, can be implemented very easily and was initially introduced for volatility derivatives pricing purpose. The main idea behind our model is to take into account some past history of the variance process in its (risk-neutral) diffusion. In a follow-up article we focus on volatility swap pricing and hedging.

The volatility process is an important concept in financial modeling as it quantifies at each time t how likely the modeled asset log-return is to vary significantly over some short immediate time period $[t, t + \epsilon]$. This process can be stochastic or deterministic, e.g. local volatility models in which the (deterministic) volatility depends on time and spot price level. In quantitative finance, we often consider the volatility process $\sqrt{V_t}$ (where V_t is the variance process) to be stochastic as it allows to fit the observed vanilla option market prices with an acceptable bias as well as to model the risk linked with the future evolution of the volatility smile (which deterministic model cannot), namely the forward smile. Many derivatives are known to be very sensitive towards the forward smile, one of the most popular examples being the cliquet options (options on future asset performance).

Heston model (Heston, 1993; Heston and Nandi, 2000) is one of the most popular stochastic volatility models in the industry as semi-closed formulas for vanilla option prices are available, few (five) parameters need to be calibrated, and it accounts for the mean-reverting feature of the volatility.

One might be willing, in the variance diffusion, to take into account not only its current state but also its past history over some interval $[t - \tau, t]$, where $\tau > 0$

---

[1] Average of the absolute differences between market and model implied BS volatilities.

is a constant and is called the delay. Starting from the discrete-time GARCH(1,1) model (Bollerslev, 1986), a first attempt was made in this direction in (Kazmerchuk, Swishchuk and Wu, 2002), where a non-Markov delayed continuous-time GARCH model was proposed (S_t being the asset price at time t, γ, θ, α some positive constants):

$$\frac{dV_t}{dt} = \gamma\theta^2 + \frac{\alpha}{\tau}\ln^2\left(\frac{S_t}{S_{t-\tau}}\right) - (\alpha+\gamma)V_t,$$

this model being inherited from its discrete-time analogue:

$$\sigma_n^2 = \tilde{\gamma}\tilde{\theta}^2 + \frac{\tilde{\alpha}}{L}\ln^2\left(\frac{S_{n-1}}{S_{n-1-L}}\right) + (1-\tilde{\alpha}+\tilde{\gamma})\sigma_{n-1}^2.$$

The parameter θ^2 (resp. γ) can be interpreted as the value of the long-range variance[2] (resp. variance mean-reversion speed) when the delay tends to 0 (we will see that introducing delay modifies the value of these two model features), and α a continuous-time equivalent of the variance ARCH(1,1) autoregressive coefficient. In fact, we can interpret the right-hand side of previous diffusion equation as the sum of two terms:

- the delay-free term $\gamma(\theta^2 - V_t)$ which accounts for the mean-reverting feature of the variance process
- $\alpha\left(\frac{1}{\tau}\ln^2\left(\frac{S_t}{S_{t-\tau}}\right) - V_t\right)$ which is a pure (noisy) delay term, i.e. that vanishes when $\tau \to 0$ and takes into account the past history of the variance (via the term $\ln\left(\frac{S_t}{S_{t-\tau}}\right)$). As we will see below, the introduction of this pure delay term modifies the value of both the long-range variance and variance mean-reversion speed of the model. The autoregressive coefficient α can be seen as the amplitude of this pure delay term.

J.C. Duan remarked the importance to incorporate the real world $P-$drift $d_P(t,\tau) := \int_{t-\tau}^t (\mu - \frac{1}{2}V_u)du$ of $\ln\left(\frac{S_t}{S_{t-\tau}}\right)$ in the model, transforming the variance dynamics into:

$$\frac{dV_t}{dt} = \gamma\theta^2 + \frac{\alpha}{\tau}\left[\ln\left(\frac{S_t}{S_{t-\tau}}\right) - d_P(t,\tau)\right]^2 - (\alpha+\gamma)V_t. \qquad (11.1)$$

The latter diffusion (11.1) was introduced in (Swishchuk, 2005) and (Kazmerchuk, Swishchuk and Wu, 2005), and the proposed model was proved to be complete and to account for the mean-reverting feature of the volatility process. This model is also non-Markov as the past history $(V_u)_{u\in[t-\tau,t]}$ of the variance appears in its diffusion equation via the term $\ln\left(\frac{S_t}{S_{t-\tau}}\right)$, as it is shown in (Swishchuk, 2005).

In the continuity of this approach, pricing of variance swaps for one-factor stochastic volatility with delay has been presented in Swishchuk (2005), for multi-factor stochastic volatility in Swishchuk (2006) and for one-factor stochastic volatility with delay and jumps in Swishchuk and Li (2011). Variance swap for local

---

[2]The long-range variance being defined as $\lim_{t\to\infty} E^Q(V_t)$.

Lévy-based stochastic volatility with delay has been calculated in Swishchuk and Malenfant (2011).

Unfortunately, the model (11.1) doesn't lead to (semi-)closed formulas for the vanilla options, making it difficult to use for practitioners willing to calibrate on vanilla market prices. Nevertheless, one can notice that the Heston model and the delayed continuous-time GARCH model (11.1) are very similar in the sense that the expected values of the variances are the same - when we make the delay tend to 0 in (11.1). As mentioned before, the Heston framework is very convenient, and therefore it is naturally tempting to adjust the Heston dynamics in order to incorporate the delay introduced in (11.1). In this way, we considered in a first approach adjusting the Heston drift by a deterministic function of time so that the expected value of the variance under the delayed Heston model is equal to the one under the delayed GARCH model (11.1). In addition to making our delayed Heston framework coherent with (11.1), this construction makes the variance process diffusion dependent not on its past history $(V_u)_{u \in [t-\tau, t]}$, but on the past history of its risk-neutral expectation $(E^Q(V_u))_{u \in [t-\tau, t]}$, preserving the Markov feature of the Heston model.

11.2 Modeling of Delayed Heston Stochastic Volatility

Throughout this chapter, we will assume constant risk-free rate r, dividend yield q and finite time-horizon T.

Assume risk-neutral $Q-$ stock price dynamics (Z_t^Q and W_t^Q being two correlated standard Brownian motions):

$$dS_t = (r - q)S_t dt + S_t \sqrt{V_t} dZ_t^Q. \tag{11.2}$$

The well-known Heston model has the following $Q-$dynamics for the variance V_t:

$$dV_t = \gamma(\theta^2 - V_t)dt + \delta\sqrt{V_t}dW_t^Q, \tag{11.3}$$

where θ^2 is the long-range variance, γ the variance mean-reversion speed, δ the volatility of the variance and c the brownian correlation coefficient ($\langle W^Q, Z^Q \rangle_t = ct$).

As explained in the introduction, delayed continuous GARCH dynamics has been introduced for the variance in Swishchuk (2005):

$$\frac{dV_t}{dt} = \gamma\theta^2 + \frac{\alpha}{\tau}\left[\int_{t-\tau}^t \sqrt{V_s}dZ_s^Q - (\mu - r)\tau\right]^2 - (\alpha + \gamma)V_t. \tag{11.4}$$

We can interpret the right-hand side of previous diffusion equation[3] as the sum of two terms:

---

[3] Note that θ^2 (resp. γ) has been defined in introduction for the delayed continuous-time GARCH model as the value of the long-range variance (resp. variance mean-reversion speed) when $\tau \to 0$, therefore it has the same meaning as the Heston long-range variance (resp. variance mean-reversion speed).

- the delay-free term $\gamma(\theta^2 - V_t)$ which accounts for the mean-reverting feature of the variance process
- $\alpha \left(\frac{1}{\tau} \left[\int_{t-\tau}^{t} \sqrt{V_s} dZ_s^Q - (\mu - r)\tau \right]^2 - V_t \right)$ which is a pure (noisy) delay term of amplitude α (continuous-time equivalent of the variance ARCH(1,1) autoregressive coefficient), i.e. that vanishes when $\tau \to 0$ and takes into account the past history of the variance via the integral $\int_{t-\tau}^{t} \sqrt{V_s} dZ_s^Q$. As we will see below, the introduction of this pure delay term modifies the value of both the long-range variance and variance mean-reversion speed of the model.

We can see that the two models are very similar. Indeed, they both give the same expected value for V_t as the delay goes to 0 in (11.6), namely $\theta^2 + (v_0 - \theta^2)e^{-\gamma t}$. The idea here is to adjust the Heston dynamics (11.5) in order to account for the delay introduced in (11.6). Our approach is to adjust the drift by a deterministic function of time so that the expected value of V_t under the adjusted Heston model is the same as under (11.6). This approach can be seen as a correction by a pure delay term of amplitude α (in the sense of (11.14)) of the Heston drift by a deterministic function in order to account for the delay.

Namely, we assume the adjusted Heston dynamics:

$$dV_t = \left[\gamma(\theta^2 - V_t) + \epsilon_\tau(t) \right] dt + \delta\sqrt{V_t} dW_t^Q, \tag{11.5}$$

$$\epsilon_\tau(t) := \alpha\tau(\mu - r)^2 + \frac{\alpha}{\tau} \int_{t-\tau}^{t} v_s ds - \alpha v_t, \tag{11.6}$$

with $v_t := E^Q(V_t)$. It was shown in (Swishchuk, 2005) that v_t solves the following:

$$\frac{dv_t}{dt} = \gamma\theta^2 + \alpha\tau(\mu - r)^2 + \frac{\alpha}{\tau} \int_{t-\tau}^{t} v_s ds - (\alpha + \gamma)v_t. \tag{11.7}$$

And we have the following expression for v_t:

$$v_t = \theta_\tau^2 + (v_0 - \theta_\tau^2)e^{-\gamma_\tau t}, \quad (\gamma_\tau \neq 0) \tag{11.8}$$

$$\theta_\tau^2 := \theta^2 + \frac{\alpha\tau(\mu - r)^2}{\gamma}. \tag{11.9}$$

The parameter θ_τ^2 can be interpreted as the adjusted long-range variance - that has been (positively) shifted from its original value θ^2 because of the introduction of delay. We have $\theta_\tau^2 \to \theta^2$ when $\tau \to 0$, which is coherent. We will see below that we can interpret the parameter γ_τ as the adjusted mean-reversion speed. This parameter is given in Swishchuk (2005) by:

$$\gamma_\tau = \alpha + \gamma + \frac{\alpha}{\gamma_\tau \tau}(1 - e^{\gamma_\tau \tau}). \tag{11.10}$$

By (11.6), (11.8), (11.9) and (11.10) we get an explicit expression for the drift adjustment:

$$\epsilon_\tau(t) = \alpha\tau(\mu - r)^2 + (v_0 - \theta_\tau^2)(\gamma - \gamma_\tau)e^{-\gamma_\tau t}. \tag{11.11}$$

The two following simple lemmas give some information about the correction term $\epsilon_\tau(t)$ and γ_τ, that will be useful for interpretation purpose and in the derivation of the semi-closed formulas for call options in the next section. Indeed, given Lemma 1 and (11.10), the parameter γ_τ can be interpreted as the adjusted variance mean-reversion speed, and we have by (11.10) that $\gamma_\tau \to \gamma$ when $\tau \to 0$, which is coherent.

Lemma 1:

$$0 < \gamma_\tau < \gamma \tag{11.12}$$

$$\gamma_\tau \text{ is unique.} \tag{11.13}$$

Proof: Let's show $\gamma_\tau \geq 0$. If $\gamma_\tau < 0$ then by (11.10) we have $\alpha + \gamma + \frac{\alpha}{\gamma_\tau \tau}(1 - e^{\gamma_\tau \tau}) < 0$, i.e. $1 - e^{\gamma_\tau \tau} + \gamma_\tau \tau > -\frac{\gamma}{\alpha}\gamma_\tau \tau$. But $\tau > 0$ so $\exists x_0 > 0$ s.t. $1 - e^{-x_0} - x_0 > \frac{\gamma}{\alpha}x_0$. A simple study shows that is impossible whenever $\frac{\gamma}{\alpha} \geq 0$, which is what we have by assumption. Therefore $\gamma_\tau \geq 0$. And we have $\gamma_\tau \neq 0$ by assumption so $\gamma_\tau > 0$.
If $\gamma \leq \gamma_\tau$ then by (11.10) $\gamma_\tau \tau + 1 - e^{\gamma_\tau \tau} \geq 0$. But $\gamma_\tau \tau > 0$ therefore $\exists x_0 > 0$ s.t. $x_0 + 1 - e^{x_0} \geq 0$. A simple study shows that is impossible.
The uniqueness comes from a similar simple study. $\qquad\square$

Lemma 2:

$$\lim_{\tau \to 0} \sup_{t \in R^+} |\epsilon_\tau(t)| = 0. \tag{11.14}$$

Proof: By (11.12) we have $\gamma_\tau > 0$. Therefore $\sup_{t \in R^+} |\epsilon_\tau(t)| \leq \alpha\tau(\mu - r)^2 + |(v_0 - \theta_\tau^2)(\gamma - \gamma_\tau)|$ and $(v_0 - \theta_\tau^2)(\gamma - \gamma_\tau) = o(1)$ by (11.10). So $\lim_{\tau \to 0} \alpha\tau(\mu - r)^2 + |(v_0 - \theta_\tau^2)(\gamma - \gamma_\tau)| = 0$. $\qquad\square$

11.3   Model Calibration

It is possible to get a semi-closed formula for the call option in our delayed Heston model. Indeed, our model is a time-dependent Heston model with time-dependent long-range variance $\tilde{\theta}^2(t) := \theta_\tau^2 + (v_0 - \theta_\tau^2)\frac{(\gamma - \gamma_\tau)}{\gamma}e^{-\gamma_\tau t}$. From Kahl and Jäckel (2005) we get equations (11.15) to (12.15) for the price of a call option with maturity T and strike K in the time-dependent long-range variance Heston model:

$$C_0 = e^{-rT}\left[\frac{1}{2}(F - K) + \frac{1}{\pi}\int_0^\infty (Fh_1(u) - Kh_2(u))du\right], \tag{11.15}$$

$$h_1(u) = \Re\left(\frac{e^{-iu\ln(K)}\varphi(u - i)}{iuF}\right), \tag{11.16}$$

$$h_2(u) = \Re\left(\frac{e^{-iu\ln(K)}\varphi(u)}{iu}\right), \tag{11.17}$$

with $F = S_0 e^{(r-q)T}$ and

$$\varphi(u) = e^{C(T,u)+v_0 D(T,u)+iu \ln(F)}. \tag{11.18}$$

By Michailov and Nöegel (2005) we have that $C(t,u)$ and $D(t,u)$ solve the following differential equations:

$$\frac{dC(t,u)}{dt} = \gamma \tilde{\theta}^2(t) D(t,u), \tag{11.19}$$

$$\frac{dD(t,u)}{dt} - \frac{\delta^2}{2} D^2(t,u) + (\gamma - iuc\delta) D(t,u) + \frac{1}{2}(u^2 + iu) = 0, \tag{11.20}$$

$$C(0,u) = D(0,u) = 0. \tag{11.21}$$

The Riccati equation for $D(t,u)$ doesn't depend on $\tilde{\theta}^2(t)$, therefore its solution is just the solution of the classical Heston model given in Kahl and Jäckel (2005):

$$D(t,u) = \frac{\gamma - ic\delta u + d}{\delta^2} \left[\frac{1 - e^{dt}}{1 - ge^{dt}} \right], \tag{11.22}$$

$$g = \frac{\gamma - ic\delta u + d}{\gamma - ic\delta u - d}, \tag{11.23}$$

$$d = \sqrt{(\gamma - ic\delta u)^2 + \delta^2 (iu + u^2)}. \tag{11.24}$$

Given $D(t,u)$, we can compute $C(t,u)$ from (11.19) and (11.21):

$$C(t,u) = \gamma \theta_\tau^2 f(t,u) + (v_0 - \theta_\tau^2)(\gamma - \gamma_\tau) \int_0^t e^{-\gamma_\tau s} D(s,u) ds. \tag{11.25}$$

Where $f(t,u) = \int_0^t D(s,u) ds$ is given in Kahl and Jäckel (2005):

$$f(t,u) = \frac{1}{\delta^2} \left((\gamma - ic\delta u + d)t - 2\ln\left(\frac{1 - ge^{dt}}{1 - g}\right) \right). \tag{11.26}$$

Unfortunately, the integral $\int_0^t e^{-\gamma_\tau s} D(s,u) ds$ in (11.25) cannot be computed directly as $\int_0^t D(s,u) ds$. The logarithm in $f(t,u)$ can be handled as suggested in Kahl and Jäckel (2005), as well as the integration of the Heston integral, namely:

$$C_0 = e^{-rT} \int_0^1 y(x) dx, \tag{11.27}$$

$$y(x) = \frac{1}{2}(F - K) + \frac{Fh_1(-\frac{\ln(x)}{C_\infty}) - Kh_2(-\frac{\ln(x)}{C_\infty})}{x\pi C_\infty}, \tag{11.28}$$

where $C_\infty > 0$ is an integration constant.

The following limit conditions are given in Kahl and Jäckel (2005):

$$\lim_{x \to 0} y(x) = \frac{1}{2}(F - K),$$ (11.29)

$$\lim_{x \to 1} y(x) = \frac{1}{2}(F - K) + \frac{FH_1 - KH_2}{\pi C_\infty},$$ (11.30)

$$H_j = \lim_{u \to 0} h_j(u) = \ln\left(\frac{F}{K}\right) + \tilde{c}_j(T) + v_0 \tilde{d}_j(T),$$ (11.31)

where:

$$\tilde{d}_1(t) = \Im\left(\frac{\partial D}{\partial u}(t, -i)\right),$$ (11.32)

$$\tilde{c}_1(t) = \Im\left(\frac{\partial C}{\partial u}(t, -i)\right),$$ (11.33)

$$\tilde{d}_2(t) = \Im\left(\frac{\partial D}{\partial u}(t, 0)\right),$$ (11.34)

$$\tilde{c}_2(t) = \Im\left(\frac{\partial C}{\partial u}(t, 0)\right).$$ (11.35)

Expressions for $\tilde{d}_1(t)$ and $\tilde{d}_2(t)$ are the same as in Kahl and Jäckel (2005) as $\tilde{\theta}^2(t)$ doesn't play any role in them. Given (11.19) and (11.21), we compute $\tilde{c}_1(T)$ and $\tilde{c}_2(T)$ in our time-dependent long-range variance Heston model by:

$$\tilde{c}_j(T) = \gamma \int_0^T \tilde{\theta}^2(t) \tilde{d}_j(t) dt.$$ (11.36)

After computations we get:

If $\gamma - c\delta \neq 0$ and $\gamma - c\delta + \gamma_\tau \neq 0$:

$$\tilde{d}_1(T) = \frac{1 - e^{-(\gamma - c\delta)T}}{2(\gamma - c\delta)},$$ (11.37)

$$\tilde{c}_1(T) = \gamma \theta_\tau^2 \frac{e^{-(\gamma - c\delta)T} - 1 + (\gamma - c\delta)T}{2(\gamma - c\delta)^2} + \frac{(v_0 - \theta_\tau^2)(\gamma - \gamma_\tau)}{2(\gamma - c\delta)}$$

$$\times \left(-\frac{e^{-\gamma_\tau T} - 1}{\gamma_\tau} + \frac{e^{-(\gamma - c\delta + \gamma_\tau)T} - 1}{\gamma - c\delta + \gamma_\tau}\right).$$ (11.38)

If $\gamma - c\delta \neq 0$ and $\gamma - c\delta + \gamma_\tau = 0$:

$$\tilde{d}_1(T) = \frac{1 - e^{-(\gamma - c\delta)T}}{2(\gamma - c\delta)},$$ (11.39)

$$\tilde{c}_1(T) = \gamma \theta_\tau^2 \frac{e^{-(\gamma - c\delta)T} - 1 + (\gamma - c\delta)T}{2(\gamma - c\delta)^2} + \frac{(v_0 - \theta_\tau^2)(\gamma - \gamma_\tau)}{2(\gamma - c\delta)}\left(-\frac{e^{-\gamma_\tau T} - 1}{\gamma_\tau} - T\right).$$ (11.40)

If $\gamma - c\delta = 0$:

$$\tilde{d}_1(T) = \frac{T}{2}, \tag{11.41}$$

$$\tilde{c}_1(T) = \gamma\theta_\tau^2 \frac{T^2}{4} + \frac{(v_0 - \theta_\tau^2)(\gamma - \gamma_\tau)}{2} \left(-\frac{Te^{-\gamma_\tau T}}{\gamma_\tau} + \frac{1 - e^{-\gamma_\tau T}}{\gamma_\tau^2} \right), \tag{11.42}$$

and

$$\tilde{d}_2(T) = \frac{e^{-\gamma T} - 1}{2\gamma}, \tag{11.43}$$

$$\tilde{c}_2(T) = \gamma\theta_\tau^2 \frac{1 - e^{-\gamma T} - \gamma T}{2\gamma^2} + \frac{(v_0 - \theta_\tau^2)(\gamma - \gamma_\tau)}{2\gamma} \left(-\frac{1 - e^{-\gamma_\tau T}}{\gamma_\tau} - \frac{e^{(-\gamma_\tau - \gamma)T} - 1}{\gamma_\tau + \gamma} \right). \tag{11.44}$$

11.4   Numerical Results

The calibration procedure is a least-squares minimization procedure that we perform via MATLAB (function *lsqnonlin* that uses a trust-region-reflective algorithm). The Heston integral (11.15) is computed via the MATLAB function *quadl* that uses a recursive adaptive Lobatto quadrature. The integral $\int_0^t e^{-\gamma_\tau s} D(s, u) ds$ in (11.25) is computed via a composite Simpson's rule with 100 points.

We perform our calibration on September 30th 2011 for underlying EURUSD on the whole volatility surface (maturities from 1M to 10Y, strikes ATM, 25D Call/Put, 10D Call/Put). The implied volatility surface, the Zero Coupon curves EUR Vs. Euribor 6M and USD Vs. Libor 3M and the spot price are taken from Bloomberg (mid prices). We use $C_\infty = 0.5$. The drift $\mu = 0.0188$ is estimated from 7.5Y of daily close prices (source: www.forexrate.co.uk).

Table 11.1   Heston Absolute Calibration Error (in bp of the BS volatility).

	ATM	25D Call	25D Put	10D Call	10D Put
1M	152	192	41	193	67
2M	114	139	15	136	81
3M	89	109	3	110	92
4M	48	61	17	67	101
6M	5	15	34	29	85
9M	59	42	63	2	85
1Y	107	83	102	31	96
1.5Y	141	116	111	42	73
2Y	166	137	127	54	68
3Y	145	124	77	52	0
4Y	96	95	18	37	66
5Y	29	47	52	7	138
7Y	39	10	112	28	186
10Y	100	67	168	58	225

Table 11.2 Delayed Heston Absolute Calibration Error (in bp of the BS volatility).

	ATM	25D Call	25D Put	10D Call	10D Put
1M	116	91	109	128	115
2M	44	24	59	54	88
3M	14	3	32	36	60
4M	18	28	1	5	29
6M	31	37	23	19	3
9M	45	45	56	37	57
1Y	51	47	82	50	104
1.5Y	29	30	79	49	129
2Y	24	23	83	47	139
3Y	11	9	29	30	90
4Y	41	28	14	17	38
5Y	76	55	59	5	16
7Y	71	49	58	1	14
10Y	26	8	18	47	24

The calibrated parameters are $(v_0, \gamma, \theta^2, \delta, c, \alpha, \tau) = (0.0343, 3.9037, 10^{-8},$ $0.808, -0.5057, 71.35, 0.7821)$ for delayed Heston and $(v_0, \gamma, \theta^2, \delta, c) = (0.0328,$ $0.5829, 0.0256, 0.3672, -0.4824)$ for Heston. The absolute calibration error (in bp of the BS volatility) for Heston model and our delayed Heston model are given below. The results show a 44% reduction of the average absolute calibration error (46bp for delayed Heston, 81bp for Heston), i.e., average of the absolute differences between market and model implied BS volatilities.

11.5 Summary

- In this Chapter, we presented an easily implementable adjustment of the Heston model, the delayed Heston model, based on the will to take into account the past history of the variance process in its risk-neutral diffusion.
- The adjustment we make to the Heston variance drift is based on the delayed continuous-GARCH model presented in Swishchuk (2005).
- Calibration results show a significant improvement of the market volatility surface fitting compared to Heston: the average absolute calibration error was reduced by 44% in the numerical example we considered with recent market data.

Bibliography

Bollerslev, T. (1986). Generalized autoregressive conditional heteroscedasticity. *Journal of Economics*, 31, 307-27.

Broadie, M. and Jain, A. (2008). The effect of jumps and discrete sampling on volatility and variance swaps. *International Journal of Theoretical and Applied Finance*, 11, 8, 761-797.

Brockhaus, O. and Long, D. (2000). Volatility swaps made simple. *Risk Magazine*, January, 92-96.

Carr, P. and Lee, R. (2009). Robust replication of volatility derivatives. Working Paper.

Carr, P. and Lee, R. (2009). Volatility derivatives. *Annual Review of Financial Economics*, 1, 1-21.

Carr, P. and Lee, R. (2007). Realized volatility and variance: Options via swaps. *Risk Magazine*, 20, 5, 76-83.

Carr, P. and Madan, D. (1998). Towards a theory of volatility trading. In Jarrow, R. (eds.) *Volatility*, 417-427, USA: Risk Book Publications.

Demeterfi, K., Derman, E., Kamal, M. and Zou, J. (1999). A guide to volatility and variance swaps. *The Journal of Derivatives*, Summer, 9-32.

Heston, S. (1993). A closed-form solution for options with stochastic volatility with applications to bond and currency options. *Review of Financial Studies*, 6, 327-343.

Heston, S. and Nandi, S. (2000). Derivatives on volatility: Some simple solutions based on observables. Federal Rserve Bank of Atlanta. Working Paper. 20 November.

Hobson, D. and Rogers, L. (1998). Complete models with stochastic volatility. *Mathematical Finance*, 8, 1, 27-48.

Howison, S., Rafailidis, A. and Rasmussen, H. (2004). On the pricing and hedging of volatility derivatives. *Applied Mathematical Finance*, 11, 4, 317-346.

Javaheri, A., Wilmott, P. and Haug, E. (2002). GARCH and volatility swaps. *Wilmott Technical Article*, January, 17 pages.

Kazmerchuk, Y., Swishchuk, A. and Wu, J. (2002). The pricing of options for security markets with delayed response. *Mathematical Finance Journal* (Submitted).

Kazmerchuk, Y., Swishchuk, A. and Wu, J. (2005). A continuous-time GARCH model for stochastic volatility with delay. *Canadian Applied Mathematics Quarterly*, 13, 2.

Mikhailov, S. and Noegel, U. (2005). Heston's Stochastic Volatility Model Implementation, Calibration and Some Extensions. *Wilmott Magazine*, 1, 74-79.

Kahl, C. and Jäckel, P. (2005). Not-so-complex logarithms in the Heston model. *Wilmott Magazine*, September, 94-103.

Swishchuk, A and Malenfant, K. (2011). Variance swap for local Lévy based stochastic volatility with delay. *International Review of Applied Financial Issues and Economics*, 3, 2, 432-441.

Swishchuk, A. and Li, X. (2011). Pricing variance swaps for stochastic volatilities with delay and jumps. *International Journal of Stochastic Analysis*, Volume 2011, 27 pages.

Swishchuk, A. and Couch, M. (2010). Volatility and variance swaps for the COGARCH(1,1) model. *Wilmott Magazine*, 2, 5, 231-246.

Swishchuk, A. (2006). Modeling and pricing of variance swaps for multi-factor stochastic volatilities with delay. *Canadian Applied Mathematicals Quarterly*, 14, 4, 439-468.

Swishchuk, A. (2005). Modeling and pricing of variance swaps for stochastic volatilities with delay. *Wilmott Magazine*, September, No. 19, 63-73.

Swishchuk, A. (2004). Modeling of variance and volatility swaps for financial markets with stochastic volatilities. *Wilmott Magazine*, September issue, No. 2, 64-72.

Chapter 12

Pricing and Hedging of Volatility Swap in the Delayed Heston Model

12.1 Introduction

Using change of time method for continuous local martingales, we derive a closed formula for the Brockhaus and Long approximation of the volatility swap price. The model we consider is a variance drift adjusted Heston model — the delayed Heston Model — that has been presented in Chapter 11 (see also Swishchuk and Vadori (February 2012)). The main idea behind this model is to take into account some past history of the variance process in its (risk-neutral) diffusion. We also consider dynamic hedging of volatility swaps using a portfolio of variance swaps.

In Chapter 11, we explained the motivation of some authors[1] to take into account in the (stochastic) volatility dynamics its past history over some interval $[t - \tau, t]$, where $\tau > 0$ is a constant and is called the delay. We presented the delayed Heston model, which makes the link between the well-known Heston model, and the delayed continuous-time GARCH model presented in Swishchuk (2005). In this Chapter 11, we focus on pricing and hedging of variance and volatility swaps in the delayed Heston model framework (see Swishchuk and Vadori (March 2012)).

Volatility and variance swaps are contracts whose payoff depend (respectively convexly and linearly) on the realized variance of the underlying asset over some specified time interval. They provide pure exposure to volatility, and therefore make it a tradable market instrument. Variance Swaps are even considered by some practitioners to be vanilla derivatives. The most commonly traded variance swaps are discretely sampled and have a payoff $P_n^V(T)$ at maturity T of the form:

$$P_n^V(T) = N \left[\frac{252}{n} \sum_{i=0}^{n} \ln^2 \left(\frac{S_{i+1}}{S_i} \right) - K_{var} \right],$$

where S_i is the asset spot price on fixing time $t_i \in [0, T]$ (usually there is one fixing time each day, but there could be more, or less), N the notional amount of the contract (in currency per unit of variance) and K_{var} the strike specified in the

---

[1]E.g. Kazmerchuk, Swishchuk and Wu (2002) or Swishchuk (2005).

contract. The corresponding volatility swap payoff $P_n^v(T)$ is given by:

$$P_n^v(T) = N \left[\sqrt{\frac{252}{n} \sum_{i=0}^{n} \ln^2 \left(\frac{S_{i+1}}{S_i} \right)} - K_{vol} \right].$$

One can also consider continuously sampled volatility and variance swaps (on which we will focus in this article), which payoffs are respectively defined as the limit when $n \to +\infty$ of their discretely sampled versions. Formally, if we denote $(V_t)_{t \geq 0}$ the stochastic volatility process of our asset, adapted to some brownian filtration $(\mathcal{F}_t)_{t \geq 0}$, then the continuously-sampled realized variance V_R from initiation date of the contract $t = 0$ to maturity date $t = T$ is given by $V_R = \frac{1}{T} \int_0^T V_s ds$. The fair variance strike K_{var} is calculated such that the initial value of the contract is 0, and therefore is given by:

$$E_0^Q \left[e^{-rT} (V_R - K_{var}) \right] = 0 \Rightarrow K_{var} = E_0^Q (V_R),$$

where we denote $E_t^Q(\cdot) := E^Q(\cdot | \mathcal{F}_t)$. In the same way, the fair volatility strike K_{vol} is given by:

$$E_0^Q \left[e^{-rT} (\sqrt{V_R} - K_{vol}) \right] = 0 \Rightarrow K_{vol} = E_0^Q (\sqrt{V_R}).$$

The volatility swap fair strike might be difficult to compute explicitly as we have to compute the expectation of a square-root. In Brockhaus and Long (2000), the following approximation — based on Taylor expansion — was proposed to compute the expected value of the square-root of an almost surely non negative random variable Z:

$$E(\sqrt{Z}) \approx \sqrt{E(Z)} - \frac{Var(Z)}{8E(Z)^{\frac{3}{2}}}.$$

We will refer to this approximation in our paper as the Brockhaus and Long approximation.

The paper by Carr and Lee (2009) provides an overview of the current market for volatility derivatives. They survey the early literature on the subject. They also provide relatively simple proofs of some fundamental results related to variance swaps and volatility swaps. Pricing of variance swaps for one-factor stochastic volatility with delay has been presented in Swishchuk (2004), for multi-factor stochastic volatility in Swishchuk (2006) and for one-factor stochastic volatility with delay and jumps in Swishchuk and Li (2011). Variance swap for local Levy-based stochastic volatility with delay has been calculated in (Swishchuk and Malenfant (2011): Variance swap for local Levy-based stochastic volatility with delay. *International Review of Applied Financial Issues and Economics (IRAFIE)*, forthcoming). Variance and volatility swaps in energy markets have been considered in Swishchuk (2011). Variance and volatility swaps in energy markets. *Journal of Energy Markets*, forthcoming). Broadie and Jain (2008) covers pricing and dynamic hedging of volatility derivatives in the Heston model.

In Section 12.2, we provide a quick recap of the main features of the delayed Heston model that has been presented in Chapter 11 (see also Swishchuk and Vadori, 2012). In Section 12.3, we compute the price process $X_t(T) := E_t^Q(V_R)$ of the floating leg of the variance swap of maturity T, as well as the Brockhaus and Long approximation of the price process $Y_t(T) := E_t^Q(\sqrt{V_R})$ of the floating leg of the volatility swap of maturity T. This leads in particular to closed formulas for the fair volatility and variance strikes. In Section 12.4, we consider dynamic hedging of volatility swaps using variance swaps. A numerical example for vanilla options on September 30th 2011 for underlying EURUSD (with parameters calibrated in Swishchuk and Vadori, 2012) is presented in Section 12.5. Section 12.6 summarizes the chapter.

12.2 Modeling of Delayed Heston Stochastic Volatility: Recap

Throughout this chapter, we will assume constant risk-free rate r, dividend yield q and finite time-horizon T.

We recap the main features of the Delayed Heston model that has been presented in Part 1 of our article.

We assume $Q-$ stock price dynamics (Z_t^Q and W_t^Q being two correlated standard Brownian motions):

$$dS_t = (r - q)S_t dt + S_t \sqrt{V_t} dZ_t^Q \qquad (12.1)$$

$$dV_t = \left[\gamma(\theta^2 - V_t) + \epsilon_\tau(t)\right] dt + \delta \sqrt{V_t} dW_t^Q \qquad (12.2)$$

where $\tau > 0$ is the delay, θ^2 (resp. γ) can be interpreted as the value of the long-range variance (resp. variance mean-reversion speed) when the delay tends to 0, δ the volatility of the variance and c the brownian correlation coefficient ($\langle W^Q, Z^Q \rangle_t = ct$).

The variance drift adjustment $\epsilon_\tau(t)$ and the adjusted long-range variance θ_τ^2 being respectively given by:

$$\epsilon_\tau(t) = \alpha\tau(\mu - r)^2 + (V_0 - \theta_\tau^2)(\gamma - \gamma_\tau)e^{-\gamma_\tau t} \qquad (12.3)$$

$$\theta_\tau^2 := \theta^2 + \frac{\alpha\tau(\mu - r)^2}{\gamma}. \qquad (12.4)$$

α is a continuous-time equivalent of the variance ARCH(1,1) autoregressive coefficient, and can also be seen as the amplitude of the pure delay adjustment $\epsilon_\tau(t)$.[2]

The adjusted variance mean-reversion speed γ_τ is the unique positive solution to:

$$\gamma_\tau = \alpha + \gamma + \frac{\alpha}{\gamma_\tau \tau}(1 - e^{\gamma_\tau \tau}) \qquad (0 < \gamma_\tau < \gamma). \qquad (12.5)$$

---

[2]Recall from Part 1 that the adjustment was defined to be $\epsilon_\tau(t) := \alpha\left[\tau(\mu - r)^2 + \frac{1}{\tau}\int_{t-\tau}^{t} v_s ds - v_t\right]$, where $v_t := E^Q(V_t)$.

Furthermore, $v_t := E^Q(V_t)$ is given by:

$$v_t = \theta_\tau^2 + (V_0 - \theta_\tau^2)e^{-\gamma_\tau t}. \tag{12.6}$$

12.3 Pricing Variance and Volatility Swaps

In this Section, we derive a closed formula for the Brockhaus and Long approximation of the volatility swap price using change of time method introduced in Swishchuk (2004), as well as the price of the variance swap. Precisely, in Brockhaus and Long (2000), the following approximation was presented to compute the expected value of the square-root of an almost surely non-negative random variable Z: $E(\sqrt{Z}) \approx \sqrt{E(Z)} - \frac{Var(Z)}{8E(Z)^{\frac{3}{2}}}$. We denote $V_R := \frac{1}{T}\int_0^T V_s ds$ the realized variance on $[0, T]$. We will also denote $E_t^Q(\cdot) := E^Q(\cdot|\mathcal{F}_t)$ and $Var_t^Q(\cdot) := Var^Q(\cdot|\mathcal{F}_t)$.

We let $X_t(T) := E_t^Q(V_R)$ (resp. $Y_t(T) := E_t^Q(\sqrt{V_R})$) the price process of the floating leg of the variance swap (resp. volatility swap) of maturity T.

Theorem 1: The price process $X_t(T)$ of the floating leg of the variance swap of maturity T in the delayed Heston model (12.1)–(12.2) is given by:

$$X_t(T) = \frac{1}{T}\int_0^t V_s ds + \frac{T-t}{T}\theta_\tau^2 + (V_t - \theta_\tau^2)\left(\frac{1 - e^{-\gamma(T-t)}}{\gamma T}\right)$$

$$+ (V_0 - \theta_\tau^2)e^{-\gamma_\tau t}\left(\frac{1 - e^{-\gamma_\tau(T-t)}}{\gamma_\tau T} - \frac{1 - e^{-\gamma(T-t)}}{\gamma T}\right). \tag{12.7}$$

Proof: By definition, $X_t(T) = E_t^Q(\frac{1}{T}\int_0^T V_s ds) = \frac{1}{T}\int_0^t V_s ds + \frac{1}{T}\int_t^T E_t^Q(V_s)ds$. Let $s \geq t$. Then we have by (12.2) that $E_t^Q(V_s - V_t) = E_t^Q(V_s) - V_t = \int_t^s \gamma(\theta^2 - E_t^Q(V_u)) + \epsilon_\tau(u)du + E_t^Q(\int_t^s \sqrt{V_u}dW_u^Q)$. But $(\sqrt{V_t})_{t\geq 0}$ is an adapted process s.t. $E^Q(\int_0^T V_u du) < +\infty$, therefore $\int_0^t \sqrt{V_u}dW_u^Q$ is a martingale and we have $E_t^Q(\int_t^s \sqrt{V_u}dW_u^Q) = 0$. Therefore $\forall s \geq t \geq 0$, the function $s \to E_t^Q(V_s)$ is a solution of $y_s' = \gamma(\theta^2 - y_s) + \epsilon_\tau(s)$ with initial condition $y_t = V_t$. Simple calculations give us $E_t^Q(V_s) = \theta_\tau^2 + (V_t - \theta_\tau^2)e^{-\gamma(s-t)} + (V_0 - \theta_\tau^2)e^{-\gamma_\tau t}(e^{-\gamma_\tau(s-t)} - e^{-\gamma(s-t)})$. A calculation of $\int_t^T E_t^Q(V_s)ds$ completes the proof.

Corollary 1: The price K_{var} of the variance swap of maturity T at initiation of the contract $t = 0$ in the delayed Heston model (12.1)–(12.2) is given by:

$$K_{var} = \theta_\tau^2 + (V_0 - \theta_\tau^2)\frac{1 - e^{-\gamma_\tau T}}{\gamma_\tau T}. \tag{12.8}$$

Proof: By definition, $K_{var} = X_0(T)$.

Now, let $x_t := -(V_0 - \theta_\tau^2)e^{(\gamma-\gamma_\tau)t} + e^{\gamma t}(V_t - \theta_\tau^2)$. Then by Ito's Lemma we get:

$$dx_t = \delta e^{\gamma t}\sqrt{(x_t + (V_0 - \theta_\tau^2)e^{(\gamma-\gamma_\tau)t})e^{-\gamma t} + \theta_\tau^2}dW_t^Q. \tag{12.9}$$

Which is of the form $dx_t = f(t, x_t)dW_t^Q$ with

$$f(t, x) := \delta e^{\gamma t}\sqrt{(x + (V_0 - \theta_\tau^2)e^{(\gamma-\gamma_\tau)t})e^{-\gamma t} + \theta_\tau^2}.$$

The process $(x_t)_{t\geq 0}$ is therefore a continuous local martingale, and even a true martingale since $E^Q(\int_0^T f^2(s, x_s)ds) < \infty$. We can use the change of time method introduced in (Swishchuk, 2004) and we get $x_t = \tilde{W}_{\phi_t}$, where $\tilde{W}_t$ is a $\mathcal{F}_{\phi_t^{-1}}-$ adapted $Q-$Brownian motion. This method basically says that every continuous local martingale can be represented as a time-changed brownian motion. The process $(\phi_t)_{t\geq 0}$ is a.e. increasing, non negative, $\mathcal{F}_t-$ adapted and is called the change of time process. This process is also equal to the quadratic variation $\langle x \rangle_t$ of the (square-integrable) continuous martingale x_t.

Expressions of ϕ_t, ϕ_t^{-1} and $\tilde{W}_t$ are given by:

$$\phi_t = \langle x \rangle_t = \int_0^t f^2(s, x_s)\,ds, \tag{12.10}$$

$$\tilde{W}_t = \int_0^{\phi_t^{-1}} f(s, x_s)dW_s^Q, \tag{12.11}$$

$$\phi_t^{-1} = \int_0^t f^{-2}\left(\phi_s^{-1}, x_{\phi_s^{-1}}\right)ds. \tag{12.12}$$

This immediately yields:

$$V_t = \theta_\tau^2 + (V_0 - \theta_\tau^2)e^{-\gamma_\tau t} + e^{-\gamma t}\tilde{W}_{\phi_t}. \tag{12.13}$$

Lemma 1:

$$E_t^Q(\tilde{W}_{\phi_s}) = \tilde{W}_{\phi_{t\wedge s}}. \tag{12.14}$$

And for $s, u \geq t$:

$$E_t^Q(\tilde{W}_{\phi_s}\tilde{W}_{\phi_u}) = x_t^2 + \delta^2\left[\theta_\tau^2\left(\frac{e^{2\gamma(s\wedge u)} - e^{2\gamma t}}{2\gamma}\right)\right.$$

$$+ (V_0 - \theta_\tau^2)\left(\frac{e^{(2\gamma-\gamma_\tau)(s\wedge u)} - e^{(2\gamma-\gamma_\tau)t}}{2\gamma - \gamma_\tau}\right)$$

$$\left. + x_t\left(\frac{e^{\gamma(s\wedge u)} - e^{\gamma t}}{\gamma}\right)\right]. \tag{12.15}$$

Proof: (12.14) comes from the fact that $x_t = \tilde{W}_{\phi_t}$ is a martingale. Let $s \geq u \geq t$. Then by iterated conditioning: $E_t^Q(\tilde{W}_{\phi_s}\tilde{W}_{\phi_u}) = E_t^Q(E_u^Q(\tilde{W}_{\phi_s}\tilde{W}_{\phi_u})) = E_t^Q(\tilde{W}_{\phi_u}E_u^Q(\tilde{W}_{\phi_s})) = E_t^Q(\tilde{W}_{\phi_u}^2)$, because $x_t = \tilde{W}_{\phi_t}$ is a martingale. Now, by definition of the quadratic variation, $x_u^2 - \langle x \rangle_u$ is a martingale and therefore $E_t^Q(\tilde{W}_{\phi_u}^2) = x_t^2 - \langle x \rangle_t + E_t^Q(\langle x \rangle_u) = x_t^2 - \phi_t + E_t^Q(\phi_u) = x_t^2 - \phi_t + \phi_t + E_t^Q(\int_t^u f^2(s, x_s) \, ds)$. By definiton of $f^2(s, x_s)$ and since x_t martingale, then we have (for $s \geq t$) $E_t^Q(f^2(s, x_s)) = f^2(s, x_t)$, and so that $E_t^Q(\tilde{W}_{\phi_s}\tilde{W}_{\phi_u}) = x_t^2 + \int_t^u f^2(s, x_t) \, ds$. A simple integration completes the proof.

The following theorem gives the expression of the Brockhaus and Long approximation of the volatility swap floating leg price process $Y_t(T)$.

Theorem 2: The Brockhaus and Long approximation of the price process $Y_t(T)$ of the floating leg of the volatility swap of maturity T in the delayed Heston model (12.1)–(12.2) is given by:

$$Y_t(T) \approx \sqrt{X_t(T)} - \frac{Var_t^Q(V_R)}{8X_t(T)^{\frac{3}{2}}} \qquad (12.16)$$

where $X_t(T)$ is given by Theorem 1 and:

$$Var_t^Q(V_R) = \frac{x_t \delta^2}{\gamma^3 T^2}\left[e^{-\gamma t}\left(1 - e^{-2\gamma(T-t)}\right) - 2(T-t)\gamma e^{-\gamma T}\right]$$

$$+ \frac{\delta^2}{2\gamma^3 T^2}\left[2\theta_\tau^2 \gamma(T-t) + 2(V_0 - \theta_\tau^2)\frac{\gamma}{\gamma_\tau}e^{-\gamma_\tau t}\right.$$

$$\left. + 4\theta_\tau^2 e^{-\gamma(T-t)} - \theta_\tau^2 e^{-2\gamma(T-t)} - 3\theta_\tau^2\right] - \frac{\delta^2(V_0 - \theta_\tau^2)}{\gamma^2 T^2(\gamma_\tau^2 + 2\gamma^2 - 3\gamma\gamma_\tau)}$$

$$\times \left[2(\gamma_\tau - 2\gamma)e^{-\gamma(T-t)-\gamma_\tau t} + (\gamma - \gamma_\tau)e^{-2\gamma(T-t)-\gamma_\tau t} + 2\frac{\gamma^2}{\gamma_\tau}e^{-\gamma_\tau T}\right]$$

$$(12.17)$$

Proof: The (conditioned) Brockhaus and Long approximation gives us: $Y_t(T) = E_t^Q(\sqrt{V_R}) \approx \sqrt{E_t^Q(V_R)} - \frac{Var_t^Q(V_R)}{8E_t^Q(V_R)^{\frac{3}{2}}} = \sqrt{X_t(T)} - \frac{Var_t^Q(V_R)}{8X_t(T)^{\frac{3}{2}}}$.

Furthermore:

$$Var_t^Q(V_R) = E_t^Q((V_R - E_t^Q(V_R))^2)$$

$$= \frac{1}{T^2}E_t^Q\left(\left(\int_0^T (V_s - E_t^Q(V_s))ds\right)^2\right).$$

From (12.13) we have $V_t = \theta_\tau^2 + (V_0 - \theta_\tau^2)e^{-\gamma_\tau t} + e^{-\gamma t}\tilde{W}_{\phi_t}$, and since $\tilde{W}_{\phi_t}$ is a martingale, $V_s - E_t^Q(V_s) = 0$ if $s \leq t$, and $V_s - E_t^Q(V_s) = e^{-\gamma s}(\tilde{W}_{\phi_s} - x_t)$ if $s > t$.

Therefore

$$Var_t^Q(V_R) = \frac{1}{T^2} E_t^Q \left(\left(\int_t^T e^{-\gamma s}(\tilde{W}_{\phi_s} - x_t)ds \right)^2 \right)$$

$$= \frac{1}{T^2} x_t^2 \left(\int_t^T e^{-\gamma s}ds \right)^2 + \frac{1}{T^2} E_t^Q \left(\left(\int_t^T e^{-\gamma s}\tilde{W}_{\phi_s}ds \right)^2 \right)$$

$$- \frac{2}{T^2} x_t \left(\int_t^T e^{-\gamma s} E_t^Q(\tilde{W}_{\phi_s})ds \right) \left(\int_t^T e^{-\gamma s}ds \right)$$

$$= -\frac{1}{T^2} x_t^2 \left(\int_t^T e^{-\gamma s}ds \right)^2 + \frac{1}{T^2} E_t^Q \left(\left(\int_t^T e^{-\gamma s}\tilde{W}_{\phi_s}ds \right)^2 \right)$$

$$= \frac{1}{T^2} \int_t^T \int_t^T e^{-\gamma(s+u)} E_t^Q(\tilde{W}_{\phi_s}\tilde{W}_{\phi_u})dsdu - \frac{1}{T^2}x_t^2 e^{-2\gamma t} \left(\frac{1-e^{-\gamma(T-t)}}{\gamma} \right)^2.$$

Lemma 1 and some straightforward computations complete the proof.

Corollary 2: The Brockhaus and Long approximation of the volatility swap price K_{vol} of maturity T at initiation of the contract $t = 0$ in the delayed Heston model (12.1)–(12.2) is given by:

$$K_{vol} \approx \sqrt{K_{var}} - \frac{Var^Q(V_R)}{8K_{var}^{\frac{3}{2}}} \tag{12.18}$$

where K_{var} is given by Corollary 1 and:

$$Var^Q(V_R) = \frac{\delta^2 e^{-2\gamma T}}{2T^2\gamma^3} \left[\theta_\tau^2 \left(2\gamma Te^{2\gamma T} + 4e^{\gamma T} - 3e^{2\gamma T} - 1 \right) \right.$$

$$+ \frac{\gamma}{2\gamma - \gamma_\tau}(V_0 - \theta_\tau^2) \left(2e^{2\gamma T} \left(2\frac{\gamma}{\gamma_\tau} - 1 \right) \right.$$

$$\left. - 4\gamma e^{\gamma T} \left(\frac{e^{(\gamma-\gamma_\tau)T} - 1}{\gamma - \gamma_\tau} \right) + 4e^{\gamma T} \left(1 - \frac{\gamma}{\gamma_\tau}e^{(\gamma-\gamma_\tau)T} \right) - 2 \right) \right]. \tag{12.19}$$

We notice that letting $\tau \to 0$ (and therefore $\gamma_\tau \to \gamma$) we get the formula of (Swishchuk, 2004).

Proof: We have by definition $K_{vol} = Y_0(T)$, and straightforward computations using theorem 2 finish the proof.

12.4   Volatility Swap Hedging

In this Section, we consider dynamic hedging of volatility swaps using variances swap. In the spirit of Broadie and Jain (2008), we consider a portfolio containing at time t one unit of volatility swap and β_t units of variance swaps, both of maturity T. Therefore the value Π_t of the portfolio at time t is:

$$\Pi_t = e^{-r(T-t)} \left[Y_t(T) - K_{vol} + \beta_t(X_t(T) - K_{var}) \right]. \tag{12.20}$$

The portfolio is self-financing, therefore:

$$d\Pi_t = r\Pi_t dt + e^{-r(T-t)} \left[dY_t(T) + \beta_t dX_t(T) \right]. \tag{12.21}$$

The price processes $X_t(T)$ and $Y_t(T)$ can be expressed, denoting $I_t := \int_0^t V_s ds$ the accumulated variance at time t (known at this time):

$$X_t(T) = E_t^Q \left[\frac{1}{T} I_t + \frac{1}{T} \int_t^T V_s ds \right] = g(t, I_t, V_t) \tag{12.22}$$

$$Y_t(T) = E_t^Q \left[\sqrt{\frac{1}{T} I_t + \frac{1}{T} \int_t^T V_s ds} \right] = h(t, I_t, V_t). \tag{12.23}$$

Letting $\tilde{\theta}_t^2 := \theta_\tau^2 + (V_0 - \theta_\tau^2) \frac{(\gamma - \gamma_\tau)}{\gamma} e^{-\gamma_\tau t}$ and noticing that $dI_t = V_t dt$, by Ito's lemma we get:

$$dX_t(T) = \left[\frac{\partial g}{\partial t} + \frac{\partial g}{\partial I_t} V_t + \frac{\partial g}{\partial V_t} \gamma(\tilde{\theta}_t^2 - V_t) + \frac{1}{2} \frac{\partial^2 g}{\partial V_t^2} \delta^2 V_t \right] dt + \frac{\partial g}{\partial V_t} \delta \sqrt{V_t} dW_t^Q \tag{12.24}$$

$$dY_t(T) = \left[\frac{\partial h}{\partial t} + \frac{\partial h}{\partial I_t} V_t + \frac{\partial h}{\partial V_t} \gamma(\tilde{\theta}_t^2 - V_t) + \frac{1}{2} \frac{\partial^2 h}{\partial V_t^2} \delta^2 V_t \right] dt + \frac{\partial h}{\partial V_t} \delta \sqrt{V_t} dW_t^Q. \tag{12.25}$$

As conditional expectations of cashflows at maturity of the contract, the price processes $X_t(T)$ and $Y_t(T)$ are by construction martingales, and therefore we should have:

$$\frac{\partial g}{\partial t} + \frac{\partial g}{\partial I_t} V_t + \frac{\partial g}{\partial V_t} \gamma(\tilde{\theta}_t^2 - V_t) + \frac{1}{2} \frac{\partial^2 g}{\partial V_t^2} \delta^2 V_t = 0 \tag{12.26}$$

$$\frac{\partial h}{\partial t} + \frac{\partial h}{\partial I_t} V_t + \frac{\partial h}{\partial V_t} \gamma(\tilde{\theta}_t^2 - V_t) + \frac{1}{2} \frac{\partial^2 h}{\partial V_t^2} \delta^2 V_t = 0. \tag{12.27}$$

The second equation, combined with some appropriate boundary conditions, was used in Broadie and Jain (2008) to compute the value of the price process $Y_t(T)$, whereas we focus on its Brockhaus and Long approximation.

Therefore we get:

$$dX_t(T) = \frac{\partial g}{\partial V_t} \delta \sqrt{V_t} dW_t^Q \tag{12.28}$$

$$dY_t(T) = \frac{\partial h}{\partial V_t} \delta \sqrt{V_t} dW_t^Q \tag{12.29}$$

and so:

$$d\Pi_t = r\Pi_t dt + e^{-r(T-t)}\left[\frac{\partial h}{\partial V_t}\delta\sqrt{V_t}dW_t^Q + \beta_t\frac{\partial g}{\partial V_t}\delta\sqrt{V_t}dW_t^Q\right]. \qquad (12.30)$$

In order to dynamically hedge a volatility swap of maturity T, one should therefore hold β_t units of variance swap of maturity T, with:

$$\beta_t = -\frac{\frac{\partial h}{\partial V_t}}{\frac{\partial g}{\partial V_t}} = -\frac{\frac{\partial Y_t(T)}{\partial V_t}}{\frac{\partial X_t(T)}{\partial V_t}}. \qquad (12.31)$$

The initial hedge ratio β_0 is given by $(Var^Q(V_R)$, K_{var} being given resp. in Corollary 2 and 1):

$$\beta_0 = -\frac{\frac{\partial Y_0(T)}{\partial V_0}}{\frac{\partial X_0(T)}{\partial V_0}} \qquad (12.32)$$

$$\frac{\partial X_0(T)}{\partial V_0} = \frac{1 - e^{-\gamma_\tau T}}{\gamma_\tau T} \qquad (12.33)$$

$$\frac{\partial Y_0(T)}{\partial V_0} \approx \frac{\frac{\partial X_0(T)}{\partial V_0}}{2\sqrt{K_{var}}} - \frac{K_{var}\frac{\partial Var^Q(V_R)}{\partial V_0} - \frac{3}{2}\frac{\partial X_0(T)}{\partial V_0}Var^Q(V_R)}{8K_{var}^{\frac{5}{2}}} \qquad (12.34)$$

$$\frac{\partial Var^Q(V_R)}{\partial V_0} = \frac{\delta^2 e^{-2\gamma T}}{T^2\gamma^3}\frac{\gamma}{2\gamma - \gamma_\tau}\left[e^{2\gamma T}\left(2\frac{\gamma}{\gamma_\tau} - 1\right) - 2\gamma e^{\gamma T}\left(\frac{e^{(\gamma-\gamma_\tau)T} - 1}{\gamma - \gamma_\tau}\right)\right.$$

$$\left. + 2e^{\gamma T}\left(1 - \frac{\gamma}{\gamma_\tau}e^{(\gamma-\gamma_\tau)T}\right) - 1\right] \qquad (12.35)$$

The hedge ratio β_t for $t > 0$ is given by $(Var_t^Q(V_R)$, $X_t(T)$ being given resp. in Theorem 2 and 1):

$$\beta_t = -\frac{\frac{\partial Y_t(T)}{\partial V_t}}{\frac{\partial X_t(T)}{\partial V_t}} \qquad (12.36)$$

$$\frac{\partial X_t(T)}{\partial V_t} = \frac{1 - e^{-\gamma(T-t)}}{\gamma T} \qquad (12.37)$$

$$\frac{\partial Y_t(T)}{\partial V_t} \approx \frac{\frac{\partial X_t(T)}{\partial V_t}}{2\sqrt{X_t(T)}} - \frac{X_t(T)\frac{\partial Var_t^Q(V_R)}{\partial V_t} - \frac{3}{2}\frac{\partial X_t(T)}{\partial V_t}Var_t^Q(V_R)}{8X_t(T)^{\frac{5}{2}}} \qquad (12.38)$$

$$\frac{\partial Var_t^Q(V_R)}{\partial V_t} = \frac{\delta^2}{\gamma^3 T^2}\left[1 - e^{-2\gamma(T-t)} - 2(T-t)\gamma e^{-\gamma(T-t)}\right]. \qquad (12.39)$$

12.5 Numerical Results

We take the parameters that have been calibrated in Part 1 of our article (vanilla options on September 30th 2011 for underlying EURUSD, maturities

from 1M to 10Y, strikes ATM, 25D Put/Call, 10D Put/Call), namely
$(v_0, \gamma, \theta^2, \delta, c, \alpha, \tau) = (0.0343, 3.9037, 10^{-8}, 0.808, -0.5057, 71.35, 0.7821)$. We plot
the naive Volatility Swap strike $\sqrt{K_{var}}$ and the adjusted Volatility Swap strike
$\sqrt{K_{var}} - \dfrac{Var^Q(V_R)}{8K_{var}^{\frac{3}{2}}}$ along the maturity dimension, as well as the convexity adjust-

ment $\dfrac{Var^Q(V_R)}{8K_{var}^{\frac{3}{2}}}$:

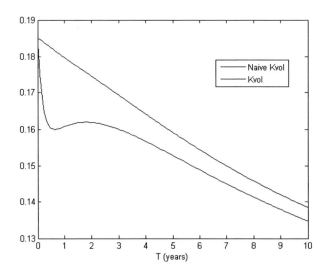

Fig. 12.1   Naive Volatility Swap Strike Vs. Adjusted Volatility Swap Strike.

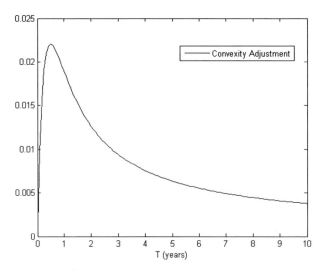

Fig. 12.2   Convexity Adjustment.

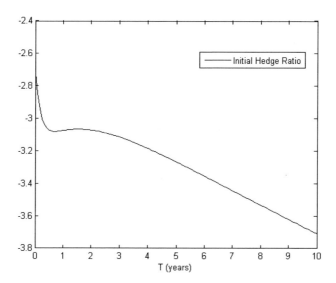

Fig. 12.3   Initial Hedge Ratio.

We also plot the initial hedge ratio β_0 along the maturity dimension.

12.6   Summary

- Following Chapter 11 that presented an easily implementable adjustment of the Heston model, the delayed Heston model, based on the will to take into account the past history of the variance process in its risk-neutral diffusion, we considered variance and volatility swap pricing and dynamic hedging in this context.
- We derived a closed formula for the variance swap fair strike, as well as for the Brockhaus and Long approximation of the volatility swap fair strike.
- Based on these results, we considered hedging of a position on a volatility swap with variance swaps. A closed formula — based on the Brockhaus and Long approximation — was derived for the number of variance swaps one should hold at each time t in order to hedge the position (hedge ratio).

Bibliography

Bollerslev, T. (1986). Generalized autoregressive conditional heteroscedasticity. *Journal of Economics*, 31, 307-27.

Broadie, M. and Jain, A. (2008). The effect of jumps and discrete sampling on volatility and variance swaps. *International Journal of Theoretical Applied Finance*, 11, 8, 761-797.

Broadie, M and Jain, A. (2008). Pricing and hedging volatility derivatives. *Journal of Derivatives*, 15, 3, 7-24.

Brockhaus, O. and Long, D. (2000). Volatility swaps made simple. *Risk Magazine*, January, 92-96.

Carr, P. and Lee, R. (2009). Robust replication of volatility derivatives. Working Paper.

Carr, P. and Lee, R. (2009). Volatility derivatives. *Annual Review of Financial Economics*, 1, 1-21.

Carr, P. and Lee, R. (2007). Realized volatility and variance: Options via swaps. *Risk Magazine*, 20, 5, 76-83.

Carr, P. and Madan, D. (1998). Towards a theory of volatility trading. In Jarrow, R. (ed.), *Volatility*, 417-427, Risk Book Publications.

Demeterfi, K., Derman, E., Kamal, M. and Zou, J. (1999). A guide to volatility and variance swaps. *The Journal of Derivatives*, Summer, 9-32.

Heston, S. and Nandi, S. (2000). Derivatives on volatility: Some simple solutions based on observables. Federal Rserve Bank of Atlanta. Working Paper. 20 November.

Heston, S. (1993). A closed-form solution for options with stochastic volatility with applications to bond and currency options. *Review of Financial Studies*, 6, 327-343.

Hobson, D. and Rogers, L. (1998). Complete models with stochastic volatility. *Mathematical Finance*, 8, 1, 27-48.

Howison, S., Rafailidis, A. and Rasmussen, H. (2004). On the pricing and hedging of volatility derivatives. *Applied Mathematical Finance*, 11, 4, 317-346.

Javaheri, A., Wilmott, P. and Haug, E. (2002). GARCH and volatility swaps. *Wilmott Technical Article*, January, 17 pages.

Kahl, C. and Jäckel, P. (2005). Not-so-complex logarithms in the Heston model. *Wilmott Magazine*, September, 94-103.

Kazmerchuk, Y., Swishchuk, A. and Wu, J. (2005). A continuous-time GARCH model for stochastic volatility with delay. *Canadian Applied Mathematics Quarterly*, 13, 2.

Mikhailov, S. and Noegel, U. (2005). Heston's stochastic volatility model implementation, calibration and some extensions. *Wilmott Magazine*, 1, 74-79.

Swishchuk, A. and Malenfant, K. (2011). Variance swaps for local levy based stochastic volatility with delay. *International Review of Applied Financial Issues and Economics*, 3, 2, 432-441.

Swishchuk, A. and Vadori, N. (2012). Delayed Heston model: Improvement of the volatility surface fitting. *Wilmott Magazine*, submitted February 2012 (Under Review).

Swishchuk, A. and Vadori, N. (2012). Pricing and hedging of volatility swap in the delayed Heston model. *Wilmott Magazine*, submitted March 2012 (Under Review).

Swishchuk, A. and Li, X. (2011). Pricing variance swaps for stochastic volatilities with delay and jumps. *International Journal of Stochastic Analysis*, Volume 2011, 27 pages.

Swishchuk, A. and Couch, M. (2010). Volatility and variance swaps for the COGARCH(1,1) model. *Wilmott Magazine*, 2, 5, 231-246.

Swishchuk, A. (2006). Modeling and pricing of variance swaps for multi-factor stochastic volatilities with delay. *Canadian Applied Mathematics Quarterly*, 14, 4, 439-468.

Swishchuk, A. (2005). Modeling and pricing of variance swaps for stochastic volatilities with delay. *Wilmott Magazine*, September, 19, 63-73.

Swishchuk, A. (2004). Modeling of variance and volatility swaps for financial markets with stochastic volatilities. *Wilmott Magazine*, September issue, No. 2, 64-72.

Chapter 13

Pricing of Variance and Volatility Swaps with Semi-Markov Volatilities

13.1 Introduction

We consider a semi-Markov modulated market consisting of a riskless asset or bond, B, and a risky asset or stock, S, whose dynamics depend on a semi-Markov process x. Using the martingale characterization of semi-Markov processes, we note the incompleteness of semi-Markov modulated markets and find the minimal martingale measure. We price variance (Theorem 1) and volatility (Theorem 2) swaps for stochastic volatilities driven by the semi-Markov processes. We will also discuss some extensions of the obtained results such as local semi-Markov volatility, Dupire formula for the local semi-Markov volatility and residual risk associated with the swap pricing.

The chapter is organized as follows. The literature review and necessarily notions is presented in Section 13.2. Martingale characterization of semi-Markov processes is considered in Section 13.3. Minimal martingale measure for stock price with semi-Markov volatility is constructed in Section 13.4. Section 13.5 contains pricing of varinace swap for stochastic volatility driven by a semi-Markov process. An example of variance swap for stochastic volatility driven by two-state continuous-time Markov chain is presented in Section 13.6. Section 13.7 is devoted to the pricing of volatility swap for stochastic volatility driven by a semi-Markov process. In Section 13.8 we consider some extensions of the obtained results: local/current semi-Markov volatility, Dupire formula for a semi-Markov local volatility, risk-minimizing strategies and residual risk associated with the swaps pricing.

13.2 Martingale Characterization of Semi-Markov Processes

13.2.1 *Markov Renewal and Semi-Markov Processes*

Let $(\Omega, \mathcal{F}, P)$ be a probability space, $(X, \mathcal{X})$ be a measurable space and $Q(x, B, t) = P(x, B)G_x(t), x \in X, B \in \mathcal{X}, t \in R_+$, be a semi-Markov kernel. Let us consider a $(X \times R_+, \mathcal{X} \otimes \mathcal{B}_+)$-valued stochastic process $(x_n, \tau_n; n \geq 0)$, with $\tau_0 \leq \tau_1 \leq ... \leq \tau_n \leq \tau_{n+1} \leq ....$ (See Korolyuk and Swishchuk (1995), Swishchuk (2000)).

Definition 1. A Markov renewal process is a two component Markov chain, $(x_n \tau_n; n \geq 0)$, homogeneous with respect to the second component with transition probabilities

$$P(x_{n+1} \in B, \tau_{n+1} - \tau_n \leq t | \mathcal{F}_n) = Q(x_n, B, t) = P(x, B)G_x(t) \quad a.s.$$

Let us define the counting process of jumps $\nu(t)$ by $\nu(t) = \sup\{n \geq 0 : \tau_n \leq t\}$, that gives the number of jumps of the Markov renewal process in the time interval $(0, t]$.

Definition 2. A stochastic process $x(t), t \geq 0$, defined by the following relation $x(t) = x_{\nu(t)}$ is called a semi-Markov process, associated to the Markov renewal process $(x_n \tau_n; n \geq 0)$.

Remark 1. Markov jump processes are special cases of semi-Markov processes with semi-Markov kernel $Q(x, B, t) = P(x, B)[1 - e^{-\lambda(x)t}]$.

Definition 3. The following auxiliary process $\gamma(t) = t - \tau_{\nu(t)}$ is called the backward recurrence time-the time period since the last renewal epoch until t (or the current life in the terminology of reliability theory, or age random variable).

Remark 2. The current life is a Markov process with generator $Qf(t) = f'(t) + \lambda(t)[f(0) - f(t)]$, where $\lambda(t) = -\frac{\bar{F}'(t)}{\bar{F}(t)}$, $\bar{F}(t) = 1 - F(t)$, $Domain(Q) = C^1(R)$.

Remark 3. If we expand the state space of the semi-Markov process to include a component that records the amount of time already spent in the current state, then this additional information in the state description makes the semi-Markov process Markovian. For example, the following process $(x(t), \gamma(t))$ is a Markov process.

Definition 4. The compensating operator Q of the Markov renewal process is defined by the following relation

$$Qf(x_0, \tau_0) = q(x_0)E[f(x_1, \tau_1) - f(x_0, \tau_0)|calF_0],$$

where $q(x) = 1/m(x), m(x) = \int_0^{+\infty} \bar{G}_x(t)dt, \mathcal{F}_t = \sigma\{x(s), \tau_{\nu(s)}; 0 \leq s \leq t\}$.

Proposition 1. The compensating operator of the Markov renewal process can be defined by the relation

$$Qf(x, t) = q(x)\left[\int_0^{+\infty} G_x(ds)\int_X P(x, dy)f(y, t + s) - f(x, t)\right].$$

This statement follows directly from Definition 3.

Proposition 2. Let (x_n, τ_n) be the Markov renewal process, Q be the compensating operator,

$$m_n := f(x_n, \tau_n) - \sum_{i=1}^{n}(\tau_i - \tau_{i-1})Qf(x_{i-1}, \tau_{i-1})),$$

and $\mathcal{F}_n = \sigma\{x_k, \tau_k; k \leq n\}$. Then the process m_n is a $\mathcal{F}_n$-martingale for any function f such that $E_x|f(x_1, \tau_1)| < +\infty$.

Let $x(t)$ be a Markov process with infinitesimal generator Q.

Proposition 3. The process

$$m(t) := f(x(t)) - f(x) - \int_0^t Qf(x(s))ds$$

is an $\mathcal{F}_t^x = \sigma\{x(s); s \leq t\}$-martingale (see Korolyuk and Swishchuk (1995)).

This statement follows from Dynkin formula.

Proposition 4. The quadratic variation of the martingale $m(t)$ is the process

$$< m(t) >= \int_0^t [Qf^2(x(s)) - 2f(x(s))Qf(x(s))]ds.$$

(See Elliott and Swishchuk (2007)).

13.2.2 *Jump Measure for Semi-Markov Process*

The *jump measure* for $x(t)$ is defined in the following way (see Korolyuk and Swishchuk, 1995):

$$\mu([0,t] \times A) = \sum_{n \geq 0} \mathbf{1}(x_n \in A, \tau_n \leq t), \quad A \in \mathcal{X}, t \geq 0.$$

It is known (see Korolyuk and Swishchuk (1995)) that predictable projection (compensator) for μ has the form:

$$\nu(dt, dy) = \sum_{n \geq 0} \mathbf{1}(\tau_n < t \leq \tau_{n+1}) \frac{P(x_n, dy) g_{x_n}(t)}{\bar{G}_{x_n}(t)} dt,$$

where $\bar{G}_x(t) = 1 - G_x(t), g_x(t) = dG_x(t)/dt, \forall x \in X$.

13.2.3 *Martingale Characterization of Semi-Markov Processes*

Let x_t be a semi-Markov process in with semi-Markov kernel $Q(x, B, t) = P(x, B)G_x(t)$ and $\gamma(t)$ be the current life.

Lemma 1. The process

$$m_t^f := f(x_t, \gamma(t)) - \int_0^t Qf(x_s, \gamma(s)))ds \tag{13.1}$$

is a martingale with respect to the filtration $F_t^* := \sigma\{x_s, \tau_{nu(s)}; 0 \leq s \leq t\}$, where Q is the infinitesimal operator of Markov process $(x_t, \gamma(t))$:

$$Qf(x, t) = \frac{df(x, t)}{dt} + \frac{g_x(t)}{\bar{G}_x(t)} \int_X P(x, dy)[f(y, t) - f(x, t)]. \tag{13.2}$$

This statement follows from Proposition 3 and the fact that $(x(t), \gamma(t))$ is a Markov process (see Remark 3).

Let us calculate the quadratic variation of the martingale m_t^f.

Lemma 2. Let Q be such that if $f \in Domain(Q)$, then $f^2 \in Domain(Q)$. The quadratic variation $\langle m_t^f \rangle$ of the martingale m_t^f in (13.5) is equal to

$$\langle m_t^f \rangle = \int_0^t [Qf^2(x_s, \gamma(s)) - 2f(x_s, \gamma(s))Qf(x_s, \gamma(s))]ds. \tag{13.3}$$

This statement follows from Proposition 4 and the fact that $(x(t), \gamma(t))$ is a Markov process (see Remark 3).

Lemma 3. Let the following condition (Novikov's condition) is satisfied

$$E^P \exp\left\{\frac{1}{2}\int_0^t [Qf^2(x_s,\gamma(s)) - 2f(x_s,\gamma(s))Qf(x_s,\gamma(s))]ds\right\} < +\infty,$$

$$\forall f^2 \in Domain(Q).$$

Then $E^P e_t^f = 1$, where

$$e_t^f := e^{m_t^f - \frac{1}{2}\langle m_t^f \rangle},$$

and e_t^f is a P-martingale (Doléans-Dade martingale).

13.3 Minimal Risk-Neutral (Martingale) Measure for Stock Price with Semi-Markov Stochastic Volatility

13.3.1 *Current Life Stochastic Volatility Driven by Semi-Markov Process (Current Life Semi-Markov Volatility)*

Let x_t be a semi-Markov process in measurable phase space $(X, \mathcal{X})$.

We suppose that the stock price S_t satisfies the following stochastic differential equation

$$dS_t = S_t(rdt + \sigma(x_t,\gamma(t))dw_t)$$

with the volatility $\sigma := \sigma(x_t,\gamma(t))$ depending on the process x_t, which is independent on standard Wiener process w_t, and the current life $\gamma(t) = t - \tau_{\nu(t)}$, $\mu \in R$. We call the volatility $\sigma(x_t,\gamma(t))$ the *current life semi-Markov volatility*.

Remark 4. We note that process (x_t,γ_t) is a Markov process on (X, R_+) with infinitesimal operator

$$Qf(x,t) = \frac{df(x,t)}{dt} + \frac{g_x(t)}{\bar{G}_x(t)}\int_X P(x,dy)[f(y,t) - f(x,t)]. \tag{13.4}$$

13.3.2 *Minimal Martingale Measure*

We consider the following (B, S)-security market (B stands for 'Bond' and S stands for 'Stock'). Let the stock price S_t satisfies the following eqaution

$$dS_t = S_t(\mu dt + \sigma(x_t,\gamma(t))dw_t), \tag{13.5}$$

where $\mu \in R$ is the appreciation rate and $\sigma(x_t,\gamma(t))$ is the semi-Markov volatility, and the bond price $B(t)$ is

$$B(t) = B_0 e^{rt}, \tag{13.6}$$

where $r > 0$ is the risk-free rate of return (interest rate).

As long as we have two sources of randomness, Brownian motion $w(t)$ and semi-Markov process x_t, the above (B, S)-security market (13.5)–(13.6) is incomplete (see Theorem 1, Elliott and Swishchuk (2007)) and there are many risk-neutral (or martingale) measures. We are going to construct the minimal martingale measure (see Föllmer and Schweitzer (1991)). With respect to this construction (see Lemma 4, Elliott and Swishchuk (2007)) the following measure P^* is the minimal martingale measure.

Using Girsanov's Theorem (see Shiryaev (1999)) we obtain the following result concerning the minimal martingale measure in the above market.

Lemma 4. Under the assumption $\int_0^T (\frac{r-\mu}{\sigma(x_t,\gamma(t))})^2 dt < +\infty, a.s.$, the following holds:

1) There is a probability measure P^* equivalent to P such that

$$\frac{dP^*}{dP} = \exp\left\{ \int_0^T \frac{r-\mu}{\sigma(x_t,\gamma(t))} dw(t) - \frac{1}{2} \int_0^T \left(\frac{r-\mu}{\sigma(x_t,\gamma(t))} \right)^2 dt \right\} \quad (13.7)$$

is its Radon-Nikodym density;

2) The discounted stock price $Z(t) = \frac{S_t}{B(t)}$ is a positive local martingale with respect to P^* and is given by

$$Z(t) = \exp\left\{ -\frac{1}{2} \int_0^t \sigma^2(x_s,\gamma(s)) ds + \int_0^t \sigma(x_s,\gamma(s)) dw^*(s) \right\},$$

where $w^*(t) = \int_0^t \frac{r-\mu}{\sigma(x_s,\gamma(s))} ds + w(t)$ is a standard Brownian motion with respect to P^*.

Remark 5. Measure P^* is called the minimal martingale measure.

Remark 6. We note that under risk-neutral measure P^* the stock price S_t satisfies the following equation:

$$dS_t = S_t(rdt + \sigma(x_t,\gamma(t)) dw^*(t)) \quad (13.8)$$

and discounted process $Z(t)$ has the presentation

$$dZ(t) = Z(t)\sigma(x_s,\gamma(s)) dw^*(t).$$

Remark 7. A sufficient condition (Novikov's condition) for the right-hand side of (7) to be a martingale is

$$E \exp\left\{ \frac{1}{2} \int_0^T \left(\frac{r-\mu}{\sigma(x_t,\gamma(t))} \right)^2 dt \right\} < +\infty.$$

13.4 Pricing of Variance Swaps for Stochastic Volatility Driven by a Semi-Markov Process

A variance swap is a forward contract on annualized variance, the square of the realized volatility. Its payoff at expiration is equal to

$$N(\sigma_R^2(x) - K_{var}),$$

where $\sigma_R^2(x)$ is the realized stock variance (quoted in annual terms) over the life of the contract,

$$\sigma_R^2(x) := \frac{1}{T} \int_0^T \sigma^2(x_s,\gamma(s)) ds, \quad (13.9)$$

K_{var} is the delivery price for variance, and N is the notional amount of the swap in dollars per annualized volatility point squared. The holder of variance swap at expiration receives N dollars for every point by which the stock's realized variance $\sigma_R^2(x)$ has exceeded the variance delivery price K_{var}.

Pricing a variance forward contract or swap is no different from valuing any other derivative security. The value of a forward contract F on future realized variance with strike price K_{var} is the expected present value of the future payoff in the risk-neutral world:

$$P_{var}(x) = E\{e^{-rT}(\sigma_R^2(x) - K_{var})\}, \tag{13.10}$$

where r is the risk-free discount rate corresponding to the expiration date T, and E denotes the expectation with respect to the minimal martingale measure P (from now on we use simplier notation P instead of P^*).

Let us show how we can calculate $EV(x)$, where $V(x) := \sigma_R^2(x)$. For that we need to calculate $E\sigma^2(x_t, \gamma(t))$.

We note (see Section 2, Lemma 1) that for $\sigma(x) \in Domain(Q)$ the following process

$$m_t^\sigma := \sigma(x_t, \gamma(t)) - \sigma(x, 0) - \int_0^t Q\sigma(x_s, \gamma(s)))ds$$

is a zero-mean martingale with respect to $F_t := \sigma\{x_s, \tau_{\nu(s)}; 0 \le s \le t\}$ and Q is the infinitesimal operator defined in (13.4).

The quadratic variation of the martingale m_t^f by Lemma 2 is equal to

$$\langle m_t^\sigma \rangle = \int_0^t [Q\sigma^2(x_s, \gamma(t)) - 2\sigma(x_s, \gamma(t))Q\sigma(x_s, \gamma(t))]ds, \quad \sigma^2(x, t) \in Domain(Q). \tag{13.11}$$

Since $\sigma(x_s, \gamma(s))$ satisfies the following stochastic differential equation

$$d\sigma(x_t, \gamma(t)) = Q\sigma(x_t, \gamma(t))dt + dm_t^\sigma$$

then we obtain from *Itô* formula (see Shiryaev (1999)) that $\sigma^2(x_t, \gamma(t))$ satisfies the following stochastic differential equation

$$d\sigma^2(x_t, \gamma(t)) = 2\sigma(x_t, \gamma(t))dm_t^\sigma + 2\sigma(x_t, \gamma(t))Q\sigma(x_t, \gamma(t))dt + d\langle m_t^\sigma \rangle, \tag{13.12}$$

where $\langle m_t^\sigma \rangle$ is defined in (13.11).

Substituting (13.11) into (13.12) and taking the expectation of both parts of (13.12) we obtain that

$$E\sigma^2(x_t, \gamma(t)) = \sigma^2(x, 0) + \int_0^t QE\sigma^2(x_s, \gamma(s))ds,$$

and solving this equation we have

$$E\sigma^2(x_t, \gamma(t)) = e^{tQ}\sigma^2(x, 0).$$

Finally, we obtain that

$$EV(x) = \frac{1}{T}\int_0^T e^{tQ}\sigma^2(x, 0)dt. \tag{13.13}$$

In this way, we have obtained the following result.

Theorem 1. The value of a variance swap for semi-Markov stochastic volatility $\sigma(x_t, \gamma(t))$ equals to

$$P_{var}(x) = e^{-rT} \left(\frac{1}{T} \int_0^T e^{tQ} \sigma^2(x, 0) dt - K_{var} \right) \tag{13.14}$$

where Q is defined in (13.4).

13.5   Example of Variance Swap for Stochastic Volatility Driven by Two-State Continuous-Time Markov Chain

Let Q be a generator of two-state continuous time Markov chain

$$Q = \begin{pmatrix} q_{11} & q_{12} \\ q_{21} & q_{22} \end{pmatrix}$$

and

$$P(t) = \begin{pmatrix} p_{11}(t) & p_{12}(t) \\ p_{21}(t) & p_{22}(t) \end{pmatrix}$$

be a Markov transition function. Thus,

$$P(t) = e^{tQ}.$$

In this case, the variance takes two values: $\sigma^2(1)$ and $\sigma^2(2)$.

From formula (85) it follows that the value of variance swap in this case is equal to

$$P(i) = e^{-rT} \left(\frac{1}{T} \int_0^T [p_{i1}(s)\sigma^2(1) + p_{i2}(s)\sigma^2(2)] ds - K_{var} \right) \tag{13.15}$$

for $i = 1, 2$.

Thus, the value of varinace swap depends on the initial state of Markov chain.

We note, that if Markov chain is stationary with ergodic distribution (p_1, p_2), then the value of variance swap is equal to

$$P = p_1 P(1) + p_2 P(2),$$

where $P(i)$, $i = 1, 2$, are defined in (13.15) (see also Elliott and Swishchuk (2007)).

13.6   Pricing of Volatility Swaps for Stochastic Volatility Driven by a Semi-Markov Process

13.6.1   *Volatility Swap*

Volatility swaps are forward contracts on future realized stock volatility.

A stock's volatility is the simplest measure of its risk less or uncertainty. Formally, the volatility $\sigma_R(S)$ is the annualized standard deviation of the stock's returns during the period of interest, where the subscript R denotes the observed or "realized" volatility for the stock S.

The easy way to trade volatility is to use volatility swaps, sometimes called realized volatility forward contracts, because they provide pure exposure to volatility (and only to volatility).

A stock *volatility swap* is a forward contract on the annualized volatility. Its payoff at expiration is equal to

$$N(\sigma_R(S) - K_{vol}),$$

where $\sigma_R(S)$ is the realized stock volatility (quoted in annual terms) over the life of contract,

$$\sigma_R(S) := \sqrt{\frac{1}{T} \int_0^T \sigma_s^2 ds}, \qquad (13.16)$$

σ_t is a stochastic stock volatility, K_{vol} is the annualized volatility delivery price, and N is the notional amount of the swap in dollar per annualized volatility point. The holder of a volatility swap at expiration receives N dollars for every point by which the stock's realized volatility σ_R has exceeded the volatility delivery price K_{vol}. The holder is swapping a fixed volatility K_{vol} for the actual (floating) future volatility σ_R. We note that usually $N = \alpha I$, where α is a converting parameter such as 1 per volatility-square, and I is a long-short index (+1 for long and −1 for short).

Pricing a variance forward contract or swap is no different from valuing any other derivative security. The value of a forward contract F on future realized variance with strike price K_{var} is the expected present value of the future payoff in the risk-neutral world:

$$P_{vol}(x) = E\{e^{-rT}(\sigma_R(x) - K_{vol})\}, \qquad (13.17)$$

where r is the risk-free discount rate corresponding to the expiration date T, and E denotes the expectation with respect to the minimal martingale measure P (from now on we use simplier notation P instead of P^*).

Thus, for calculating variance swaps we need to know only $E\{\sigma_R^2(S)\}$, namely, mean value of the underlying variance.

To calculate volatility swaps we need more. From Brockhaus and Long (2000) approximation (which is used the second order Taylor expansion for function $\sqrt{x}$) we have (see also Javaheri *et al.* (2002), p.16):

$$E\{\sqrt{\sigma_R^2(S)}\} \approx \sqrt{E\{V\}} - \frac{Var\{V\}}{8E\{V\}^{3/2}},$$

where $V := \sigma_R^2(S)$ and $\frac{Var\{V\}}{8E\{V\}^{3/2}}$ is the convexity adjustment.

Thus, to calculate volatility swaps we need both $E\{V\}$ and $Var\{V\}$.

13.6.2   *Pricing of Volatility Swap*

From Brockhaus and Long (2000) approximation we have (see also Javaheri *et al.* (2002), p. 16):

$$E\{\sqrt{\sigma_R^2(S)}\} \approx \sqrt{E\{V\}} - \frac{Var\{V\}}{8E\{V\}^{3/2}}, \qquad (13.18)$$

where $V := \sigma_R^2(S)$ and $\frac{Var\{V\}}{8E\{V\}^{3/2}}$ is the convexity adjustment.

As we can see from (13.18), to calculate volatility swaps we need both $E\{V\}$ and $Var\{V\}$.

We have already calculated $E\{\sigma_R^2(S)\} = E[V]$ (see (13.13)). Let us calculate $Var(V) := E[V]^2 - (E[V])^2$. In this way, we need $E[V]^2 = E\sigma_R^4(S)$. Taking into account the expression for $V = \sigma_R^2(S)$ we have:

$$E[V]^2 = \frac{1}{T^2}\int_0^T \int_0^T E[\sigma^2(x_t,\gamma(t))\sigma^2(x_s,\gamma(s))dtds]. \qquad (13.19)$$

In this way, the variance of V, $Var[V]$, is

$$Var[V] = E[V^2] - (E[V])^2$$

$$= \frac{1}{T^2}\int_0^T \int_0^T E\sigma^2(x_t,\gamma(t))\sigma^2(x_s,\gamma(s))dtds$$

$$- \left(\frac{1}{T}\int_0^T e^{tQ}\sigma^2(x,0)dt\right)^2. \qquad (13.20)$$

Taking into account (13.16)–(13.20) we obtain:

$$P_{vol}(x) = e^{-rT}[E\sigma_R(S) - K_{vol}]$$

$$= e^{-rT}\left[E\sqrt{\frac{1}{T}\int_0^T \sigma^2(x_s,\gamma(s))dt} - K_{vol}\right]$$

$$\approx e^{-rT}\left[\sqrt{EV} - \frac{Var(V)}{8(EV)^{3/2}} - K_{vol}\right]$$

$$= e^{-rT}\left\{\sqrt{\frac{1}{T}\int_0^T e^{tQ}\sigma^2(x,0)dt}\right.$$

$$- \left[\frac{1}{T^2}\int_0^T \int_0^T E\sigma^2(x_t,\gamma(t))\sigma^2(x_s,\gamma(s))dtds\right.$$

$$\left.\left.- \left(\frac{1}{T}\int_0^T e^{tQ}\sigma^2(x,0)dt\right)^2\right]\middle/\left(8\left(\frac{1}{T}\int_0^T e^{tQ}\sigma^2(x,0)dt\right)^{3/2}\right) - K_{vol}\right\}. \qquad (13.21)$$

Summarizing, we have

Theorem 2. The value of volatility swap for semi-Markov stochastic volatility $\sigma(x_t, \gamma(t))$ equals to

$$
P_{vol}(x) \approx e^{-rT} \left\{ \sqrt{\frac{1}{T} \int_0^T e^{tQ} \sigma^2(x,0) dt } \right.
$$

$$
- \left[\frac{1}{T^2} \int_0^T \int_0^T E\sigma^2(x_t, \gamma(t)) \sigma^2(x_s, \gamma(s)) dt ds \right.
$$

$$
\left. \left. - \left(\frac{1}{T} \int_0^T e^{tQ} \sigma(x,0) dt \right)^2 \right] \middle/ \left(8 \left(\frac{1}{T} \int_0^T e^{tQ} \sigma^2(x,0) dt \right)^{3/2} \right) - K_{vol} \right\}.
$$

$$(13.22)$$

13.7   Discussions of Some Extensions

13.7.1   *Local Current Stochastic Volatility Driven by a Semi-Markov Process (Local Current Semi-Markov Volatility)*

We suppose that the stock price S_t satisfies the following stochastic differential equation

$$
dS_t = S_t(rdt + \sigma_{loc}(S_t, x_t, \gamma(t)) dw_t)
$$

with the volatility $\sigma := \sigma_{loc}(S_t, x_t, \gamma(t))$ depending on the process x_t, which is independent on standard Wiener process w_t, stock price S_t, and the current life $\gamma(t) = t - \tau_{\nu(t)}$.

Remark 8. We note that process (S_t, x_t, γ_t) is a Markov process on (R_+, X, R_+) with infinitesimal operator

$$
Qf(s,x,t) = \frac{\partial f(s,x,t)}{\partial t} + \frac{g_x(t)}{\bar{G}_x(t)} \int_X P(x, dy)[f(s,y,t) - f(s,x,t)]
$$

$$
+ rS \frac{\partial f(s,x,t)}{\partial s} + \frac{1}{2} \sigma^2(s,x,0) S^2 \frac{\partial^2 f(s,x,t)}{\partial s^2}.
$$

$$(13.23)$$

Using the same procedure as in Section 4 with infinitesimal operator Q in (13.23) we get the following result

Theorem 3. The value of a variance swap for Markov stochastic volatility $\sigma(x_t)$ equals to

$$
P(x) = e^{-rT} \left(\frac{1}{T} \int_0^T e^{tQ} \sigma^2(s,x,0) dt - K_{var} \right),
$$

where Q is defined in (13.23).

13.7.2   *Local Stochastic Volatility Driven by a Semi-Markov Process (Local Semi-Markov Volatility)*

We suppose that the stock price S_t satisfies the following stochastic differential equation

$$dS_t = S_t(rdt + \sigma(S_t, x_t, t)dw_t)$$

with the volatility $\sigma := \sigma(x_t, t)$ depending on the process x_t, which is independent on standard Wiener process w_t, stock price S_t and current time t.

Suppose that $\sigma(S, x, t)$ is differentiable function by t with bounded derivative.

Then we can reduce the problem to the previous one by the following expansion:

$$\sigma(S_t, x_t, t) = \sigma(S_t, x_t, \gamma(t)) + \tau_{\nu(t)} \frac{d\sigma(S_t, x_t, t)}{dt} + o(\tau_{\nu(t)})$$

The error of estimation will be

$$E[\sigma(S_t, x_t, t) - \sigma(S_t, x_t, \gamma(t))]^2 \le E[\tau_{\nu(t)}]^2 \times C,$$

where $C = \max_{0 \le t \le T} E[\frac{d\sigma(S_t, x_t, t)}{dt}]^2$.

13.7.3   *Dupire Formula for Semi-Markov Local Volatility*

Unlike the implied volatility produced by applying the Black-Scholes formulae to market prices, the local volatility is the volatility implied by the market prices and the one factor Black-Scholes. In 1994, Dupire showed (see Dupire (1994)) that if the spot price follows a risk-neutral random walk of the form:

$$\frac{dS}{S} = (r - D)dt + \sigma(t, S)dW$$

and if no-arbitrage market prices for European vanilla options are available for all strikes K and expires T, then $\sigma(t, S)$ can be extracted analytically from these option prices. If C denotes the price of a European call with strike K and expiry T, we obtain Dupire's famous equation:

$$\frac{\partial C}{\partial T} = \sigma_{local}^2(K, T) \frac{K^2}{2} \frac{\partial^2 C}{\partial K^2} - (r - D)K \frac{\partial C}{\partial K} - DC,$$

where r and D are interest and dividend rates, respectively. After rearranging this equation, we obtain the direct expression to calculate the local volatility (Dupire formulae):

$$\sigma_{local}(K, T) = \sqrt{\frac{\frac{\partial C}{\partial T} + (r - D)K \frac{\partial C}{\partial K} + DC}{\frac{K^2}{2} \frac{\partial^2 C}{\partial K^2}}}. \tag{13.24}$$

We suppose that the stock price S_t satisfies the following stochastic differential equation

$$dS_t = S_t((r - D)dt + \sigma_{loc}(S_t, x_t, t)dw_t)$$

with the volatility $\sigma := \sigma(S_t, x_t, \gamma(t))$ depending on the process x_t, which is independent on standard Wiener process w_t, stock price S_t, and the current life $\gamma(t) = t - \tau_{\nu(t)}$. Here D is the dividend rate.

Let C denotes the price of a European call with strike K and expiry T for our (B, S)-security market with semi-Markov volatility. Then C satisfies the following equation:

$$\frac{\partial C}{\partial T} + (r - D)K\frac{\partial C}{\partial K} - \frac{1}{2}\sigma_{loc}(K, x, T)\frac{\partial^2 C}{\partial K^2} - DC + QC = 0$$

where Q is defined by

$$Qf(x, t) = \frac{g_x(t)}{\bar{G}_x(t)}\int_X P(x, dy)[f(y, t) - f(x, t)]. \tag{13.25}$$

Then the semi-Markov local volatility can be calculated using the following expression

$$\sigma_{loc}(K, x, T) = \sqrt{\frac{\frac{\partial C}{\partial T} + (r - D)K\frac{\partial C}{\partial K} + DC - QC}{\frac{K^2\partial^2 C}{2\partial K^2}}}, \tag{13.26}$$

where Q is defined in (13.25). Of course, if σ does not depend on x (and C does not depend as well), then $QC = 0$ in (13.26) and (13.26) coincides with (13.24).

13.7.4   *Risk-Minimizing Strategies (or Portfolios) and Residual Risk*

Let $\pi(t) := (\alpha(t), \beta(t))$ is a *portfolio or strategy* (predictable processes), $V_t(\pi) = \alpha(t) + \beta(t)S_t$ is the *value process*, where S_t satisfies

$$dS_t = S_t\sigma(S_t, x_t, \gamma(t))dw^*(t). \tag{13.27}$$

We suppose that $r = 0$ just for simplicity. Let's define the *cost process*

$$C_t(\pi) = V_t(\pi) - \int_0^t \beta(u)dS_u.$$

The *residual risk* is defined by the formula (see Föllmer and Schweizer (1991))

$$R_t(\pi) = E[(C_T(\pi) - C_t(\pi))^2|\mathcal{F}_t].$$

Portfolio $\pi(t)$ is said to be H admissible if $V_T(\pi) = H$. Also $\pi(t)$ is *self-financing* if the cost process is a martingale. The H-admissible self-financing portfolio π^* is called a *risk-minimizing* if, for any H admissible π for any t

$$R_t(\pi^*) \leq R_t(\pi).$$

Proposition 5 (see Swishchuk (1995)). The risk-minimizing H-admissible strategy $\pi^* = (\alpha^*, \beta^*)$ is given by the following formula:

$$\beta^*(t) = u_s(S_t, x_t, t)$$

and

$$\alpha^*(t) = V_t(\pi^*) - \alpha^*(t)S_t,$$

where

$$V_t(\pi^*) = E^* f(S_T) + \int_0^t u_s(S_r, x(r), r) dS_r$$

$$+ \int_0^t \int_X \phi(r, y)(\mu - \nu)(dr, dy),$$

$\phi(r, y) := u(S_r, y, r) - u(S_r, x(r-), r)$, and function $u(z, x, t)$ satisfies the following Cauchy problem:

$$\begin{cases} u_t(z, x, t) + \frac{1}{2}\sigma^2(z, x, t)z^2 u_{zz}(z, x, t) + Qu(z, x, t) = 0 \\ \qquad\qquad\qquad\qquad\qquad\qquad\qquad u(z, xT) = f(z), \end{cases} \tag{13.28}$$

with operator Q as

$$Qu(z, x, t) = \frac{g_x(t)}{\bar{G}_x(t)} \int_X P(x, dy)[u(z, y, t) - u(z, x, t)], \tag{13.29}$$

and measures μ and ν have defined in Section 13.2.

Proposition 6 (see Swishchuk (1995)). The residual risk process has the form

$$R_t(\pi^*) = E\left(\int_t^T [Qu^2(S_r, x(r), r) - 2u(S_r, x(r), r)Qu(S_r, x(r), r)]ds \Big| \mathcal{F}_t\right).$$

In particular, the total residual risk on the interval $[0, T]$ at the moment $t = 0$ is equal to

$$R_0(\pi^*) = E\left(\int_0^T [Qu^2(S_r, x(r), r) - 2u(S_r, x(r), r)Qu(S_r, x(r), r)]ds\right). \tag{13.29}$$

Remark 9. We note, that process

$$m_t := u(S_t, x(t), \gamma(t)) - u(z, x, 0) - \int_0^t \left(Q + \frac{d}{dr}\right) u(S_r, x(r), \gamma(r)) dr \tag{13.30}$$

is an $\mathcal{F}_t$-martingale, where $\mathcal{F}_t = \sigma\{x(s), w(s); 0 \le s \le t\}$.

Proposition 7. The total residual risk on the interval $[0, T]$ at time $t = 0$ is the risk-neutral expectation of quadratic variation (or characteristic) $\langle m_t \rangle$ of the martingale m_t in (13.30):

$$R_0(\pi^*) = E^*\langle m_t \rangle. \tag{13.31}$$

Proof. Follows from the definition of the residual risk, (13.29) and the following relationship:

$$\langle m_t \rangle = \int_0^T \left[\left(Q + \frac{d}{dr}\right) u^2(S_r, x(r), r) - 2u(S_r, x(r), r)\left(Q + \frac{d}{dr}\right) u(S_r, x(r), r)\right] ds$$

$$= \int_0^T [Qu^2(S_r, x(r), r) - 2u(S_r, x(r), r)Qu(S_r, x(r), r)]ds,$$

hence,

$$R_0(\pi^*) = E^* \langle m_t \rangle,$$

and (13.31) follows.

Remark 10. In the paper Swishchuk and Manca (2012) we studied a semi-Markov modulated security market consisting of a riskless asset or bond with constant interest rate and risky asset or stock, whose dynamics follow gemoetric Brownian motion with volatility that depends on semi-Markov process. Two cases for semi-Markov volatilities are studied: local current and local semi-Markov volatilities. Using the martingale characterization of semi-Markov processes, we find the minimal martingale measure for this incomplete market. Then we model and price variance and volatility swaps for local semi-Markov stochastic volatilities.

13.8   Summary

- We considered in this Chapter a semi-Markov modulated market consisting of a riskless asset or bond, B, and a risky asset or stock, S, whose dynamics depend on a semi-Markov process x.
- Using the martingale characterization of semi-Markov processes, we note the incompleteness of semi-Markov modulated markets and find the minimal martingale measure. We price variance (Theorem 1) and volatility (Theorem 2) swaps for stochastic volatilities driven by the semi-Markov processes.
- We also discuss some extensions of the obtained results such as local semi-Markov volatility, Dupire formula for the local semi-Markov volatility and residual risk associated with the swap pricing.

Bibliography

Brockhaus, O. and Long, D. (2000). Volatility swaps made simple. *Risk Magazine*, January, 92-96.

Dupire, B. (1994). Pricing with a smile. *Risk Magazine*, 7, 1, 18-20.

Elliott, R. and Swishchuk, A. (2007). Pricing options and variance swaps in Markov-modulated Brownian markets. In Mamon, R. and Elliot, R. (eds.), *Hidden Markov Models in Finance*. New York: Springer.

Föllmer, H. and Schweizer, M. (1991). Hedging of contingent claims under incomplete information. In *Applied Stochastic Analysis*, 389-414. New-York-London: Gordon and Beach.

Javaheri, A., Wilmott, P. and Haug, E. (2002). GARCH and volatility swaps. *Wilmott Technical Article*, January, 17 pages.

Korolyuk, V. and Swishchuk, A. (1995). *Semi-Markov Random Evolutions*. Dordrecht, The Netherlands: Kluwer Academic Publishers.

Shiryaev, A. (1999). *Essentials of Stochastic Finance: Facts, Models, Theory*. Singapore: World Scientific.

Swishchuk, A. (1995). Hedging of options under mean-square criterion and with semi-Markov volatility. *Ukrainian Mathematics Journal*, 47, 7, 1119-1127.

Swishchuk, A. (2000). *Random Evolutions and Their Applications. New Trends.* Dordrecht, The Netherlands: Kluwer Academic Publishers.

Swishchuk, A. (2004). Modeling of variance and volatility swaps for financial markets with stochastic volatilities. *Wilmott Magazine*, September Issue, Technical article No. 2, pp. 64-72.

Swishchuk, A. (2005). Modeling and pricing of variance swaps for stochastic volatilities with delay. *Wilmott Magazine*, Issue 19, September 2005, 63-73.

Swishchuk, A. and Manca, R. (2010). Modeling and pricing of variance and volatility swaps for local semi-Markov volatilities in financial engineering. *Mathematical Problems in Engineering*, Volume 2010, 17 pages.

Chapter 14

Covariance and Correlation Swaps for Markov-Modulated Volatilities

14.1 Introduction

In this Chapter, we price covariance and correlation swaps for financial markets with Markov-modulated volatilities. As an example, we consider stochastic volatility driven by two-state continuous Markov chain. In this case, numerical example is presented for VIX and VXN volatility indeces ($S\&P500$ and NASDAQ-100, respectively, since January 2004 to June 2012). We also use VIX (January 2004 to June 2012) to price variance and volatility swaps for the two-state Markov-modulated volatility and to present a numerical result in this case.

Among of the recent and new financial products are covariance and correlation swaps, which are useful for volatility hedging and speculation using two different financial underlying assets.

For example, option dependent on exchange rate movements, such as those paying in a currency different from the underlying currency, have an exposure to movements of the correlation between the asset and the exchange rate, this risk may be eliminated by using covariance swap.

A *covariance swap* is a covariance forward contact of the underlying rates S^1 and S^2 which payoff at expiration is equal to

$$N(Cov_R(S^1, S^2) - K_{cov}),$$

where K_{cov} is a strike price, N is the notional amount, $Cov_R(S^1, S^2)$ is a covariance between two assets S^1 and S^2.

A *correlation swap* is a correlation forward contract of two underlying rates S^1 and S^2 which payoff at expiration is equal to:

$$N(Corr_R(S^1, S^2) - K_{corr}),$$

where $Corr(S^1, S^2)$ is a realized correlation of two underlying assets S^1 and S^2, K_{corr} is a strike price, N is the notional amount.

Pricing covariance swap, from a theoretical point of view, is similar to pricing variance swaps, since

$$Cov_R(S^1, S^2) = 1/4\{\sigma_R^2(S^1 S^2) - \sigma_R^2(S^1/S^2)\}$$

where S^1 and S^2 are given two assets, $\sigma_R^2(S)$ is a variance swap for underlying assets, $Cov_R(S^1, S^2)$ is a realized covariance of the two underlying assets S^1 and S^2.

Thus, we need to know variances for $S^1 S^2$ and for S^1/S^2 (see Section 6.3, Chapter 6 for details). Correlation $Corr_R(S^1, S^2)$ is defined as follows:

$$Corr_R(S^1, S^2) = \frac{Cov_R(S^1, S^2)}{\sqrt{\sigma_R^2(S^1)}\sqrt{\sigma_R^2(S^2)}},$$

where $Cov_R(S^1, S^2)$ is defined above and $\sigma_R^2(S^1)$ in Section 6.3, Chapter 6.

Given two assets S_t^1 and S_t^2 with $t \in [0, T]$, sampled on days $t_0 = 0 < t_1 < t_2 < ... < t_n = T$ between today and maturity T, the log-return each asset is:
$$R_i^j := \log(\frac{S_{t_i}^j}{S_{t_{i-1}}^j}), \quad i = 1, 2, ..., n, \quad j = 1, 2.$$

Covariance and correlation can be approximated by

$$Cov_n(S^1, S^2) = \frac{n}{(n-1)T} \sum_{i=1}^{n} R_i^1 R_i^2$$

and

$$Corr_n(S^1, S^2) = \frac{Cov_n(S^1, S^2)}{\sqrt{Var_n(S^1)}\sqrt{Var_n(S^2)}},$$

respectively.

The literature devoted to the volatility derivatives is growing. We give here a short overview of the latest development in this area. The Non-Gaussian Ornstein-Uhlenbeck stochastic volatility model was used by Benth *et al.* (2007) to study volatility and variance swaps. Broadie and Jain (2008) evaluated price and hedging strategy for volatility derivatives in the Heston square root stochastic volatility model. In another paper, Broadie and Jain (2008) compare result from various model in order to investigate the effect of jumps and discrete sampling on variance and volatility swaps. Pure jump process with independent increments return models were used by Carr *et al.* (2005) to price derivatives written on realized variance, and subsequent development by Carr and Lee (2007). We also refer to Carr and Lee (2009) for a good survey on volatility derivatives. Da Foneseca *et al.* (2009) analyzed the influence of variance and covariance swap in a market by solving a portfolio optimization problem in a market with risky assets and volatility derivatives. Correlation swap price has been investigated by Bossu (2005) and Bossu (2005) for component of an equity index using statistical method. The paper Drissien, Maenhout and Vilkov (2009) discusses the price of correlation risk for equity options. Pricing volatility swaps under Heston's model with regime-switching and pricing options under a generalized Markov-modulated jump-diffusion model are discussed in two papers by Elliott, Siu and Chan (2007). The paper Howison, Rafailidis and Rasmussen (2004) considers the pricing of a range of volatility derivatives, including volatility and variance swaps and swaptions. The pricing options on realized variance in the Heston model with jumps in returns and volatility is studied in Sepp (2008). An analytical closed-forms pricing of pseudo-variance, pseudo-volatility, pseudo-covariance and pseudo-correlation swaps is studied in Swishchuk *et al.* (2002). The paper Windcliff, Forsyth and Vetzal (2006) investigates the behaviour and hedging of discretely observed volatility derivatives.

14.2 Martingale Representation of Markov Processes

Let $(\Omega, \mathcal{F}, (\mathcal{F}_t)_{t \in R_+}, P)$ be a filtered probability space, with a right-continuous filtration $(\mathcal{F}_t)_{t \in R_+}$ and probability P. Let $(X, \mathcal{X})$ be a measurable space and $(x_t)_{t \in R_+}$ be a $(X, \mathcal{X})$-valued Markov process with generator Q. The following two results allow us to associate a martingale to this process and to obtain its quadratic variation, we refer to Elliott and Swishchuk (2007) for the proofs.

Proposition 14.1. (Elliott and Swishchuk (2007)) Let $(x_t)_{t \in R_+}$ be a Markov process with generator Q and $f \in$ Domain (Q), then

$$m_t^f := f(x_t) - f(x_0) - \int_0^t Qf(x_s)ds \qquad (14.1)$$

is a zero-mean martingale with respect to $\mathcal{F}_t := \sigma\{y(s); 0 \le s \le t\}$.

Let us evaluate the quadratic variation of this martingale.

Proposition 14.2. (Elliott and Swishchuk (2007)) Let $(x_t)_{t \in R_+}$ be a Markov process with generator Q, $f \in$ Domain (Q) and $(m_t^f)_{t \in R_+}$ its associated martingale, then

$$\langle m^f \rangle_t := \int_0^t [Qf^2(x_s) - 2f(x_s)Qf(x_s)]ds \qquad (14.2)$$

is the quadratic variation of m^f.

In what follows it will be useful to consider the quadratic covariation of two martingales associated to a generic couple of functions of a Markov process.

Proposition 14.3. Let $(x_t)_{t \in R_+}$ be a Markov process with generator Q, $f, g \in$ Domain (Q) such that $fg \in$ Domain (Q). Denote by $(m_t^f)_{t \in R_+}, (m_t^g)_{t \in R_+}$ their associated martingale. Then

$$\langle f(x.), g(x.) \rangle_t := \int_0^t \{Q(f(x_s)g(x_s)) - [g(x_s)Qf(x_s) + f(x_s)Qg(x_s)]\}ds \qquad (14.3)$$

is the quadratic covariation of f and g.

Proof. First of all, we note that

$$m_t^f m_t^g = f(x_t)g(x_t) - f(x_t)\int_0^t Qg(x_s)ds - g(x_t)\int_0^t Qf(x_s)ds$$

$$+ \int_0^t Qf(x_s)ds \int_0^t Qg(x_s)ds$$

$$= f(x_t)g(x_t) - m_t^f \int_0^t Qg(x_s)ds - m_t^g \int_0^t Qf(x_s)ds$$

$$- \int_0^t Qf(x_s)ds \int_0^t Qg(x_u)du \tag{14.4}$$

Moreover,

$$d\left[m_t^f \int_0^t Qg(x_s)ds + m_t^g \int_0^t Qf(x_s)ds + \int_0^t Qf(x_s)ds \int_0^t Qg(x_u)du \right]$$

$$= \left(\int_0^t Qg(x_s)ds \right) dm_t^f + \left(\int_0^t Qf(x_s)ds \right) dm_t^g + m_t^f Qg(x_t)dt + m_t^g Qf(x_t)dt$$

$$+ \left(\int_0^t Qg(x_s)ds \right) Qf(x_t)dt + \left(\int_0^t Qf(x_s)ds \right) Qg(x_t)dt,$$

now using the expression for m^f and m^g (cf. Proposition 14.1) we obtain

$$d\left[m_t^f \int_0^t Qg(x_s)ds + m_t^g \int_0^t Qf(x_s)ds + \int_0^t Qf(x_s)ds \int_0^t Qg(x_u)du \right]$$

$$= \left(\int_0^t Qg(x_s)ds \right) dm_t^f + \left(\int_0^t Qf(x_s)ds \right) dm_t^g + f(x_t)Qg(x_t)dt + g(x_t)Qf(x_t)dt. \tag{14.5}$$

Using (14.5) the equation (14.4) becomes

$$m_t^f m_t^g = f(x_t)g(x_t) - \left[\int_0^t \left(\int_0^s Qg(x_u)du \right) dm_s^f + \int_0^t \left(\int_0^s Qf(x_u)du \right) dm_s^g \right]$$

$$- \int_0^t [f(x_s)Qg(x_s) + g(x_s)Qf(x_s)]\, ds.$$

Adding and subtracting $\int_0^t Q(f(x_s)g(x_s))ds$ on the right hand side of the previous equation, we have

$$m_t^f m_t^g = f(x_t)g(x_t) - \int_0^t Q(f(x_s)g(x_s))ds$$

$$- \left[\int_0^t \left(\int_0^s Qg(x_u)du \right) dm_s^f + \int_0^t \left(\int_0^s Qf(x_u)du \right) dm_s^g \right]$$

$$+ \int_0^t [Q(f(x_s)g(x_s)) - f(x_s)Qg(x_s) - g(x_s)Qf(x_s)]\, ds. \tag{14.6}$$

Since $fg \in$ Domain (Q), then (cf. Proposition 14.1)

$$f(x_t)g(x_t) - \int_0^t Q(f(x_s)g(x_s))ds, \qquad t \in R_+$$

is a martingale. The term in square bracket on the right hand side of (14.6) is a martingale too. Therefore

$$m_t^f m_t^g - \int_0^t \left[Q(f(x_s)g(x_s)) - f(x_s)Qg(x_s) - g(x_s)Qf(x_s) \right] ds, \qquad t \in R_+$$

is a martingale and we have that

$$\langle f(x.), g(x.) \rangle_t = \int_0^t \left[Q(f(x_s)g(x_s)) - f(x_s)Qg(x_s) - g(x_s)Qf(x_s) \right] ds.$$

$\square$

Now, we are able to evaluate the expectation of a generic function of a Markov process.

Proposition 14.4. (Elliott and Swishchuk (2007)) Let $(x_t)_{t \in R_+}$ be a Markov process with generator Q and $f \in$ Domain (Q), then

$$E\{f(x_t)\} = e^{tQ}f(x_0).$$

Proof. From Proposition 14.1 we have

$$E\{f(x_t)\} = f(x_0) + \int_0^t QE\{f(x_s)\}ds,$$

and solving this differential equation we obtain the result. $\square$

Proposition 14.5. Let $(x_t)_{t \in R_+}$ be a Markov process with generator Q, $f, g \in$ Domain (Q) such that $fg \in$ Domain (Q) and denote by $(m_t^f)_{t \in R_+}, (m_t^g)_{t \in R_+}$ their associated martingale, then

$$E\{f(x_t)g(x_t)\} = e^{tQ}f(x_0)g(x_0).$$

Proof. Applying the Ito's lemma to the product fg we obtain

$$d(f(x_t)g(x_t)) = f(x_t)dg(x_t) + g(x_t)df(x_t) + d\langle f(x.), g(x.) \rangle_t,$$

here using Propositions 14.1 and 14.3 we have

$$d(f(x_t)g(x_t)) = Q\left[f(x_t)g(x_t) \right] dt + f(x_t)dm_t^g + g(x_t)dm_t^f.$$

This stochastic differential equation can be written in integral form, taking the expectation of its integral representation we get

$$E\{f(x_t)g(x_t)\} = f(x_0)g(x_0) + \int_0^t Q\left[f(x_s)g(x_s) \right] ds.$$

Solving this differential equation we finally obtain

$$E\{f(x_t)g(x_t)\} = e^{tQ}f(x_0)g(x_0).$$

$\square$

14.3   Variance and Volatility Swaps for Financial Markets with Markov-Modulated Stochastic Volatilities

Let consider a financial market with only two securities, the risk-free bond and the stock. Let us suppose that the stock price $(S_t)_{t \in R_+}$ satisfies the following stochastic differential equation

$$dS_t = S_t(rdt + \sigma(x_t)dw_t)$$

where w is a standard Wiener process independent of the Markov process $(x_t)_t$. In this model the volatility is stochastic then it is interesting to study the property of σ and in particular how to price future contracts on realized variance and volatility. The following results concern the expectation of variance and they are simple application of Propositions 14.1 and 14.4.

Proposition 14.6. Suppose that $\sigma \in$ Domain (Q). Then

$$E\{\sigma^2(x_t)|\mathcal{F}_u\} = \sigma^2(x_u) + \int_u^t QE\{\sigma^2(x_s)|\mathcal{F}_u\}ds$$

for any $0 \le u \le t$.

The conditional expectation of the variance can be expressed as

$$E\{\sigma^2(x_t)|\mathcal{F}_u\} = e^{(t-u)Q}\sigma^2(x_u)$$

for any $0 \le u \le t$.

Proposition 14.7. Suppose that $\sigma^2 \in$ Domain (Q), then

$$E\{\sigma^4(x_t)\} = e^{tQ}\sigma^4(x) \qquad \text{for } t \in R_+,$$

here we have denoted $x_0 =: x$.

Proof. If $\sigma^2 \in$ Domain (Q), then from Proposition 14.1 we have that

$$m_t^{\sigma^2} := \sigma^2(x_t) - \sigma^2(x) - \int_0^t Q\sigma^2(x_s)ds$$

is a zero-mean martingale with respect to $\mathcal{F}_t := \sigma\{x_s; 0 \le s \le t\}$, where Q is the infinitesimal generator of the Markov process $(x_t)_{t \in [0,T]}$. Then σ^2 satisfies the following stochastic differential equation

$$d\sigma^2(x_t) = Q\sigma^2(x_t)dt + dm_t^{\sigma^2}.$$

By applying the Ito's Lemma we obtain

$$d\sigma^4(x_t) = 2\sigma^2(x_t)d\sigma^2(x_t) + d\langle m_\cdot^{\sigma^2}\rangle_t.$$

We note that, the quadrating variation of m^{σ^2} is (see Propositon 14.2) given by

$$\langle m^{\sigma^2}\rangle_t := \int_0^t [Q\sigma^4(x_s) - 2\sigma^2(x_s)Q\sigma^2(x_s)]ds.$$

Substituting the expression for the quadratic variation $\langle m^{\sigma^2} \rangle$ and for σ^2, we obtain that σ^4 satisfies the following stochastic differential equation

$$d\sigma^4(x_t) = Q\sigma^4(x_t)dt + 2\sigma^2(x_t)dm_t^{\sigma^2}.$$

By tacking the expectation of both sides of these equation we obtain

$$E\{\sigma^4(x_t)\} = \sigma^4(x) + \int_0^t QE\{\sigma^4(x_s)\}ds.$$

Solving this differential equation we finally get

$$E\{\sigma^4(x_t)\} = e^{tQ}\sigma^4(x).$$

□

This Markov-modulated financial market is incomplete (see Elliott and Swishchuk (2007)), in order to price the swaps we will use the minimal martingale measure and we will denote it by P for simplicity. We do not discuss here the details regarding minimal martingale measure, let us instead focus on the pricing of future contract on variance and volatility.

14.3.1 *Pricing Variance Swaps*

Let us start from the more straightforward variance swap. Variance swaps are forward contract on future realized level of variance. The payoff of a variance swap with expiration date T is given by

$$N(\sigma_R^2(x) - K_{var})$$

here $\sigma_R^2(x)$ is the realized stock variance over the life of the contract defined by

$$\sigma_R^2(x) := \frac{1}{T}\int_0^T \sigma^2(x_s)ds,$$

while K_{var} is the strike price for variance and N is the notional amounts of dollars per annualized variance point. Without loss of generality, we can assume $N = 1$. The price of the variance swap is the expected present value of the payoff in the risk-neutral world

$$P_{var}(x) = E\{e^{-rT}(\sigma_R^2(x) - K_{var})\}.$$

The following result concern the evaluation of the variance swap. We refer to Elliott and Swishchuk (2007) for a complete discussion and proof.

Theorem 14.1. (Elliott and Swishchuk (2007)) The present value of a variance swap for Markov stochastic volatility is

$$P_{var}(x) = e^{-rT}\left\{\frac{1}{T}\int_0^T (e^{tQ}\sigma^2(x) - K_{var})dt\right\}.$$

□

14.3.2   *Pricing Volatility Swaps*

Volatility swaps are forward contract on future realized level of volatility. The payoff of a volatility swap with maturity date T is given by

$$N(\sigma_R(x) - K_{vol})$$

where $\sigma_R(x)$ is the realized stock volatility over the life of the contract defined by

$$\sigma_R(x) := \sqrt{\frac{1}{T} \int_0^T \sigma^2(x_s)ds},$$

while K_{vol} is the strike price for volatility and N is the notional amounts of dollars per annualized volatility point. We will assume, as before, that $N = 1$ for the sake of simplicity. The price of the volatility swap is the expected present value of the payoff in the risk-neutral world

$$P_{vol}(x) = E\{e^{-rT}(\sigma_R(x) - K_{vol})\}.$$

In order to evaluate the volatility swaps we need to know the expected value of the square root of the variance, but unfortunately, we are not able to evaluate analytically this expected value. Then in order to obtain a closed formula for the price of volatility swaps we have to make an approximation. Following the approach of Brockhaus and Long (2000) (see also Javaheri *et al.* (2002)), from the second order Taylor expansion we have

$$E\{\sqrt{\sigma_R^2(x)}\} \approx \sqrt{E\{\sigma_R^2(x)\}} - \frac{Var\{\sigma_R^2(x)\}}{8E\{\sigma_R^2(x)\}^{3/2}}.$$

Then, in order to evaluate the volatility swap price we have to know both expectation and variance of $\sigma_R^2(x)$.

Theorem 14.2. The value of a volatility swap for Markov-modulated stochastic volatility is

$$P_{vol}(x) \approx e^{-rT} \left\{ \sqrt{\frac{1}{T} \int_0^T e^{tQ}\sigma^2(x)dt} - \frac{Var\{\sigma_R^2(x)\}}{8\left(\frac{1}{T}\int_0^T e^{tQ}\sigma^2(x)dt\right)^{3/2}} - K_{vol} \right\}$$

where the variance is given by

$$Var\{\sigma_R^2(x)\} = \frac{2}{T^2} \int_0^T \int_0^t e^{sQ}\left[\sigma^2(x)e^{(t-s)Q}\sigma^2(x)\right] dsdt - \left[\int_0^T e^{tQ}\sigma^2(x)dt\right]^2.$$

Proof. We have already obtained the expectation of the realized variance,

$$E\{\sigma_R^2(x)\} = \frac{1}{T} \int_0^T e^{tQ}\sigma^2(x)dt$$

then it remains to prove that

$$Var\{\sigma_R^2(x)\} = \frac{2}{T^2} \int_0^T \int_0^t e^{sQ} \left[\sigma^2(x)e^{tQ}\sigma^2(x)\right] dsdt - \left[\int_0^T e^{tQ}\sigma^2(x)dt\right]^2.$$

The variance is, from the definition, given by

$$Var\{\sigma_R^2(x)\} = E\{[\sigma_R^2(x) - E\{\sigma_R^2(x)\}]^2\},$$

using the definition of realized variance, and Fubini theorem, we have

$$Var\{\sigma_R^2(x)\} = E\left\{\left[\frac{1}{T}\int_0^T \sigma^2(x_t)dt - \frac{1}{T}\int_0^T E\{\sigma^2(x_t)\}dt\right]^2\right\}$$

$$= E\left\{\left[\frac{1}{T}\int_0^T \left(\sigma^2(x_t) - E\{\sigma^2(x_t)\}\right) dt\right]^2\right\}$$

$$= \frac{1}{T^2}\int_0^T \int_0^T E\left\{[\sigma^2(x_t) - E\{\sigma^2(x_t)\}][\sigma^2(x_s) - E\{\sigma^2(x_s)\}]\right\} dsdt,$$

and then solving the product

$$Var\{\sigma_R^2(x)\} = \frac{1}{T^2}\int_0^T \int_0^T E\left\{\sigma^2(x_t)\sigma^2(x_s)\right\} dsdt - \left[\frac{1}{T}\int_0^T E\{\sigma^2(x_t)\}dt\right]^2.$$

The second term on the right hand side is known, it follows directly from Proposition 14.6. Moreover, we observe that the integrand on the first term is invariant in the exchange of s and t. Then, we can divide the integration set in two zone above and below the line $t = s$, thanks to the symmetry the contribution on the two parts is the same. We can rewrite the variance as

$$Var\{\sigma_R^2(x)\} = \frac{2}{T^2}\int_0^T \int_0^t E\left\{\sigma^2(x_t)\sigma^2(x_s)\right\} dsdt - \left[\frac{1}{T}\int_0^T e^{tQ}\sigma^2(x)dt\right]^2.$$

We stress that, in this form, the integration set of the first term is such that the inequality $s \leq t$ holds true. Using the properties of conditional expectation we have

$$Var\{\sigma_R^2(x)\} = \frac{2}{T^2}\int_0^T \int_0^t E\{\sigma^2(x_s)E\{\sigma^2(x_t)|\mathcal{F}_s\}\}dsdt - \left[\frac{1}{T}\int_0^T e^{tQ}\sigma^2(x)dt\right]^2.$$

Using the Markov property and Corollary 14.3, the conditional expected value in the integrand, can be viewed as a function of the process at time s, that is

$$E\{\sigma^2(x_t)|\mathcal{F}_s\} = e^{(t-s)Q}\sigma(x_s) =: g(x_s).$$

Thus, the variance can be expressed as

$$Var\{\sigma_R^2(x)\} = \frac{2}{T^2}\int_0^T \int_0^t E\{\sigma^2(x_s)g(x_s)\}dsdt - \left[\frac{1}{T}\int_0^T e^{tQ}\sigma^2(x)dt\right]^2.$$

Now, using Proposition 14.5 we can solve the expectation on the integrand and we finally obtain

$$Var\{\sigma_R^2(x)\} = \frac{2}{T^2} \int_0^T \int_0^t e^{sQ} \left(\sigma^2(x)g(x)\right) dsdt - \left[\frac{1}{T} \int_0^T e^{tQ} \sigma^2(x)dt\right]^2.$$

Moreover we observe that function g, evaluated in x, becomes

$$g(x) = e^{(t-s)Q} \sigma^2(x).$$

Then, substituting in the previous formula we can expressed the variance as

$$Var\{\sigma_R^2(x)\} = \frac{2}{T^2} \int_0^T \int_0^t e^{sQ} \left[\sigma^2(x)e^{(t-s)Q}\sigma^2(x)\right] dsdt - \left[\frac{1}{T} \int_0^T e^{tQ} \sigma^2(x)dt\right]^2.$$

$\square$

14.4 Covariance and Correlation Swaps for a Two Risky Assets for Financial Markets with Markov-Modulated Stochastic Volatilities

Let's consider a market model with two risky assets and a risk-free bond. Let's assume that the risky assets are satisfying the following stochastic differential equations

$$\begin{cases} dS_t^{(1)} = S_t^{(1)}(\mu_t^{(1)} dt + \sigma^{(1)}(x_t)dw_t^{(1)}) \\ dS_t^{(2)} = S_t^{(2)}(\mu_t^{(2)} dt + \sigma^{(2)}(x_t)dw_t^{(2)}) \end{cases}$$

where $\mu^{(1)}, \mu^{(2)}$ are deterministic functions of time and $(w_t^{(1)})_t$ and $(w_t^{(2)})_t$ are standard Wiener processes with quadratic covariance given by

$$d[w_t^{(1)}, w_t^{(2)}] = \rho_t dt.$$

Here, ρ_t is a deterministic function of time and $(w_t^{(1)})_t, (w_t^{(2)})_t$ are supposed to be independent of the Markov process $(x_t)_t$. This model allow us to study the covariance and correlation structure of two risky assets and how it is possible to price future contract on them.

14.4.1 *Pricing Covariance Swaps*

A covariance swap is a covariance forward contract on the realized covariance between two risky assets which payoff at maturity is equal to

$$N(Cov_R(S^{(1)}, S^{(2)}) - K_{cov})$$

where K_{cov} is a strike reference value, N is the notional amount and $Cov_R(S^{(1)}, S^{(2)})$ is the realized covariance of the two assets $S^{(1)}$ and $S^{(2)}$ defined by

$$Cov_R(S^{(1)}, S^{(2)}) = \frac{1}{T}[\ln S_T^{(1)}, \ln S_T^{(2)}] = \frac{1}{T} \int_0^T \rho_t \sigma^{(1)}(x_t)\sigma^{(2)}(x_t)dt.$$

The price of the covariance swap is the expected present value of the payoff in the risk-neutral world

$$P_{cov}(x) = E\{e^{-rT}(Cov_R(S^{(1)}, S^{(2)}) - K_{cov})\},$$

where we assumed that $N = 1$.

Theorem 14.3. The value of a covariance swap for Markov-modulated stochastic volatility is

$$P_{cov}(x) = e^{-rT}\left\{\frac{1}{T}\int_0^T \rho_t e^{tQ}[\sigma^{(1)}(x)\sigma^{(2)}(x)]dt - K_{cov}\right\}.$$

Proof. To evaluate the price of covariance swap we only need to know

$$E\{Cov_R(S^{(1)}, S^{(2)})\} = \frac{1}{T}\int_0^T \rho_t E\{\sigma^{(1)}(x_t)\sigma^{(2)}(x_t)\}dt.$$

Then, it remains to show that

$$E\{\sigma^{(1)}(x_t)\sigma^{(2)}(x_t)\} = e^{tQ}[\sigma^{(1)}(x)\sigma^{(2)}(x)].$$

By applying the Ito's lemma we have

$$d(\sigma^{(1)}(x_t)\sigma^{(2)}(x_t)) = \sigma^{(1)}(x_t)d\sigma^{(2)}(x_t) + \sigma^{(2)}(x_t)d\sigma^{(1)}(x_t) + d\langle\sigma^{(1)}(x.), \sigma^{(2)}(x.)\rangle_t.$$

Using Proposition 14.3, we can express the covariation as

$$d\langle\sigma^{(1)}(x.), \sigma^{(2)}(x.)\rangle_t = Q[\sigma^{(1)}(x_t)\sigma^{(2)}(x_t)]dt$$
$$- [\sigma^{(1)}(x_t)Q\sigma^{(2)}(x_t) + \sigma^{(2)}(x_t)Q\sigma^{(1)}(x_t)]dt.$$

Furthermore, from Proposition 14.1, we have

$$d\sigma^{(i)}(x_t) = Q\sigma^{(i)}(x_t)dt + dm^{\sigma^{(i)}} \qquad i = 1, 2.$$

Therefore, we get

$$d(\sigma^{(1)}(x_t)\sigma^{(2)}(x_t)) = Q[\sigma^{(1)}(x_t)\sigma^{(2)}(x_t)]dt + \sigma^{(1)}(x_t)dm^{\sigma^{(2)}} + \sigma^{(2)}(x_t)dm^{\sigma^{(1)}}.$$

Taking the expectation on both side of the above equation we obtain

$$E\{\sigma^{(1)}(x_t)\sigma^{(2)}(x_t)\} = \sigma^{(1)}(x)\sigma^{(2)}(x) + \int_0^t QE\{\sigma^{(1)}(x_s)\sigma^{(2)}(x_s)\}dt.$$

Now, we can solve this differential equation and we get

$$E\{\sigma^{(1)}(x_t)\sigma^{(2)}(x_t)\} = e^{tQ}[\sigma^{(1)}(x)\sigma^{(2)}(x)],$$

this conclude the proof. □

14.4.2   *Pricing Correlation Swaps*

A correlation swap is a forward contract on the correlation between the underlying assets S^1 and S^2 which payoff at maturity is equal to

$$N(Corr_R(S^1, S^2) - K_{corr})$$

where K_{corr} is a strike reference level, N is the notional amount and $Corr_R(S^1, S^2)$ is the realized correlation defined by

$$Corr_R(S^1, S^2) = \frac{Cov_R(S^1, S^2)}{\sqrt{\sigma_R^{(1)^2}(x)}\sqrt{\sigma_R^{(2)^2}(x)}},$$

where the realized variance is given by

$$\sigma_R^{(i)^2}(x) = \frac{1}{T}\int_0^T (\sigma^{(i)}(x_t))^2 dt \qquad i = 1, 2.$$

The price of the correlation swap is the expected present value of the payoff in the risk-neutral world

$$P_{corr}(x) = E\{e^{-rT}(Corr_R(S^1, S^2) - K_{corr})\}$$

where we set $N = 1$ for simplicity. Unfortunately the expected value of $Corr_R(S^1, S^2)$ is not known analytically. In order to obtain an explicit formula for the correlation swap price we have to introduce some approximation.

14.4.3   *Correlation Swap Made Simple*

First of all, let introduce the following notations

$$X = Cov_R(S^1, S^2)$$

$$Y = \sigma_R^{(1)^2}(x)$$

$$Z = \sigma_R^{(2)^2}(x),$$

in what follows with the subscript 0 we will denote the expected value of the above random variables. Starting from the approach we have used for the volatility swap, we would like to approximate the square root of Y and Z at the first order as follows

$$\sqrt{Y} \approx \sqrt{Y_0} + \frac{Y - Y_0}{2\sqrt{Y_0}}$$

$$\sqrt{Z} \approx \sqrt{Z_0} + \frac{Z - Z_0}{2\sqrt{Z_0}}.$$

The realized correlation can now be approximated by

$$Corr_R(S^1, S^2) \approx \frac{X}{\left(\sqrt{Y_0} + \frac{Y-Y_0}{2\sqrt{Y_0}}\right)\left(\sqrt{Z_0} + \frac{Z-Z_0}{2\sqrt{Z_0}}\right)} = \frac{\frac{X}{\sqrt{Y_0}\sqrt{Z_0}}}{\left(1 + \frac{Y-Y_0}{2Y_0}\right)\left(1 + \frac{Z-Z_0}{2Z_0}\right)}.$$

Solving the product in the denominator on the right hand side last term and keeping only the terms up to the first order in the increment, we have

$$Corr_R(S^1, S^2) \approx \frac{\frac{X}{\sqrt{Y_0}\sqrt{Z_0}}}{1 + \left(\frac{Y - Y_0}{2Y_0} + \frac{Z - Z_0}{2Z_0}\right)} \approx \frac{X}{\sqrt{Y_0}\sqrt{Z_0}} \left[1 - \left(\frac{Y - Y_0}{2Y_0} + \frac{Z - Z_0}{2Z_0}\right)\right].$$

Finally we obtain the following approximation for the correlation

$$Corr_R(S^1, S^2) \approx \frac{X}{\sqrt{Y_0}\sqrt{Z_0}} - \frac{X}{\sqrt{Y_0}\sqrt{Z_0}} \left(\frac{Y - Y_0}{2Y_0} + \frac{Z - Z_0}{2Z_0}\right),$$

we are going to evaluate the expectation only on the first term on the right hand side, which represent the zero order of approximation and the most intuitive part. We discuss the first order correction in the appendix. In what follows we are going to approximate the realized correlation as

$$Corr_R(S^1, S^2) \approx \frac{X}{\sqrt{Y_0}\sqrt{Z_0}},$$

here substituting X, Y and Z we obtain

$$Corr_R(S^1, S^2) \approx \frac{1}{\sqrt{E\{\sigma_R^{(1)^2}(x)\}}\sqrt{E\{\sigma_{2R}^{(2)^2}(x)\}}} \frac{1}{T} \int_0^T \rho_t \sigma^{(1)}(x_t) \sigma^{(2)}(x_t) dt$$

where (cf. Theorem 14.4), we have

$$E\left\{\sigma_{(i)R}^2(x)\right\} = E\left\{\frac{1}{T} \int_0^T (\sigma^{(i)}(x_t))^2 dt\right\} = \frac{1}{T} \int_0^T e^{tQ} (\sigma^{(i)}(x))^2 dt,$$

for $i = 1, 2$. In order to price a correlation swap we have to be able to evaluate the expectation of both side of the last equation. From Proposition 14.5 the expectation of the integrand in the right hand side of the last equation is given by

$$E\{\sigma^{(1)}(x_t)\sigma^{(2)}(x_t)\} = e^{tQ}\sigma^{(1)}(x)\sigma^{(2)}(x).$$

We can summarize the previous result in the following statement.

Theorem 14.4. The value of a correlation swap for a Markov-modulated stochastic volatility is

$$P_{corr}(x) \approx e^{-rT} \left[\frac{\int_0^T \rho_t e^{tQ} \sigma^{(1)}(x)\sigma^{(2)}(x) dt}{\sqrt{\int_0^T e^{tQ}(\sigma^{(1)}(x))^2 dt}\sqrt{\int_0^T e^{tQ}(\sigma^{(2)}(x))^2 dt}} - K_{corr}\right].$$

$\square$

14.5   Example: Variance, Volatility, Covariance and Correlation Swaps for Stochastic Volatility Driven by Two-State Continuous Markov Chain

Let $(x_t)_t$ be a two state continuous time Markov chain, let denote the states as "Up" (u) and "Down" (d). Let Q be the generator of this Markov chain

$$Q = \begin{pmatrix} q_{uu} & q_{ud} \\ q_{du} & q_{dd} \end{pmatrix}$$

and let

$$P(t) = \begin{pmatrix} p_{uu}(t) & p_{ud}(t) \\ p_{du}(t) & p_{dd}(t) \end{pmatrix}$$

be its transition function, such that

$$P(t) = e^{tQ}.$$

In this simple model the volatility takes only two values: σ_u and σ_d, thus we can easily express the swap prices of the futures contract so far studied.

The variance swap price in this model is given by

$$P_{var}(i) = e^{-rT} \left\{ \frac{1}{T} \int_0^T (p_{iu}(t)\sigma_u^2 + p_{id}(t)\sigma_d^2) dt - K_{var} \right\},$$

where $i = u, d$ is the initial state of the Markov chain. If we are uncertain about the initial state and we have only a probability distribution, let say (p_u, p_d) such that $p_u + p_d$, then the price is going be

$$P_{var} = p_u P_{var}(u) + p_d P_{var}(d).$$

If we assume that initial distribution is actually the stationary distribution of Markov chain the price simply becomes

$$P_{var} = e^{-rT} \left\{ \pi_u \sigma_u^2 + \pi_d \sigma_d^2 - K_{var} \right\},$$

where (π_u, π_d) is the stationary distribution.

The volatility swap price in this model can be approximateted by

$$P_{vol}(i) \approx e^{-rT} \left\{ \sqrt{\frac{1}{T} \int_0^T [p_{iu}(t)\sigma_u^2 + p_{id}(t)\sigma_d^2] dt} \right.$$

$$\left. - \frac{Var\{\sigma_R^2(i)\}}{8 \left(\frac{1}{T} \int_0^T [p_{iu}(t)\sigma_u^2 + p_{id}(t)\sigma_d^2] dt \right)^{3/2}} - K_{vol} \right\}$$

for $i = u, d$ and where

$$Var\{\sigma_R^2(i)\}$$

$$= \frac{2}{T^2} \int_0^T \int_0^t \{p_{iu}(s)p_{uu}(t-s)\sigma_u^4 + [p_{iu}(s)p_{ud}(t-s) + p_{id}(s)p_{du}(t-s)]\sigma_d^2\sigma_u^2$$

$$+ p_{id}(s)p_{dd}(t-s)\sigma_d^4\}dsdt - \left[\frac{1}{T}\int_0^T [p_{iu}(t)\sigma_u^2 + p_{id}(t)\sigma_d^2]dt\right]^2.$$

Let now consider a two risky asset market model with volatility modulated by this two state Markov chain, in this setting the covariance swap price is

$$P_{cov}(i) = e^{-rT}\left\{\frac{1}{T}\int_0^T \rho_t[p_{iu}(t)\sigma_u^{(1)}\sigma_u^{(2)} + p_{id}(t)\sigma_d^{(1)}\sigma_d^{(2)}]dt - K_{cov}\right\},$$

for $i = u, d$ representing the initial state of the chain.

The correlation swap in this model can be approximated by

$$P_{corr}(i)$$

$$\approx e^{-rT}\left[\frac{\int_0^T \rho_t[p_{iu}(t)\sigma_u^{(1)}\sigma_u^{(2)} + p_{id}(t)\sigma_d^{(1)}\sigma_d^{(2)}]dt}{\sqrt{\int_0^T [p_{iu}(t)(\sigma_u^{(1)})^2 + p_{id}(t)(\sigma_d^{(1)})^2]dt}\sqrt{\int_0^T [p_{iu}(t)(\sigma_u^{(2)})^2 + p_{id}(t)(\sigma_d^{(2)})^2]dt}} - K_{corr}\right]$$

for $i = u, d$ stand for the initial state of the chain.

14.6 Numerical Example

14.6.1 *S&P500: Variance and Volatility Swaps*

In this Section, we use the two states continuous Markov chain model for stochastic volatility to price variance and volatility swap on the volatility of the $S\&P500$ index, we used the CBOE Volatility Index (VIX), since January 2004 to June 2012. The daily datas were used to construct the one step transition matrix for the two states Markov chain, in particular we use the interpolation between the highest and the lowest daily data. We took a mean of the interpolated datas and we divided the datas in up, higher of the mean, and down, lower or equal to the mean value. In the table below we shows the one step transition probability matrix.

Table 14.1   One Step Transition Probability Matrix.

Transition Matrix		
	Up	Down
Up	0.957	0.043
Down	0.026	0.974

Using this probability matrix, given the initial state of the chain, we are able to evaluate the price of covariance and correlation swap as described in Section 14.5. In Figures 14.1 and 14.2 the variance and volatility swap prices as a function of maturity are shown. The rate of convergence of the prices depend on the rate of convergence of the transition probability matrix to the stationary distribution.

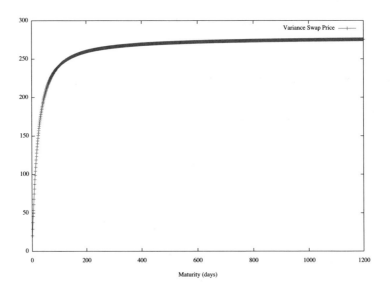

Fig. 14.1   *S&P*500 Index Variance Swap.

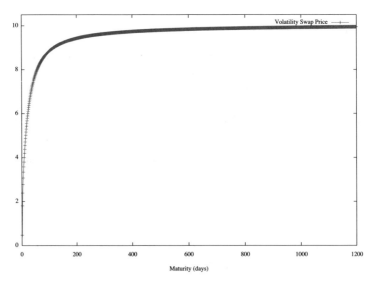

Fig. 14.2   *S&P*500 Index Volatility Swap.

14.6.2   *S&P500 and NASDAQ-100: Covariance and Correlation Swaps*

In this Section, we use the two states continuous Markov chain model for stochastic volatility to price covariance and correlation swap. The underlying assets are the *S&P*500 index, CBOE Volatility Index (VIX), and NASDAQ-100 index, CBOE NASDAQ-100 Volatility Index (VXN), since January 2004 to June 2012.

The daily datas were used to construct the one step transition matrix for the two states Markov chain, in particular we use the interpolation between the highest and the lowest daily data. We took a mean of the interpolated datas and we divided the datas in up, higher of the mean, and down, lower or equal to the mean value. In the table below we shows the one step transition probability matrix built using VIX and VXN indexes.

Table 14.2   One Step Transition Probability Matrix.

Transition Matrix		
	Up	Down
Up	0.950	0.050
Down	0.030	0.970

Using this probability matrix, given the initial state of the chain, we are able to evaluate the price of covariance and correlation swap as described in Section 14.5. In the numerical evaluation we assumed that the correlation between the brownian motion ρ was constant.

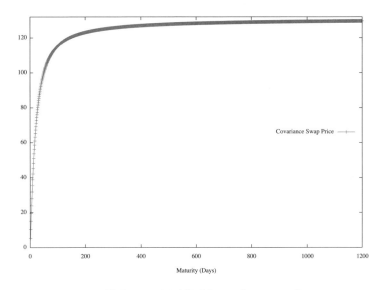

Fig. 14.3   *S&P*500 and NASDAQ-100 Covariance Swap.

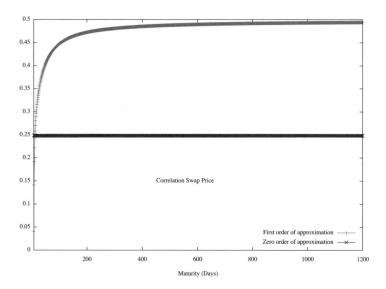

Fig. 14.4 $S\&P500$ and NASDAQ-100 Correlation Swap.

In Figures 14.3 and 14.4 the covariance and correlation swap prices as a function of maturity are shown. The rate of convergence of the prices depend on the rate of convergence of the transition probability matrix to the stationary distribution. For the correlation swap price we show the result up to the zero order of approximation, as described in Section 14.5, and up to the first order of approximation, as described in the appendix. The zero order of approximation price is constant, indeed at that order we only take into account for the correlation of the assets noise, that is *rho* the brownian motions correlation. In the first order the interaction of the volatilities due to the Markov process is taken in to account and the price changes as a function of maturity.

14.7   Correlation Swaps: First Order Correction

We would like to obtain an approximation for the realized correlation between two risky assets

$$Corr_R(S^1, S^2) = \frac{Cov_R(S^1, S^2)}{\sqrt{\sigma_R^{(1)^2}(x)}\sqrt{\sigma_R^{(2)^2}(x)}}.$$

In Section 14.4.3 we have already obtained the following approximated expression

$$Corr_R(S^1, S^2) \approx \frac{\frac{X}{\sqrt{Y_0}\sqrt{Z_0}}}{1 + \left(\frac{Y-Y_0}{2Y_0} + \frac{Z-Z_0}{2Z_0}\right)} \approx \frac{X}{\sqrt{Y_0}\sqrt{Z_0}}\left[1 - \left(\frac{Y-Y_0}{2Y_0} + \frac{Z-Z_0}{2Z_0}\right)\right].$$

where

$$X = Cov_R(S^1, S^2)$$

$$Y = \sigma_R^{(1)^2}(x)$$

$$Z = \sigma_R^{(2)^2}(x),$$

and with the pedix 0 we have denoted the expected values. We have already evaluated the expectation of the zero order approximation, now we would like to evaluate the first order.

Substituting X, Y and Z in equation above we obtain

$$Corr_R(S^1, S^2) \approx \frac{1}{\sqrt{E\{\sigma_R^{(1)^2}(x)\}}\sqrt{E\{\sigma_{2R}^{(2)^2}(x)\}}} \frac{1}{T} \int_0^T \rho_t \sigma^{(1)}(x_t) \sigma^{(2)}(x_t) dt$$

$$- \frac{1}{2T^2 (E\{\sigma_R^{(1)^2}(x)\})^{3/2} (E\{\sigma_R^{(2)^2}(x)\})^{3/2}} \int_0^T \rho_t \sigma^{(1)}(x_t) \sigma^{(2)}(x_t) dt$$

$$\times \left\{ E\{\sigma_R^{(2)^2}(x)\} \int_0^T [(\sigma^{(1)}(x_s))^2 - E\{(\sigma^{(1)}(x_s))^2\}] ds \right.$$

$$\left. + E\{\sigma_R^{(1)^2}(x)\} \int_0^T [(\sigma^{(2)}(x_u))^2 - E\{(\sigma^{(2)}(x_u))^2\}] du \right\},$$

where we have

$$E\left\{ \sigma_{(i)R}^2(x) \right\} = E\left\{ \frac{1}{T} \int_0^T (\sigma^{(i)}(x_t))^2 dt \right\} = \frac{1}{T} \int_0^T e^{tQ} (\sigma^{(i)}(x))^2 dt,$$

for $i = 1, 2$. We have to evaluate the expectation of the right hand side of the equation above. In Section 14.3 we calculated the expectation of the first term. Then we will focus now on the other terms. First of all, let us rewrite them as follows

$$\int_0^T \int_0^T \rho_t \sigma^{(1)}(x_t) \sigma^{(2)}(x_t) \left(E\{\sigma_R^{(2)^2}(x)\} [(\sigma^{(1)}(x_s))^2 - E\{(\sigma^{(1)}(x_s))^2\}] \right.$$

$$\left. + E\{\sigma_R^{(1)^2}(x)\} [(\sigma^{(2)}(x_s))^2 - E\{(\sigma^{(2)}(x_s))^2\}] \right) ds dt,$$

we have four different contributions in the integrals, the expectation of the terms

$$\int_0^T \int_0^T \rho_t \sigma^{(1)}(x_t, \gamma(t)) \sigma^{(2)}(x_t) E\{\sigma^{(i)^2}(x_s)\} E\{\sigma^{(-i)^2}(x_s)\} ds dt$$

for $i = 1, 2$, can be evaluate using Theorem 14.4. Then, in order to evaluate the expectation of the approximated realized correlation, it only remains to calculate

$$E\left\{ \int_0^T \int_0^T \rho_t \sigma^{(1)}(x_t) \sigma^{(2)}(x_t) \sigma^{(i)^2}(x_s) ds dt \right\} \qquad i = 1, 2.$$

To this end, let's first divide the range of integration in two intervals as follows

$$E\left\{\int_0^T\int_0^t \rho_t\sigma^{(1)}(x_t)\sigma^{(2)}(x_t)\sigma^{(i)^2}(x_s)dsdt + \int_0^T\int_t^T \rho_t\sigma^{(1)}(x_t)\sigma^{(2)}(x_t)\sigma^{(i)^2}(x_s)dsdt\right\}$$

for $i = 1, 2$. We notice that the first integral set is such that $t > s$ while the second has $t < s$. We can now use the property of conditional expectation to obtain

$$E\left\{\int_0^T\int_0^t \rho_t E\{\sigma^{(1)}(x_t)\sigma^{(2)}(x_t)|\mathcal{F}_s\}\sigma^{(i)^2}(x_s)dsdt \right.$$

$$\left. + \int_0^T\int_t^T \rho_t\sigma^{(1)}(x_t)\sigma^{(2)}(x_t)E\{\sigma^{(i)^2}(x_s)|\mathcal{F}_t\}dsdt\right\}.$$

Using the Markov property, we can express the conditional expectations as

$$E\{\sigma^{(1)}(x_t)\sigma^{(2)}(x_t)|\mathcal{F}_s\} = e^{(t-s)Q}\sigma^{(1)}(x_s)\sigma^{(2)}(x_s) =: h(x_s)$$

for $t > s$ and

$$E\{\sigma^{(i)^2}(x_s)|\mathcal{F}_t\} = e^{(s-t)Q}\sigma^{(i)^2}(x_t) =: g^{(i)}(x_t)$$

for $s > t$. Therefore, the first term under consideration can be expressed (cf. Proposition 14.5) as

$$E\left\{\int_0^T\int_0^t \rho_t h(x_s)\sigma^{(i)^2}(x_s)dsdt\right\} = \int_0^T\int_0^t \rho_t e^{sQ}[h(x)\sigma^{(i)^2}(x)]dsdt,$$

while the second as

$$E\left\{\int_0^T\int_t^T \rho_t\sigma^{(1)}(x_t)\sigma^{(2)}(x_t)g^{(i)}(x_t)dsdt\right\}$$

$$= \int_0^T\int_t^T \rho_t e^{tQ}[\sigma^{(1)}(x)\sigma^{(2)}(x)g^{(i)}(x)]dsdt.$$

Now, we evaluate the functions h and g at x obtaining

$$h(x) = e^{(t-s)Q}[\sigma^{(1)}(x)\sigma^{(2)}(x)]$$

and

$$g^{(i)}(x) = e^{(s-t)Q}[\sigma^{(i)^2}(x)].$$

We can summarize the previous result in the following statement which gives the correlation swap price up to the first order of approximation.

Theorem 14.5. The value of the correlation swap for a Markov-modulated volatility is

$$P_{corr}(x) = e^{-rT}\left(E\{Corr_R(S^1, S^2)\} - K_{corr}\right)$$

where the realized correlation can be approximated by

$$E\{Corr_R(S^1, S^2)\} \approx \frac{2\int_0^T \rho_t e^{tQ} \sigma^{(1)}(x)\sigma^{(2)}(x)dt}{\sqrt{\int_0^T e^{tQ}(\sigma^{(1)}(x))^2 dt}\sqrt{\int_0^T e^{tQ}(\sigma^{(2)}(x))^2 dt}}$$

$$-\frac{\int_0^T \rho_t(\int_0^t e^{sQ}\{e^{(t-s)Q}[\sigma^{(1)}(x)\sigma^{(2)}(x)]\sigma^{(1)^2}(x)\}ds + \int_t^T e^{tQ}\{\sigma^{(1)}(x)\sigma^{(2)}(x)e^{(u-t)Q}[\sigma^{(1)^2}(x)]\}du)dt}{2\left(\int_0^T e^{tQ}(\sigma^{(1)}(x))^2 dt\right)^{3/2}\left(\int_0^T e^{tQ}(\sigma^{(2)}(x))^2 dt\right)^{1/2}}$$

$$-\frac{\int_0^T \rho_t(\int_0^t e^{sQ}\{e^{(t-s)Q}[\sigma^{(1)}(x)\sigma^{(2)}(x)]\sigma^{(2)^2}(x)\}ds + \int_t^T e^{tQ}\{\sigma^{(1)}(x)\sigma^{(2)}(x)e^{(u-t)Q}[\sigma^{(2)^2}(x)]\}du)dt}{2\left(\int_0^T e^{tQ}(\sigma^{(1)}(x))^2 dt\right)^{1/2}\left(\int_0^T e^{tQ}(\sigma^{(2)}(x))^2 dt\right)^{3/2}}.$$

□

Remark. In the paper by Salvi and Swishchuk (2012), we modeled financial markets with semi-Markov volatilities and price covarinace and correlation swaps for this markets. Numerical evaluations of varinace, volatility, covarinace and correlations swaps with semi-Markov volatility have been presented as well. The novelty of the chapter lies in pricing of volatility swaps in closed form, and pricing of covariance and correlation swaps in a market with two risky assets with semi-Markov volatility.

14.8 Summary

– In this Chapter, we priced covariance and correlation swaps for financial markets with Markov-modulated volatilities.
– As an example, we consider stochastic volatility driven by two-state continuous Markov chain. In this case, numerical example is presented for VIX and VXN volatility indeces (*S&P*500 and NASDAQ-100, respectively, since January 2004 to June 2012).
– We also use VIX (January 2004 to June 2012) to price variance and volatility swaps for the two-state Markov-modulated volatility and to present a numerical result in this case.

Bibliography

Benth, F. E., Groth, M. and Kufakunesu, R. (2007). Valuing volatility and variance swaps for a non-Gaussian Ornstein-Uhlenbeck stochastic volatility model. *Applied Mathematical Finance*, 14, 4, 347-363.
Bossu, S. (2005). Arbitrage pricing of equity correlation waps. JPMorgan Equity Derivatives. Working Paper.
Bossu, S. (2007). A new approach for modelling and pricing correlation swaps. Equity Structuring — ECD London. Working Paper.
Broadie, M. and Jain, A. (2008). Pricing and hedging volatility derivatives. *The Journal of Derivatives*, 15, 3, 7-24.

Broadie, M. and Jain, A. (2008). The effect of jumps and discrete sampling on volatility and variance swaps. *International Journal of Theoretical and Applied Finance*, 11, 8. 761-797.

Brockhaus, O. and Long, D. (2000). Volatility swaps made simple. *Risk Magazine*, January, 92-96.

Carr, P., Geman, H., Madan, D. B. and Yor, M. (2005). Pricing options on realized variance. *Finance and Stochastics*, 9, 453-475.

Carr, P. and Lee, R. (2007). *Realized Volatility and Variance: Options via Swaps*. Bloomberg LP and University of Chicago. Available at: http://math.uchicago.edu/ ~rl/OVSwithAppendices.pdf.

Carr, P. and Lee, R. (2009). Volatility derivatives. *Annual Review of Financial Economics*, 1, 319-39.

Da Fonseca, J., Ielpo, F. and Grasselli, M. (2009). Hedging (Co)Variance Risk with Variance Swaps. Available at SSRN: http://ssrn.com/abstract=1341811.

Demeterfi, K., Derman, E., Kamal, M. and Zou, J. (1999). A guide to volatility and variance swaps. *The Journal of Derivatives*, 6, 9-32.

Drissien, J., Maenhout, P. J. and Vilkov, G. (2009). The price of correlation risk: Evidence from equity options. *The Journal of Finance*, 64, 3, 1377-1406.

Elliott, R. J., Siu, T. K.n and Chan, L. (2007). Pricing volatility swaps under Heston's stochastic volatility model with regime switching. *Applied Mathematical Finance*, 14, 1, 41-62.

Elliott, R. J., Siu, T. K.n and Chan, L. (2007). Pricing options under a generalized Markov-modulated jump-diffusion model. *Stochastic Analysis and Applications*, 25, 821-843.

Elliott, R. and Swishchuk, A. V. (2007). Pricing options and variance swaps in Markov-modulated Brownian markets. In Mamon, R. and Elliot, R. (eds.), *Hidden Markov Model in Finance*. New York: Springer.

Howison, S., Rafailidis, A. and Rasmussen, H. (2004). On the pricing and hedging of volatility derivatives. *Applied Mathematical Finance*, 11, 317-346.

Javaheri, A., Wilmott, P. and Haug, E. (2002). GARCH and volatility swaps. *Wilmott Technical Article*, January, 17 pages.

Salvi, G. and Swishchuk, A. V. (2012). Pricing of variance, volatility, covariance and correlation swaps in a Markov-modulated volatility model. Preprint.

Sepp, A. (2008). Pricing options on realized variance in the Heston model with jumps in returns and volatility. *Journal of Computational Finance*, 11, 4, 33-70.

Swishchuk, A. V. (2010). Pricing of variance and volatility swaps with semi-Markov volatilities. *Canadian Applied Mathematics Quarterly*. 18, 4, Winter.

Swishchuk, A. V., Cheng, R., Lawi, S., Badescu, A., Mekki, H. B., Gashaw, A. F., Hua, Y., Molyboga, M., Neocleous, T. and Petrachenko, Y. (2002). Price pseudo-variance, pseudo-covariance, pseudo-volatility, and pseudo-correlation swaps-in analytical closed-forms. *Proceedings of the Sixth PIMS Industrial Problems Solving Workshop*, PIMS IPSW 6, 45-55. May 24-31, University of British Columbia, Vancouver, Canada.

Salvi, G. and Swishchuk, A. (2012). Modeling and pricing of covariance and correlation swaps for financial markets with semi-Markov volatilities. *Canadian Applied Mathematics Quarterly* (Submitted).

Windcliff, H., Forsyth, P. A. and Vetzal, K. R. (2006). Pricing methods and hedging strategies for volatility derivatives. *Journal of Banking & Finance*, 30, 409-431.

Chapter 15

Volatility and Variance Swaps for the COGARCH(1,1) Model

15.1 Introduction

In this Chapter, we present volatility and variance swaps valuations for the COG-ARCH(1,1) model introduced by Klüppelberg $et\ al.$ (2004). We consider two numerical examples: for compound Poisson COGARCH(1,1) and for variance gamma COGARCH(1,1) processes. Also, we demonstrate two different situations for the volatility swaps: with and without convexity adjustment to show the difference in values.

The following time series models as ARCH and GARCH (see Bollerslev (1986)), popular in financial econometrics, have been designed to capture some of the distinctive features of asset prices, exchange rates, and other series. The financial returns data is characterized as heavy-tailed, uncorrelated (but not independent), with time-varying volatility (so-called stylized facts) and a long range dependence effect evident in volatility ('persistence in volatility' effect). Many and various attempts have been made to capture these features in a continuous time models, for example, a natural extension being given by diffusion approximation to the discrete time GARCH as in Nelson (1990), Duan (1997) and de Haan and Karandikar (1989). This model has separate, independent Brownian motions driving the volatility and asset price. Kazmerchuk $et\ al.$ (2002) developed a form of continuous time GARCH(1,1) (by rewriting the GARCH(1,1) equations as an SDE); this model inherits all of the parameters of the GARCH(1,1) model and maintains the form of the GARCH(1,1) equations. In this way, these lead to the following stochastic volatility models

$$\begin{cases} dX_t = \sigma_t dB_t^1 \\ d\sigma_t^2 = \theta(\gamma - \sigma_t^2)dt + \rho\sigma_t^2 dB_t^2, \end{cases}$$

with two independent Brownian motions B_t^1, B_t^2.

See Drost $et\ al.$ (1996) for a review of continuous time GARCH models.

Various related models have been proposed and studied, and many generalizations have been based on Lévy processes replacing the Brownian motions and on relaxing the independent property. See, for example, Barndorff-Nielsen and Shephard (2001), Anh *et al.* (2002). The main difference between model (∗) and the original GARCH setup is the fact that in the GARCH modeling one single source of randomness suffices (as in Kazmerchuk *et al.* (2002), for example); all stylized features are then captured by the dependence structure of the model.

Klüppelberg *et al.* (2004) adopted this idea of a single noise process and suggest a new continuous time GARCH model, so-called COGARCH(1,1) model, which captures all the stylized facts as the discrete time GARCH does. As noise process, any Lévy process is possible, its increments replacing the innovations in the discrete time GARCH model. The volatility process is modeled by a stochastic differential equation, whose solution displays the 'feedback' and 'autoregressive' aspect of the recursion formula for the discrete time GARCH model.

Haug *et al.* (2007) developed method of moments estimation for the COGARCH(1,1) model. Kallsen and Vesenmayer (2009)) showed that any COGARCH process can be represented as the limit in law of a sequence of GARCH(1,1) processes.

In this Chapter, we present variance and volatility swaps valuations for the COGARCH(1,1) model introduced by Klüppelberg *et al.* (2004). GARCH and volatility swaps for a continuous-time GARCH model were presented in Javaheri *et al.* (2002). Swishchuk (2004)) priced variance, volatility, covariance and correlations swaps for Heston model and Swishchuk (2005) priced variance swap for stochastic volatilities with delay.

15.2 Lévy Processes

We present some basic facts about Lévy processes (see, e.g., Applebaum (2004), Schoutens (2003)).

Definition. A càdlàg (right continuous with left limits), adapted, real valued stochastic process $L = \{L_t, t \geq 0\}$ with $L_0 = 0$ a.s. is called a Lévy process if the following are satisfied: L has *independent increments*, i.e. $L_t - L_s$ is independent of $\mathcal{F}_s$ for any $0 \leq s < t \leq T$, L has *stationary increments*, i.e. for any $s, t \geq 0$ the distribution of $L_{t+s} - L_t$ does not depend on t, and L is *stochastically continuous*, i.e. for all $t > 0$ and $\epsilon > 0$:

$$\lim_{s \to t} \mathbb{P}(|L_t - L_s| > \epsilon) = 0.$$

The characteristic function of L_t is given by $\phi(u) = e^{t\,\eta(u)}$ ($t > 0$ and $u \in \mathbb{R}$), where $\eta(u)$ is the characteristic exponent of the process. The characteristic exponent of L_t can be be expressed as

$$\eta(u) = i\gamma u - \frac{1}{2}\sigma^2 u^2 + \int_{\mathbb{R}}(e^{iux} - 1 - iux\,\mathbf{1}_{\{|x|<1\}})\nu(dx).$$

Lévy processes are often represented by their Lévy triplet (γ, σ^2, ν), where $\nu(dx)$ is a Lévy measure.

The Lévy-Ito decomposition for L_t has the following form:

$$L_t = \gamma t + \sigma B_t + \int_{|x|<1} x \tilde{N}(t, dx) + \int_{|x| \geq 1} x N(t, dx),$$

where $N(t, dx)$ is a Poisson jump measure, $\tilde{N}(t, dx) = N(t, dx) - t\nu(dx)$ and B_t is a Brownian motion.

15.3 The COGARCH Process of Klüppelberg *et al.*

15.3.1 *The COGARCH(1,1) Equations*

The COGARCH(1,1) equations (as described in Klüppelberg *et al.* (2004)) have the following form:

$$dG_t = \sigma_{t-} dL_t$$

$$d\sigma_{t-}^2 = (\beta - \eta \sigma_{t-}^2)dt + \phi \sigma_{t-}^2 d[L, L]_t \qquad (15.1)$$

where L_t is the driving Lévy process and $[L, L]_t$ is the quadratic variation of the driving Lévy process.

15.3.2 *Informal Derivation of COGARCH(1,1) Equation*

For the completeness of presentation and to illustrate the connection with discrete time GARCH we give here an informal derivation of COGARCH(1,1) equation (see Klüppelberg *et al.* (2004)).

The GARCH(1,1) model is defined as $Y_i = \sigma_i \epsilon_i$ with

$$\sigma_i^2 = \omega_0 + \lambda Y_{i-1}^2 + \delta \sigma_{i-1}^2$$

for i=1,2,3,... Iterating the volatility in the previous expression we obtain

$$\sigma_i^2 = \omega_0 \sum_{k=0}^{i-1} \prod_{j=k+1}^{i-1} (\delta + \lambda \epsilon_j^2) + \sigma_0^2 \prod_{j=0}^{i-1}(\delta + \lambda \epsilon_j^2)$$

$$= \omega_0 \int_{k=0}^{i-1} exp \left\{ \sum_{j=\lfloor u \rfloor + 1}^{i-1} log(\delta + \lambda \epsilon_j^2) \right\} + \sigma_0^2 exp \left\{ \sum_{j=0}^{i-1} log(\delta + \lambda \epsilon_j^2) \right\}$$

The COGARCH(1,1) Model

$$= \left[\omega_0 \int_{k=0}^{i} exp \left\{ \eta(\lfloor u \rfloor + 1) - \sum_{j=0}^{\lfloor u \rfloor} log(1 + \phi \epsilon_j^2) \right\} + \sigma_0^2 \right]$$

$$\times exp \left\{ -\eta i + \sum_{j=0}^{i-1} log(1 + \psi \epsilon_j^2) \right\}$$

where $\eta = -log(\delta)$ and $\phi = \frac{\lambda}{\delta}$ Replacing the noise terms ϵ_i with the jumps of a Lévy process $\Delta L_t = L_t - L_{t-}$ we obtain the following expression:

$$\sigma_t^2 = (\omega_0 \int_0^t e^{X_s}ds + \sigma_0^2)e^{-X_{t-}}$$

$$X_t = -t\eta - \sum_{0<s\leq t} \log(1 + \phi(\Delta L_s)^2)$$

The COGARCH(1,1) process is then defined to be $dG_t = \sigma_t dL_t$

Remark 1. In Theorem 2.2 (Brockwell *et al.* (2006)) it is shown that the volatility process σ^2 can also be defined as the solution to the SDE

$$d\sigma_{t+}^2 = (\alpha\eta - \eta\sigma_t^2)dt + \phi\sigma_t^2 d[L, L]_t^d$$

where $[L, L]_t^d = \sum_{0\leq s\leq t}(\Delta L_s)^2$. If we set $\alpha\eta = \beta = \omega_0$ this leads to the equations as presented in (15.1)

Remark 2. The *auxiliary process* $X_t = -t\eta - \sum_{0<s\leq t}\log(1 + \phi(\Delta L_s)^2)$ and, in particular the Laplace transform of the auxillary process,

$$\Psi(s) = -\eta s + \int_R ((1 + \phi x^2)^s - 1)\nu_L(dx),$$

will be useful in calculating the realized variance of the COGARCH process.

15.3.3 *The Second Order Properties of the Volatility Process σ_t*

Suppose that the Lévy process (L_t) is such that $E(L_t)^2 0 < \infty$ and that $\sigma_0^2 < \infty$. Then (see Klüppelberg *et al.* (2004), Haug (2006)):

$$E\sigma_t^2 = \frac{\beta}{-\Psi(1)} + \left(E\sigma_0^2 + \frac{\beta}{\Psi(1)}\right)e^{t\Psi(1)}. \qquad (15.2)$$

In addition, if $E(L_t)^4 0 < \infty$ and $\sigma_0^4 < \infty$ then:

$$E\sigma_t^4 = \frac{2\beta^2}{\Psi(1)\Psi(2)} + \left(\frac{e^{t\Psi(2)}}{\Psi(2)} - \frac{e^{t\Psi(1)}}{\Psi(1)}\right)\frac{2\beta^2}{\Psi(2) - \Psi(1)}$$

$$+ 2\beta E\sigma_0^2\left(\frac{e^{t\Psi(2)} - e^{t\Psi(1)}}{\Psi(2) - \Psi(1)}\right) + E\sigma_0^4 e^{t\Psi(2)}, \qquad (15.3)$$

and also:

$$E(\sigma_{t+h}^2\sigma_t^2) = (E\sigma_t^4 - E\sigma_t^2 E\sigma_0^2)e^{h\Psi(1)} + E\sigma_t^2 E\sigma_h^2. \qquad (15.4)$$

15.4 Pricing Variance and Volatility Swaps under the COGARCH(1,1) Model

Let $(\Omega, \mathcal{F}, \mathcal{P})$ be probability space with filtration $\mathcal{F}_t, t \in [0, T]$. We will take the approach of modeling the asset volatility directly, using the σ_t process as defined in (15.1). We will assume, throughout the following sections, that we are in the risk-neutral world, and we assume a constant rate of interest r.

15.4.1   Variance Swaps

Recall that the value of a variance swap is given by

$$P(\sigma^2) = Ne^{-rT}(E\sigma_R^2(S) - K_{var}),$$

where $E\sigma_R^2(S)$ is the expected annualized variance.

Thus the expected variance of the log returns of our asset is given by

$$E\sigma_R^2(s) = \frac{1}{T}\int_0^T \sigma_s^2 ds$$

$$= \frac{1}{T}\int_0^T \frac{\beta}{-\Psi(1)} + \left(E\sigma_0^2 + \frac{\beta}{\Psi(1)}\right)e^{s\Psi(1)}ds$$

$$= \frac{1}{-\Psi(1)}\left(\beta - \frac{1}{T}\left(E\sigma_0^2 + \frac{\beta}{\Psi(1)}\right)(e^{T\Psi(1)} - 1)\right). \qquad (15.5)$$

Substituting (15.5) into $P(\sigma^2)$ we obtain the variance swap price:

$$P(\sigma^2) = Ne^{-rT}\left(\frac{1}{-\Psi(1)}\left(\beta - \frac{1}{T}\left(E\sigma_0^2 + \frac{\beta}{\Psi(1)}\right)(e^{T\Psi(1)} - 1)\right) - K_{var}\right), \quad (15.6)$$

where $\Psi(s) = -\eta s + \int_R((1 + \phi x^2)^s - 1)\nu_L(dx)$.

15.4.1.1   Numerical Examples: Variance Swaps for Compound Poisson and Variance Gamma COGARCH(1,1) Processes

Here, we present two numerical examples: valuation of variance swaps for compound Poisson and variance gamma COGARCH(1,1) processes.

Compound Poisson COGARCH(1,1) Process

$L_t = \sum_{k=1}^{N_t} Y_k$ where N_t is Poisson process with jump rate $\lambda > 0$, and $(Y_k), k = 1, 2, 3, ...$ are i.i.d. random variables, independent of N_t. The Lévy measure of L has the representation

$$\nu_L(dx) = \lambda F_Y(dx).$$

Thus in the case of the compound Poisson COGARCH(1,1) process we have

$$\Psi(s) = -\eta s + \lambda \int_R ((1 + \phi y^2)^s - 1)F_Y(dy)$$

$$\Psi(1) = -\eta + \lambda\phi EY^2.$$

Thus the variance swap price is (see (15.6))

$$P(\sigma^2) = Ne^{-rT}\left(\frac{1}{\eta - \lambda\phi EY^2}\left(\beta - \frac{1}{T}\left(E\sigma_0^2 + \frac{\beta}{-\eta + \lambda\phi EY^2}\right)(e^{T(-\eta + \lambda\phi EY^2)} - 1)\right) - K_{var}\right),$$

$$(15.7)$$

If we assume $N(0, 1)$ jumps with a jump rate of $\lambda = 1$ initial volatility level $E\sigma_0^2 = 0.018$, and model parameters $\phi = 0.05$, $\eta = 1$ and $\beta = 0.5$, then applying the

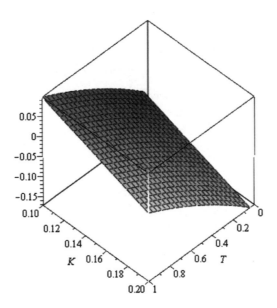

Fig. 15.1 Compound Poisson COGARCH(1,1) Variance Swap Prices $\phi = 0.05$, $\eta = 1$ and $\beta = 0.5$.

analytical solution above for a swap maturity T of 0.91 years, we find the following value $E\sigma_R^2(S) = 0.1860$. The Figure 15.1 depicts variance swap values for compound poisson COGARCH(1,1) process with $\phi = 0.05$, $\eta = 1$, $\beta = 0.5$.

Variance Gamma COGARCH(1,1) Process

If we assume that L is variance gamma with $E(L) = 0$ and $Var[L] = 1$, then the Lévy measure of L has the Lebesgue density

$$\nu_L(dx) = \frac{C}{|x|} exp(-(2C)^{1/2}|x|)dx$$

for $x \neq 0$,

$$\Psi(1) = -\eta + \phi.$$

Thus the variance swap value is (see (15.6))

$$P(\sigma^2) = Ne^{-rT}\left(\frac{1}{\eta - \phi}\left(\beta - \frac{1}{T}\left(E\sigma_0^2 + \frac{\beta}{-\eta + \phi}\right)(e^{T(-\eta+\phi)} - 1)\right) - K_{var}\right).$$
(15.8)

If we assume model parameters $C = 1$, $\phi = 0.05$, $\eta = 1$ and $\beta = 0.5$, initial volatility $E\sigma_0^2 = 0.018$, then applying the analytical solution above for a swap maturity T of 0.91 years, we find the following value $E\sigma_R^2(S) = 0.1860$. Figure 15.2 depicts the variance swap values for variance gamma COGARCH(1,1) with $C = 1$, $\phi = 0.05$, $\eta = 1$ and $\beta = 0.5$.

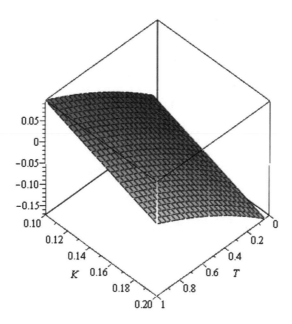

Fig. 15.2 Variance gamma COGARCH(1,1) Variance Swap Prices $C = 1$, $\phi = 0.05$, $\eta = 1$ and $\beta = 0.5$.

15.4.2 Volatility Swaps

Recall that the value of a volatility swap is given by

$$P(\sigma) = Ne^{-rT}(E\sigma_R(S) - K_{vol}),$$

where r is the risk-free discount rate corresponding to the expiration date T. We use the Brockhaus *et al.* (2000) approximation

$$E\sqrt{\sigma_R^2} \approx \sqrt{E\sigma_R^2} - \frac{Var[\sigma_R^2]}{8(E\sigma_R^2)^{3/2}}. \tag{15.9}$$

We note that

$$Var[\sigma_R^2(S)] = E(\sigma_R^2(S))^2 - (E\sigma_R^2(S))^2$$

and

$$\left(E\sigma_R^2(S)\right)^2 = \left(\frac{1}{-\Psi(1)}\left(\beta - \frac{1}{T}\left(E\sigma_0^2 + \frac{\beta}{\Psi(1)}\right)(e^{T\Psi(1)} - 1)\right)^2\right) \tag{15.10}$$

by (15.5). Also,

$$E(\sigma_R^2(S))^2 = E\left(\frac{1}{T}\int_0^T \sigma_t^2 dt\right)^2 = \frac{1}{T^2}\int_0^T\int_0^T E(\sigma_s^2\sigma_t^2)dsdt.$$

By (15.3) and (15.4) we have:

$$E\sigma_s^2\sigma_t^2 = \left(2\frac{\beta^2}{P1\Psi(2)} + 2\beta^2\left(\frac{e^{t\Psi(2)}}{\Psi(2)} - \frac{e^{tP1}}{P1}\right)\left(\Psi(2) - P1\right)^{-1}\right)e^{|s-t|}$$

$$+ \left(2\frac{\beta E\sigma_0^2(e^{t\Psi(2)} - e^{t\Psi(1)})}{\Psi(2) - \Psi(1)} + E\sigma_0^4 e^{t\Psi(2)}\right)$$

$$- \left(-\frac{\beta}{\Psi(1)} + \left(E\sigma_0^2 - \frac{\beta}{\Psi(1)}\right)e^{t\Psi(1)}\right)E\sigma_0^2\right)e^{|s-t|}$$

$$+ \left(-\frac{\beta}{\Psi(1)} + \left(E\sigma_0^2 - \frac{\beta}{\Psi(1)}\right)e^{t\Psi(1)}\right)$$

$$\times \left(-\frac{\beta}{\Psi(1)} + \left(E\sigma_0^2 - \frac{\beta}{\Psi(1)}\right)e^{|s-t|\Psi(1)}\right) \tag{15.11}$$

We may thus calculate the convexity correction by integrating (15.11) and substituting the (rather long) expression that results into (15.9). Applying the convexity correction to (15.10), we obtain the following expression for the realized volatility:

$$E\sigma_R(S) = \frac{-0.0625(\xi_1 + \xi_2)}{\sqrt{\frac{-1E\sigma_0^2\Psi(1)-1\beta-1\beta\Psi(1)T+E\sigma_0^2e^{T\Psi(1)}\Psi(1)+\beta e^{T\Psi(1)}}{T\Psi(1)^2}}}$$

$$\times \left(-1E\sigma_0^2\Psi(1) - 1\beta - 1\beta\Psi(1)T + E\sigma_0^2 e^{T\Psi(1)}\Psi(1) + \beta e^{T\Psi(1)}\right)^{-1}$$

$$\times \left(\Psi(2) + \Psi(1)\right)^{-2}\left(-1\Psi(2) + \Psi(1)\right)^{-2}\Psi(2)^{-2}T^{-1}\Psi(1)^{-2}$$

where ξ_1 and ξ_2 are defined in the Appendix.

15.4.2.1 Numerical Examples: Volatility Swaps for Compound Poisson and Variance Gamma COGARCH(1,1) Processes

Here, we present two numerical examples: valuation of volatility swaps for compound Poisson and variance gamma COGARCH(1,1) processes.

Compound Poisson COGARCH(1,1)

For the compound Poisson process we have the following:

$$\Psi(2) = -2\eta + 2\phi\lambda E[Y^2] + \phi^2\lambda E[Y^4].$$

If we assume $N(0,1)$ jumps with a jump rate of $\lambda = 1$ and model parameters $\phi = 0.05$, $\eta = 1$ and $\beta = 0.5$, then applying the analytical solution above for a swap maturity T of 0.91 years, we obtain a realized volatility of $E\sqrt{\sigma_R^2} \approx 0.4031$. Figure 15.3 depicts volatility swap values for compound Poisson COGARCH(1,1) with $\phi = 0.05$, $\eta = 1$ and $\beta = 0.5$, using the Brockhaus-Long approximation (left) and values without the convexity correction/adjustment (right).

Variance Gamma COGARCH(1,1) Process

Similarly, in the variance gamma COGARCH(1,1) process

$$\Psi(2) = -2\eta + 2\phi - 3\phi^2 C^{-1},$$

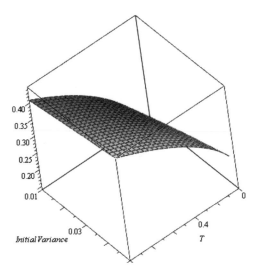

Fig. 15.3 Realized Volatility for the Compound Poisson COGARCH(1,1) Model (Varying Time and Initial Variance Level.

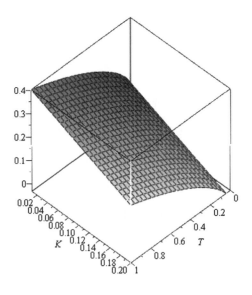

Fig. 15.4 Compound Poisson COGARCH(1,1) Volatility Swap Prices, $\phi = 0.05$, $\eta = 1$ and $\beta = 0.5$.

and we may thus calculate the value of the volatility swap. For example, if we assume $C = 1$ and model parameters $\phi = 0.05$, $\eta = 1$ and $\beta = 0.5$, then applying the analytical solution above for a swap maturity T of 0.91 years, we obtain a realized volatility of $E\sqrt{\sigma_R^2} \approx 0.4031$. Figure 15.4 depicts volatility swap values for variance gamma COGARCH(1,1) with $\phi = 0.05$, $\eta = 1$ and $\beta = 0.5$, using the Brockhaus-Long approximation (left) and values without the convexity correction/adjustment (right).

15.5   Formula for ξ_1 and ξ_2

ξ_1 and ξ_2 are defined as follows:

$$\xi_1 = 4\Psi(2)^3\Psi(1)^3\beta^2 e^{T(\Psi(2)+\Psi(1))}$$

$$+ 2E\sigma_0^4\Psi(1)^6\Psi(2)^2 e^{T(\Psi(2)+\Psi(1))}$$

$$+ 2E\sigma_0^4\Psi(1)^5\Psi(2)^3 e^{T(\Psi(2)+\Psi(1))}$$

$$+ 2E\sigma_0^4\Psi(1)^4\Psi(2)^4 e^{T(\Psi(2)+\Psi(1))}$$

$$+ 4\Psi(1)^5\beta^2 e^{T(\Psi(2)+\Psi(1))}\Psi(2) + 2E\sigma_0^4\Psi(1)^3\Psi(2)^5 e^{T(\Psi(2)+\Psi(1))}$$

$$+ 6\Psi(1)^2\beta E\sigma_0^2\Psi(2)^5 + 6\Psi(1)^4\beta E\sigma_0^2\Psi(2)^3$$

$$+ 51\Psi(2)^2 E\sigma_0^2\Psi(1)^5\beta$$

$$+ 33\beta E\sigma_0^2\Psi(2)^6\Psi(1) + 76\Psi(1)^3\beta E\sigma_0^2\Psi(2)^4 + 8E\sigma_0^2\Psi(1)^6\beta\Psi(2)$$

$$+ 34E\sigma_0^2\Psi(1)^2\beta T\Psi(2)^6$$

$$+ 34\Psi(2)^2 E\sigma_0^2\Psi(1)^6\beta T + 68\beta E\sigma_0^2\Psi(2)^6 e^{T\Psi(1)}\Psi(1) + 8E\sigma_0^2\Psi(1)^4\beta e^{\Psi(2)T}\Psi(2)^3$$

$$+ 8E\sigma_0^2\Psi(1)^6\beta e^{\Psi(2)T}\Psi(2) + 8\Psi(1)^2\beta E\sigma_0^2\Psi(2)^5 e^{T\Psi(1)}$$

$$+ 8\Psi(1)^4\beta E\sigma_0^2\Psi(2)^3 e^{T\Psi(1)}$$

$$+ 4\Psi(1)^3\beta E\sigma_0^2\Psi(2)^4 e^{\Psi(2)T} + 4\Psi(1)^5\beta E\sigma_0^2\Psi(2)^2 e^{\Psi(2)T} + 68\Psi(2)^4 E\sigma_0^2\Psi(1)^4\beta T$$

$$+ 148\Psi(1)^3\beta E\sigma_0^2\Psi(2)^4 e^{T\Psi(1)} + 80\Psi(1)^5\beta E\sigma_0^2\Psi(2)^2 e^{T\Psi(1)}$$

$$+ 76\beta^2 e^{T\Psi(1)}\Psi(2)^4\Psi(1)^3 T$$

$$+ 4\Psi(1)^2\beta^2 e^{T\Psi(1)}\Psi(2)^5 T + 4\Psi(1)^4\beta^2 e^{T\Psi(1)}\Psi(2)^3 T + 40\beta^2 e^{T\Psi(1)}\Psi(2)^2\Psi(1)^5 T$$

$$+ 35\beta E\sigma_0^2\Psi(2)^6 e^{2T\Psi(1)}\Psi(1)$$

$$+ 37\Psi(2)^2 E\sigma_0^2\Psi(1)^5\beta e^{2T\Psi(1)}$$

$$+ 2\Psi(1)^2\beta E\sigma_0^2\Psi(2)^5 e^{2T\Psi(1)} + 2\Psi(1)^4\beta E\sigma_0^2\Psi(2)^3 e^{2T\Psi(1)} + 32\beta E\sigma_0^2\Psi(1)^6\Psi(2)^2 e^{T\Psi(1)}T$$

$$+ 4\beta E\sigma_0^2\Psi(1)^3\Psi(2)^5 e^{T\Psi(1)}T + 4\beta E\sigma_0^2\Psi(1)^5\Psi(2)^3 e^{T\Psi(1)}T$$

$$+ 68\Psi(2)^4 E\sigma_0^2\Psi(1)^4\beta T e^{T\Psi(1)}$$

$$\begin{aligned}
\xi_2 = {}& 14\Psi(2)^5\beta^2\Psi(1) + 16\beta^2\Psi(2)^4\Psi(1)^2 + 3\Psi(2)^2\Psi(1)^4\beta^2 + 26\Psi(2)^3\beta^2\Psi(1)^3 \\
& + 12\Psi(1)^5\beta^2\Psi(2) + 2E\sigma_0^4\Psi(1)^3\Psi(2)^5 \\
& + 6E\sigma_0^4\Psi(1)^5\Psi(2)^3 + 4E\sigma_0^4\Psi(1)^7\Psi(2) + 2E\sigma_0^4\Psi(1)^4\Psi(2)^4 \\
& + 2E\sigma_0^4\Psi(1)^6\Psi(2)^2 + 72\Psi(2)^4E\sigma_0^2e^{2T\Psi(1)}\Psi(1)^3\beta \\
& + 18\Psi(2)^6E\sigma_0^{2^2}\Psi(1)^2 + 36\Psi(2)^4E\sigma_0^{2^2}\Psi(1)^4 \\
& + 18\Psi(2)^2E\sigma_0^{2^2}\Psi(1)^6 + 36\Psi(2)^2E\sigma_0^{2^2}\Psi(1)^6e^{T\Psi(1)} \\
& + 72\Psi(2)^4E\sigma_0^{2^2}\Psi(1)^4e^{T\Psi(1)} + 36\Psi(2)^6E\sigma_0^{2^2}\Psi(1)^2e^{T\Psi(1)} \\
& + 11\beta^2\Psi(2)^6 \\
& + 8\Psi(1)^6\beta^2 + 16\Psi(1)^4\beta^2\Psi(2)^3T \\
& + 8\Psi(1)^6\beta^2\Psi(2)T + 16\Psi(2)^5\beta^2e^{T\Psi(1)}\Psi(1) \\
& + 28\Psi(2)^3\beta^2e^{T\Psi(1)}\Psi(1)^3 \\
& + 48\Psi(2)^4\Psi(1)^2\beta^2e^{T\Psi(1)} + 4\Psi(1)^5\beta^2e^{T\Psi(1)}\Psi(2) + 12\Psi(1)^4\beta^2e^{T\Psi(1)}\Psi(2)^2 \\
& + 2\Psi(1)^3E\sigma_0^4\Psi(2)^5e^{T\Psi(1)} \\
& + 2\Psi(1)^4E\sigma_0^4\Psi(2)^4e^{T\Psi(1)} + 2\Psi(1)^5E\sigma_0^4\Psi(2)^3e^{T\Psi(1)} + 2\Psi(1)^6E\sigma_0^4\Psi(2)^2e^{T\Psi(1)} \\
& + 8\Psi(1)^2\Psi(2)^5\beta^2T \\
& + 16\beta^2T^2\Psi(1)^2\Psi(2)^6 + 32\beta^2T^2\Psi(1)^4\Psi(2)^4 \\
& + 16\Psi(2)^2\beta^2T^2\Psi(1)^6 \\
& + 30\beta^2T\Psi(2)^6\Psi(1) + 60\beta^2T\Psi(2)^4\Psi(1)^3 \\
& + 30\Psi(2)^2\beta^2T\Psi(1)^5 \\
& + 2E\sigma_0^4\Psi(1)^3\Psi(2)^5e^{\Psi(2)T} + 6E\sigma_0^4\Psi(1)^5\Psi(2)^3e^{\Psi(2)T} \\
& + 4E\sigma_0^4\Psi(1)^7e^{\Psi(2)T}\Psi(2) \\
& + 4\Psi(1)^3\beta^2e^{\Psi(2)T}\Psi(2)^3 + 12\Psi(1)^5\beta^2e^{\Psi(2)T}\Psi(2) + 2\Psi(1)^4E\sigma_0^4\Psi(2)^4e^{\Psi(2)T} \\
& + 2\Psi(1)^6E\sigma_0^4\Psi(2)^2e^{\Psi(2)T} + 32\Psi(2)^4\Psi(1)^2\beta^2e^{2T\Psi(1)} \\
& + 15\Psi(2)^2\Psi(1)^4\beta^2e^{2T\Psi(1)} \\
& + 2\Psi(2)^5\beta^2e^{2T\Psi(1)}\Psi(1) + 2\Psi(2)^3\beta^2e^{2T\Psi(1)}\Psi(1)^3 \\
& + 8\Psi(1)^4\beta E\sigma_0^2\Psi(2)^3e^{T(\Psi(2)+\Psi(1))} \\
& + 4\Psi(1)^3\beta E\sigma_0^2\Psi(2)^4e^{T(\Psi(2)+\Psi(1))} + 4\Psi(2)^2E\sigma_0^2\Psi(1)^5\beta e^{T(\Psi(2)+\Psi(1))} \\
& + 36\Psi(2)^6\beta^2\Psi(1)Te^{T\Psi(1)} \\
& + 18\Psi(2)^2E\sigma_0^{2^2}e^{2T\Psi(1)}\Psi(1)^6 + 36\Psi(2)^4E\sigma_0^{2^2}e^{2T\Psi(1)}\Psi(1)^4 \\
& + 18\Psi(2)^6E\sigma_0^{2^2}e^{2T\Psi(1)}\Psi(1)^2 + 28\beta^2\Psi(2)^6e^{T\Psi(1)} \\
& + 8\Psi(1)^6\beta^2e^{\Psi(2)T}17\beta^2\Psi(2)^6e^{2T\Psi(1)} + 36\Psi(2)^6\beta\Psi(1)^2TE\sigma_0^2e^{T\Psi(1)}.
\end{aligned}$$

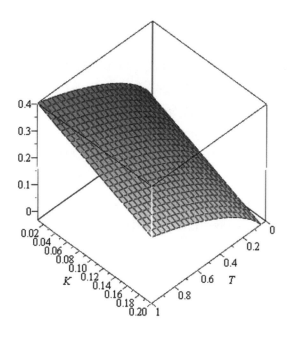

Fig. 15.5   Variance Gamma COGARCH(1,1) Volatility Swap Prices $C = 1$, $\phi = 0.05$, $\eta = 1$ and $\beta = 0.5$.

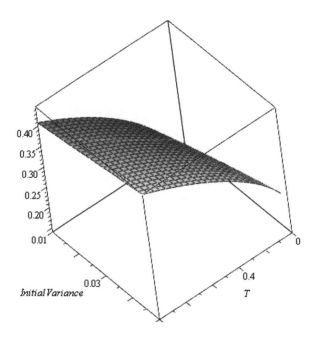

Fig. 15.6   Realized Volatility for the Variance Gamma COGARCH(1,1) Model (Varying Time and Initial Variance Level).

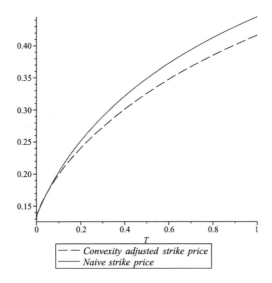

Fig. 15.7   Convexity Adjustment for the Variance Gamma Case.

15.6   Summary

– In this Chapter, we presented variance and volatility swaps valuations for the COGARCH(1,1) model introduced by Kluppelberg *et al.* (2004).
– We considered two numerical examples: for compound Poisson COGARCH(1,1) and for variance gamma COGARCH(1,1) processes.
– Also, we demonstrate two different situations for the volatility swaps: with and without convexity adjustment to show the difference in values. In future work, it would also be interesting to value covariance and correlation swaps for the COGARCH(1,1) model.

Bibliography

Anh, V. V., Heyde, C. C. and Leonenko, N. N. (2002). Dynamic models of long-memory processes driven by Lévy noise. *Journal of Applied Probability*, 39, 4, 730-747.

Applebaum, D. (2004). *Lévy Processes and Stochastic Calculus*, Volume 93 of Cambridge Studies in Advanced Mathematics. Cambridge: Cambridge University Press.

Barndorff-Nielsen, O. E. and Shephard, N. (2001). Modeling by Lévy processes for financial econometricsy. In Barndoff-Nielsen, O., Mikosch, T. and Resnick, S. (eds.) *Lévy Processes, Theory and Applications*, 283-318. Boston: Birkhauser.

Bollerslev, T. (1986). Generalized autoregressive conditional heteroskedasticity. *Journal of Econometrics*, 31, 307-327.

Brockwell, P., Chadraa, E. and Lindner, A. (2006). Continuous-time GARCH processes. *The Annals of Applied Probability*, 16, 2, 790-826.

de Haan, L. and Karandikar, R. L. (1989). Embedding a stochastic difference equation into a continuous-time process. *Stochastic Processes and Their Applications*, 32, 2, 225-235.

Duan, J.-C. (1997). Augmented GARCH (p,q) process and its diffusion limit. *Journal of Econometrics*, 79, 1, 97-127.

Haug, S. (2006). Exponential COGARCH and other continuous time models with applications to high frequency data, Ph.D. thesis, Center for Mathematical Sciences, Munich University of Technology, 85747 Garching bei Munchen.

Haug, S., Klüppelberg, C., Lindner, A. and Zapp, M. (2007). Method of moment estimation in the COGARCH(1,1) model. *Econometrics*, 10, 2, 320-341.

Javaheri, A., Wilmott, P. and Haug, E. G. (2002). Method of moment estimation in the COGARCH(1,1) model. Available at: http://www.wilmott.com/pdfs/020117_garch.pdf.

Kallsen, J. and Vesenmayer, B. (2009). COGARCH as a continuous-time limit of GARCH(1,1). *Stochastic Processes and Their Applications* 119, 1, 74-98.

Kazmerchuk, Y., Swishchuk, A. and Wu, J. (2002). A continuous-time GARCH model for stochastic volatility with delay. *Canadian Applied Mathematics Quarterly*, 13, 2, 123-150.

Klüppelberg, C., Linder, A. and Maller, R. (2004). A continuous-time GARCH process driven by a Lévy process: Stationarity and second-order behaviour. *Journal of Applied Probability.* 41, 3, 601-622.

Nelson, D. B. (1990). ARCH models as diffusion approximations. *Journal of Econometrics*, 45, 1-2, 7-38.

Schoutens, W. (2003). *Lévy Processes in Finance: Pricing Financial Derivatives.* England: John Wiley and Sons.

Swishchuk, A. (2004). Modeling of variance and volatility swaps for financial markets with stochastic volatilities. *Wilmott Magazine*, September 2004, Technical Article No. 2, pp. 64-72.

Chapter 16

Variance and Volatility Swaps for Volatilities Driven by Fractional Brownian Motion

16.1 Introduction

In this Chapter, we study financial markets with stochastic volatilities driven by fractional Brownian motion with Hurst index $H > 1/2$. Our models for stochastic volatility include new fractional versions of Ornstein-Uhlenbeck, Vasićek, geometric Brownian motion and continuous-time GARCH models. We price variance and volatility swaps for the above-mentioned models. Since pricing volatility swaps needs approximation formula, we analyze when this approximation is satisfactory. Also, we present asymptotic results for pricing variance swaps when time horizon increases.

We mention, that Klüppelberg and Kühn (2002) gave an economic justification for fractional Brownian motion to be used in modelling dynamics of the stock price. Breidt, Crato, Lima *et al.* (1998) proposed a new time series representation of persistence in conditional variance called a long memory stochastic volatility (LMSV) model. An empirical example with a long time series of stock prices demonstrates the superiority of the LMSV model over existing (short-memory) volatility models.

Remark. We study in this Chapter the pricing of variance and volatility swaps for stochastic models in finance driven by fractional Brownian motion (fBM). Of course, the model with only fractional Brownian motion (fBM) is not arbitrage-free. However, we could add a Brownian motion to the fBM noise and make it arbitrage-free (see, e.g., Mishura's *Stochastic Calculus for fBM and Related Processes*, Springer, 2008; the book contains application of fBM in finance for so-called mixed Brownian fractional Brownian models, see Chapter 5 from the book). The pricing results for financial models with only Brownian motion are available. In this way, someone may consider mixed fractional Brownian models to get further results in this direction for arbitrage-free models.

16.2   Variance and Volatility Swaps

We recall here the main definitions for varinace and volatility swaps (see also Chapter 3).

A stock *volatility swap* is a forward contract on the annualized volatility. Its payoff at expiration is equal to

$$N(\sigma_R(S) - K_{vol}),$$

where $\sigma_R(S)$ is the realized stock volatility (quoted in annual terms) over the life of contract,

$$\sigma_R(S) := \sqrt{\frac{1}{T} \int_0^T \sigma_s^2 ds},$$

σ_t is a stochastic stock volatility, K_{vol} is the annualized volatility delivery price, and N is the notional amount of the swap in dollar per annualized volatility point. The holder of a volatility swap at expiration receives N dollars for every point by which the stock's realized volatility σ_R has exceeded the volatility delivery price K_{vol}. The holder is swapping a fixed volatility K_{vol} for the actual (floating) future volatility σ_R. We note that usually $N = \alpha I$, where α is a converting parameter such as 1 per volatility-square, and I is a long-short index ($+1$ for long and -1 for short).

A *variance swap* is a forward contract on annualized variance, the square of the realized volatility. Its payoff at expiration is equal to

$$N(\sigma_R^2(S) - K_{var}),$$

where $\sigma_R^2(S)$ is the realized stock variance (quoted in annual terms) over the life of the contract,

$$\sigma_R^2(S) := \frac{1}{T} \int_0^T \sigma_s^2 ds,$$

K_{var} is the delivery price for variance, and N is the notional amount of the swap in dollars per annualized volatility point squared. The holder of variance swap at expiration receives N dollars for every point by which the stock's realized variance $\sigma_R^2(S)$ has exceeded the variance delivery price K_{var}.

Therefore, pricing the variance swap reduces to calculating the realized volatility square.

Valuing a variance forward contract or swap is no different from valuing any other derivative security. The value of a forward contract P on future realized variance with strike price K_{var} is the expected present value of the future payoff in the risk-neutral world:

$$P_{var} = e^{-rT} E\{\sigma_R^2(S) - K_{var}\}, \tag{16.1}$$

where r is the risk-free discount rate corresponding to the expiration date T, and E denotes the expectation.

Thus, for calculating variance swaps we need to know only $E\{\sigma_R^2(S)\}$, namely, mean value of the underlying variance.

To calculate volatility swaps we need more. From Brockhaus-Long (2000) approximation (which is obtained via the second order Taylor expansion for function $\sqrt{x}$) we have (see also Javaheri *et al.* (2002), p. 16):

$$E\left\{\sqrt{\sigma_R^2(S)}\right\} \approx \sqrt{E\{V\}} - \frac{Var\{V\}}{8E\{V\}^{3/2}}, \qquad (16.2)$$

where $V := \sigma_R^2(S)$ and $\frac{Var\{V\}}{8E\{V\}^{3/2}}$ is the convexity adjustment.

Thus, to calculate volatility swaps we need both $E\{V\}$ and $Var\{V\}$ and the price of volatility swap is:

$$P_{vol} = e^{-rT}E\{\sigma_R(S) - K_{vol}\} = e^{-rT}E\left\{\sqrt{\sigma_R^2(S)} - K_{vol}\right\}, \qquad (16.3)$$

where $E\{\sqrt{\sigma_R^2(S)}\}$ is defined in (16.2). Of course, this approximation is good in the neighborhood of $E\{V\}$ and we must check if the variance $Var\{V\}$ is sufficiently small, in comparison with $(E\{V\})^2$.

16.3 Fractional Brownian Motion and Financial Markets with Long-Range Dependence

16.3.1 *Definition and Some Properties of Fractional Brownian Motion*

Fractional Brownian motion is a popular model in applied probability, in particular in tele-traffic modeling and to some extent in finance.

Fractional Brownian motion is a self-similar Gaussian process with stationary increments. More exactly, we say that a continuous square integrable centered process $B^H = (B_t^H)_{t \geq 0}$ with $B_0^H = 0$ is a *fractional Brownian motion* with self-similarity index $H \in (0, 1)$ if it is a Gaussian process with zero mean and covariance function

$$E\left(B_s^H B_t^H\right) = \frac{1}{2}\left(t^{2H} + s^{2H} - |t - s|^{2H}\right) \quad s, t \geq 0. \qquad (16.4)$$

If B^H is a continuous Gaussian process with covariance (16.4), then obviously B^H has stationary increments and B^H is self-similar with index H. Mandelbrot named the Gaussian process B^H with covariance function (16.4) as *fractional Brownian motion*. An important representation result for fractional Brownian motion in terms of standard Brownian motion was proved in Mandelbrot *et al.* (1968). For results concerning fractional Brownian motion before Mandelbrot we refer to Molchan (2002). The overview of results concerning stochastic calculus for fBm are contained in the book by Mihura (2008). Fractional Brownian motion (fBm) is not a semimartingale nor a Markov process unless $H = \frac{1}{2}$, when it coincides with standard Brownian motion. The behavior of trajectories of fBm is the following that they are Hölder up to order H. Therefore, the quadratic variation of fBm is zero for $H > \frac{1}{2}$ and infinite for $H < \frac{1}{2}$. Moreover, there is the crucial difference between these two domains of H: for $H > \frac{1}{2}$ the increments of fBm are positively correlated, for $H < \frac{1}{2}$ they are negatively correlated; furthermore, for any $n \in \mathbf{N}$, the auto-covariance function is given by

$$r(n) := EB_1^H(B_{n+1}^H - B_n^H) = 2\alpha H \int_0^1 \int_n^{n+1} (u - v)^{2\alpha - 1} \, du \, dv$$

$$\sim 2\alpha H n^{2\alpha - 1}, \quad n \to \infty.$$

Here and it what follows $\alpha := H - 1/2$.

If $H \in (0, \frac{1}{2})$, then $\sum_{n=1}^{\infty} |r(n)| \sim \sum_{n=1}^{\infty} n^{2\alpha - 1} < \infty$.

If $H \in (\frac{1}{2}, 1)$, then $\sum_{n=1}^{\infty} |r(n)| \sim \sum_{n=1}^{\infty} |n|^{2\alpha - 1} = \infty$. In this case we say that fBm B^H has the property of long-range dependence.

16.3.2   *How to Model Long-Range Dependence on Financial Market*

When we deal with the model of the stock price on financial market, it is better to consider this model itself in the spirit of Black-Scholes, since such model is arbitrage free and the changes in price can be modulated with some degree of adequacy by standard Brownian motion. However, the stochastic volatility in this model (if there are some reasons to think that the volatility is stochastic) can vary slowly than standard Brownian motion and can have long-range dependence, or long memory. Arguments in favour of long-range dependent models and in particular of such dependence for stochastic volatility are contained in the book by Doukhan (2003) (see, for instance, also Taqqu (2003), Deo *et al.* (2003)), and in the book by Kallianpur *et al.* (2000). In this case it is reasonable to model it as a function of fBm with Hurst index $H > \frac{1}{2}$. Some of such models are considered in Section 16.4. All these models of stochastic volatility appear as the solutions of corresponding stochastic differential equations involving fBm. The stochastic integrals with respect to fBm in these equations are considered as pathwise ones (for the theory of path-wise integration w.r.t. fBm see the papers Zähle (1998, 2001)). The general conditions of existence and uniqueness of solution of SDE involving pathwise integral w.r.t. fBm were considered in Nualart *et al.* (2000). Note that all our equations in Section 16.5 satisfy these conditions and have unique solutions which we present in explicit form, therefore we shall not discuss this question in what follows. We only mention that the Wiener integral w.r.t. fBm, i.e., the integral with deterministic integrand, $I(T) = \int_0^T f(s) dB_s^H$ exists for f satisfying the condition

$$\int_0^T \int_0^T f(s)f(u)|s - u|^{2H-2} ds \, du < \infty, \tag{16.5}$$

it is Gaussian random variable with zero mean and variance

$$E(I(T))^2 = H(2H - 1) \int_0^T \int_0^T f(s)f(u)|s - u|^{2H-2} ds\, du. \qquad (16.6)$$

All these facts are contained, for example, in the paper Norros *et al.* (1999). At last, let B^H have parameter $H > 1/2$, the function $F \in C^2(R)$, the process X_t have the form

$$X_t = X_0 + \int_0^t a(s)ds + \int_0^t f(s)dB_s^H,$$

where $a \in L^1([0,t])$ for any $t > 0$ and $f \in C^1(R)$ (hence f satisfies assumption (16.5)). Then the Itô formula for X has a form

$$F(X_t) = F(X_0) + \int_0^t F'(X_s)a(s)ds + \int_0^t F'(X_s)f(s)dB_s^H.$$

Here, the integral $\int_0^t F'(X_s)f(s)dB_s^H$ exists in a path-wise sense because for almost all $\omega \in \Omega$ the function $g(s) := F'(X_s)f(s)$ is Hölder one up to the order H (see [46]). Note that such Itô formula does not contain any term with $F''(X_t)$, the reason of this is zero quadratic variation of fBm with $H > 1/2$. We use this Itô formula when we deduce the solutions of equations (16.8), (16.10), (16.12) and (16.14).

Note that the most part of calculations below is based on the formula (16.6), the fact that for fBm B^H we have $E(B_t^H - B_s^H)^2 = |t - s|^{2H}$, and on the formula $E \exp\{\rho\xi\} = \exp\{\frac{1}{2}\rho^2 D^2\}$ for random variable $\xi \sim N(0, D^2)$.

16.4 Modeling of Financial Markets with Stochastic Volatilities Driven by Fractional Brownian Motion (fBm)

In this Section we consider four models of stochastic volatility σ_t involving fBm, and present the solutions of the corresponding stochastic differential equations (SDE). The models (16.7), (16.10) (Ornstein-Uhlenbeck process and its generalization, Vasićek process) on one hand, and models (16.12), (16.14) (geometric Brownian motion and fractional GARCH process), on the other hand, are different in the following sense: Ornstein-Uhlenbeck and Vasićek processes are Gaussian ones, but geometric Brownian motion and fractional GARCH process contain exponents of Gaussian processes and are non-negative. Therefore, it is natural to suppose that the volatility itself satisfies corresponding SDE in the first pair of cases, but in the second pair of cases we suppose that the square of volatility, σ_t^2, satisfies corresponding SDE and σ_t is simply a square root from its solution. we suppose everywhere that the volatility is independent of the "main" standard Brownian motion. In general, all further calculations do not depend on this fact; however, if we say that E is the mathematical expectation in the risk-neutral world, it is much more natural, from financial point of view, to suppose such independence.

16.4.1   Markets with Stochastic Volatility Driven by Fractional Ornstein-Uhlenbeck Process

Let the stock price S_t satisfy the following equation in risk-neutral world:
$$dS_t = rS_t dt + \sigma_t S_t dW_t, \qquad (16.7)$$
where $r > 0$ is an interest rate, W_t is a standard Brownian motion and volatility σ_t satisfies the following equation:
$$d\sigma_t = -a\sigma_t dt + \gamma dB_t^H, \qquad (16.8)$$
where $a > 0$ is a mean-reverting speed, $\gamma > 0$ is a volatility coefficient of this stochastic volatility, B_t^H is a fractional Brownian motion with Hurst index $H > 1/2$, independent of W_t.

Note that the solution of the equation (16.7) has the following form:
$$\sigma_t = \sigma_0 e^{-at} + \gamma e^{-at} \int_0^t e^{as} dB_s^H. \qquad (16.9)$$
Evidently, the Wiener integral w.r.t. fBm exists since the function $f(s) = e^{as}$ satisfies the condition (16.5).

Moreover, σ_t is the continuous Gaussian process with the second moment $E\sigma_t^2$, which is bounded on any finite interval (we present all the calculations in the next two sections), therefore the unique solution of the equation (16.7) has a form
$$S_t = S_0 \exp[rt - \frac{1}{2} \int_0^t \sigma^2(s)ds + \int_0^t \sigma(s)dW(s)],$$
where the stochastic integral $\int_0^t \sigma(s)dW(s)$ exists and is a square-integrable martingale. The process S_t itself is locally square-integrable martingale. This situation will repeat and we will not mention this again in what follows.

16.4.2   Markets with Stochastic Volatility Driven by Fractional Vasiček Process

Let the stock price S_t satisfy equation (16.7) and the volatility σ_t satisfy the following equation:
$$d\sigma_t = a(b - \sigma_t)dt + \gamma dB_t^H, \qquad (16.10)$$
where $a > 0$ is a mean-reverting speed, $b \geq 0$ is an equilibrium (or mean-reverting) level, $\gamma > 0$ is a volatility coefficient of this stochastic volatility, B_t^H is a fractional Brownian motion with Hurst index $H > 1/2$ independent of W_t. Since the limit case $b = 0$ corresponds the fractional Ornstein-Uhlenbeck process, we suppose that for fractional Vasiček process the parameter b is positive, $b > 0$. In this sense the model (16.10) is the generalization of the model (16.7).

Note that the solution of the equation (16.10) has the following form:
$$\sigma_t = \sigma_0 e^{-at} + b(1 - e^{-at}) + \gamma e^{-at} \int_0^t e^{as} dB_s^H. \qquad (16.11)$$
Evidently, this Wiener integral w.r.t. fBm exists, and fractional Vasiček process is Gaussian. As we have mentioned above, both fractional Ornstein-Uhlenbeck and Vasiček volatilities get both positive and negative values.

16.4.3 Markets with Stochastic Volatility Driven by Geometric Fractional Brownian Motion

Let the stock price S_t satisfy equation (16.7) and the square σ_t^2 of volatility σ_t satisfy the following equation:

$$d\sigma_t^2 = a\sigma_t^2 dt + \gamma \sigma_t^2 dB_t^H, \qquad (16.12)$$

where $a > 0$ is a drift, $\gamma > 0$ is a volatility of σ_t^2, B_t^H is a fractional Brownian motion with Hurst index $H > 1/2$, independent of W_t.

Note that the solution of the equation (16.12) has the following form:

$$\sigma_t^2 = \sigma_0^2 e^{at + \gamma B_t^H}, \qquad (16.13)$$

and it is evidently the positive process, so, we can put $\sigma_t = \sigma_0 \exp\{\frac{a}{2}t + \frac{1}{2}\gamma B_t^H\}$.

16.4.4 Markets with Stochastic Volatility Driven by Fractional Continuous-Time GARCH Process

Let the stock price S_t satisfy equation (16.7) and the square σ_t^2 of volatility σ_t satisfy the following equation:

$$d\sigma_t^2 = a(b - \sigma_t^2)dt + \gamma \sigma_t^2 dB_t^H, \qquad (16.14)$$

where $a > 0$ is a mean-reverting speed, b is a mean-reverting level, $\gamma > 0$ is a volatility of σ_t^2, B_t^H is a fractional Brownian motion with Hurst index $H > 1/2$, independent of W_t. Note that the solution of the equation (16.14) has the following form:

$$\sigma_t^2 = \sigma_0^2 e^{-at + \gamma B_t^H} + abe^{-at + \gamma B_t^H} \int_0^t e^{as - \gamma B_s^H} ds. \qquad (16.15)$$

Note also that fractional GARCH process is a generalization of fractional geometric Brownian motion in the same sense that fractional Vasiček process is a generalization of fractional Ornstein-Uhlenbeck process.

16.5 Pricing of Variance Swaps

We start with the pricing of variance swaps. Introduce the following notations (note that $\alpha = H - \frac{1}{2} > 0$ consequently $2\alpha - 1 > -1$):

$$I_{st}(a) := \int_0^s \int_0^t e^{au + av} |u - v|^{2\alpha - 1} du\, dv,$$

$$J_t(a) := \int_0^t \left(\int_0^u + \int_0^{t-u} \right) e^{-az} z^{2\alpha - 1} dz\, dv, \quad C_H := H(2H - 1) = 2H\alpha.$$

Note that the integrals $I_{st}(a)$ and $J_t(a)$ cannot be calculated explicitly, but numerically.

16.5.1   Variance Swaps for Markets with Stochastic Volatility Driven by Fractional Ornstein-Uhlenbeck Process

Let consider the model (16.7)–(16.8) and use formula (16.1) to price variance swap for market with stochastic volatility driven by fractional Ornstein-Uhlenbeck process. It follows from the formulae (16.6) and (16.9) that

$$\sigma_t^2 = (\sigma_0 e^{-at} + \gamma e^{-at} \int_0^t e^{as} dB_s^H)^2,$$

$$E\sigma_t^2 = \sigma_0^2 e^{-2at} + \gamma^2 e^{-2at} C_H I_{tt}(a),$$

whence

$$E\{V\} = \frac{1}{T} \int_0^T E\sigma_t^2 dt = \frac{\sigma_0^2}{2aT}(1 - e^{-2aT}) + \frac{\gamma^2 C_H}{2aT}\Big(J_{TT}(a) - e^{-2aT} I_{TT}(a)\Big).$$
(16.16)

Now, it follows from (16.1) and (16.16) that the price P_{var} of variance swap for the market with stochastic volatility driven by fractional Ornstein-Uhlenbeck process equals.

Proposition 16.1.

$$P_{var} = e^{-rT} E\{\sigma_R^2(S) - K_{var}\}$$

$$= e^{-rT}\bigg\{ \frac{\sigma_0^2}{2aT}(1 - e^{-2aT}) + \frac{\gamma^2 C_H}{2aT}\Big(J_{TT}(a) - e^{-2aT} I_{TT}(a)\Big) - K_{var}\bigg\}.$$
(16.17)

16.5.2   Variance Swaps for Markets with Stochastic Volatility Driven by Fractional Vasićek Process

Now consider the model (16.7), (16.10) and use formula (16.1) to price variance swap for market with stochastic volatility driven by fractional Vasićek process. With the help of formulae (16.6) and (16.11) we obtain, similarly to previous subsection, that

$$E\{V\} = \frac{1}{T} \int_0^T E\sigma_t^2 dt$$

$$= \frac{(\sigma_0 - b)^2}{2aT}(1 - e^{-2aT}) + b^2 + \frac{2b(\sigma_0 - b)}{aT}(1 - e^{-aT})$$

$$+ \frac{\gamma^2 C_H}{2aT}\Big(J_{TT}(a) - e^{-2aT} I_{TT}(a)\Big).$$
(16.18)

Now, it follows from (16.1) and (16.18) that the price of variance swap for market with stochastic volatility driven by fractional Vasićek process equals.

Proposition 16.2.

$$P_{var} = e^{-rT} E\{\sigma_R^2(S) - K_{var}\}$$

$$= e^{-rT} \left\{ \frac{(\sigma_0 - b)^2}{2aT}(1 - e^{-2aT}) + b^2 + \frac{2b(\sigma_0 - b)}{aT}(1 - e^{-aT}) \right.$$

$$\left. + \frac{\gamma^2 C_H}{2aT}\left(J_{TT}(a) - e^{-2aT} I_{TT}(a) \right) - K_{var} \right\}. \qquad (16.19)$$

16.5.3 *Variance Swaps for Markets with Stochastic Volatility Driven by Geometric fBm*

We consider the model (16.7), (16.12) and use formula (16.1) to price variance swap for market with stochastic volatility driven by geometric fBm. With the help of (16.13) we obtain that

$$E\{V\} = \frac{1}{T} \int_0^T E\sigma_t^2 dt = \frac{\sigma_0^2}{T} \int_0^T e^{at + \gamma^2 t^{2H}/2} dt. \qquad (16.20)$$

Now, it follows from (16.1) and (16.20) that the price of variance swaps for markets with SV driven by geometric fBm equals.

Proposition 16.3.

$$P_{var} = e^{-rT} E\{\sigma_R^2(S) - K_{var}\} = e^{-rT} \left\{ \frac{\sigma_0^2}{T} \int_0^T e^{at + \gamma^2 t^{2H}/2} dt - K_{var} \right\}. \qquad (16.21)$$

Note that the integral in the right-hand side of (16.21) cannot be calculated explicitly unless $H = \frac{1}{2}$.

16.5.4 *Variance Swaps for Markets with Stochastic Volatility Driven by Fractional Continuous-Time GARCH Process*

Now we consider model (16.7), (16.14) and use formula (16.1) to price variance swap for market with stochastic volatility driven by fractional continuous-time GARCH process. With the help of formula (16.15) we obtain that

$$E\{V\} = \frac{1}{T} \int_0^T E\sigma_t^2 dt$$

$$= \frac{\sigma_0^2}{T} \int_0^T e^{-at + \gamma^2 t^{2H}/2} dt + \frac{ab}{T} \int_0^T e^{-at} \int_0^t e^{as + \gamma^2 (t-s)^{2H}/2} ds\, dt, \qquad (16.22)$$

consequently, in this case the price of variance swaps for markets with stochastic volatility driven by fractional continuous-time GARCH process equals.

Proposition 16.4.

$$P_{var} = e^{-rT} E\{\sigma_R^2(S) - K_{var}\}$$

$$= e^{-rT} \left\{ \frac{\sigma_0^2}{T} \int_0^T e^{-at + \gamma^2 t^{2H}/2} dt + \frac{ab}{T} \int_0^T e^{-at} \int_0^t e^{as + \gamma^2 (t-s)^{2H}/2} ds\, dt - K_{var} \right\}.$$

16.6   Pricing of Volatility Swaps

For pricing volatility swaps we use formula (16.2). It means that in addition to calculation of $E\{V\}$, we need to know $Var\{V\}$. Of course, we can calculate separately $E\{V^2\}$ and use previous calculations of $E\{V\}$; however, in Ornstein-Uhlenbeck and Vasiček cases it will be more convenient to use the following fact: if the stochastic process σ_t can be presented as a sum of two processes, $\sigma_t = \xi_t + B_t$, where ξ_t is Gaussian process with continuous in t fourth moment and B_t is nonrandom continuous function, then for $V = \frac{1}{T}\int_0^T \sigma_s^2 ds$ we have that

$$Var\{V\} = \frac{1}{T^2}\int_0^T \int_0^T \left(E\xi_s^2\xi_t^2 - E\xi_s^2 E\xi_t^2 + 4E\xi_s\xi_t B_s B_t \right) ds\, dt. \qquad (16.23)$$

Denote $r(s,t) = E\xi_s\xi_t$, $\theta_t = E\xi_t^2$, then $E\xi_s^2\xi_t^2 = \theta_s\theta_t + 2r^2(s,t)$, and (16.23) can be rewritten as

$$Var\{V\} = \frac{1}{T^2}\int_0^T \int_0^T (2r^2(s,t) + 4r(s,t)B_s B_t)ds\, dt. \qquad (16.24)$$

If we want to use the approximation (16.2), we must be sure, at least, that its right-hand side is positive. So, we must check the inequality

$$Var\{V\} \leq 8(E\{V\})^2. \qquad (16.25)$$

If $\sigma_t = \xi_t + B_t$, then the inequality (16.25) is transformed to

$$\int_0^T \int_0^T (r^2(s,t) + 2r(s,t)B_s B_t)ds\, dt \leq 4\left(\int_0^T \theta_t dt \right)^2 + 8\int_0^T \theta_t dt \int_0^T B_t^2 dt$$

$$+ 4\left(\int_0^T B_t^2 dt \right)^2,$$

and the latter inequality evidently holds, since

$$\int_0^T \int_0^T (r^2(s,t) + 2r(s,t)B_s B_t)ds\, dt \leq \int_0^T \int_0^T (2r^2(s,t) + (B_s B_t)^2)ds\, dt$$

$$\leq 2\left(\int_0^T \theta_t dt \right)^2 + \left(\int_0^T B_t^2 dt \right)^2.$$

However, the approximation (16.2) is satisfactory in the neighborhood of $E\{V\}$, when $Var\{V\}$ is comparatively small, and for Gaussian models (Ornstein-Uhlenbeck and Vasiček processes, see subsections 16.6.1 and 16.6.2 below) this approximation is better when the drift term $\sigma_0 e^{-at}$ or $\sigma_0 e^{-at} + b(1 - e^{-at})$, correspondingly, exceeds significantly the diffusion term $\gamma e^{-at}\int_0^t e^{as}dB_s^H$. The behavior of approximation (2) for fractional geometric Brownian motion and fractional GARCH process will be analyzed in subsections 16.6.3 and 16.6.4.

16.6.1 *Volatility Swaps for Markets with Stochastic Volatility Driven by Fractional Ornstein-Uhlenbeck Process*

Consider again the model (16.7)–(16.8) and use formulae (16.2)–(16.3) to price volatility swap for market with stochastic volatility driven by fractional Ornstein-Uhlenbeck process. We can substitute from (16.9) into (16.24) $\xi_t = \gamma e^{-at} \int_0^t e^{as} dB_s^H$, $B_t = \sigma_0 e^{-at}$ and obtain, after some calculations, that

$$Var\{V\} = \frac{2\gamma^4 C_H^2}{T^2} \int_0^T \int_0^T e^{-2as-2at} I_{st}^2 ds\, dt$$

$$+ \frac{4C_H \sigma_0^2}{T^2} \int_0^T \int_0^T e^{-2as-2at} I_{st} ds\, dt. \tag{16.26}$$

Now, using formula (16.2), (16.16) and (16.26) we have:

$$E\{\sqrt{\sigma_R^2(S)}\} \approx \sqrt{E\{V\}} - \frac{Var\{V\}}{8E\{V\}^{3/2}}$$

$$\approx \sqrt{\frac{\sigma_0^2}{2aT}(1 - e^{-2aT}) + \frac{\gamma^2 C_H}{2aT}\left(J_{TT}(a) - e^{-2aT} I_{TT}(a)\right)}$$

$$- \left(\frac{2\gamma^4 C_H^2}{T^2} \int_0^T \int_0^T e^{-2as-2at} I_{st}^2 ds\, dt\right.$$

$$+ \frac{2C_H \sigma_0^2}{T^2} \int_0^T \int_0^T e^{-2as-2at} I_{st} ds\, dt \left.\right) \left(8\left(\frac{\sigma_0^2}{2aT}(1 - e^{-2aT})\right.\right.$$

$$+ \frac{\gamma^2 C_H}{2aT}\left(J_{TT}(a) - e^{-2aT} I_{TT}(a))\right)^{3/2}\right)^{-1}. \tag{16.27}$$

Substituting (16.27) into formula (16.3) we have the price of volatility swap for markets with stochastic volatility driven by fractional Ornstein-Uhlenbeck process.

Proposition 16.5.

$$P_{vol} = E\{e^{-rT}(\sigma_R(S) - K_{vol})\} = e^{-rT} E\left\{\sqrt{\sigma_R^2(S)} - K_{vol}\right\}$$

$$= e^{-rT}\left(\sqrt{\frac{\sigma_0^2}{2aT}(1 - e^{-2aT}) + \frac{\gamma^2 C_H}{2aT}\left(J_{TT}(a) - e^{-2aT} I_{TT}(a)\right)}\right.$$

$$- \left(\frac{2\gamma^4 C_H^2}{T^2} \int_0^T \int_0^T e^{-2as-2at} I_{st}^2 ds\, dt\right.$$

$$+ \frac{2C_H \sigma_0^2}{T^2} \int_0^T \int_0^T e^{-2as-2at} I_{st} ds\, dt \left.\right) \left(8\left(\frac{\sigma_0^2}{2aT}(1 - e^{-2aT})\right.\right.$$

$$+ \frac{\gamma^2 C_H}{2aT}\left(J_{TT}(a) - e^{-2aT} I_{TT}(a))\right)^{3/2}\right)^{-1} - K_{vol}\right).$$

16.6.2 Volatility Swaps for Markets with Stochastic Volatility Driven by Fractional Vasićek Process

We return to the model (16.7), (16.10). This model differs from the previous one only by additional term $b(1 - e^{-at})$ in B_t (see formula $\sigma_t = \xi_t + B_t$ for what B_t means and compare (16.9) and (16.11)). Therefore, using (16.24) and (16.26), it is easy to price volatility swap for market with stochastic volatility driven by fractional Vasićek process. The additional terms appear both in $E\{V\}$ (see (16.18) for $E\{V\}$) and in $Var\{V\}$:

$$Var\{V\} = \frac{2\gamma^4 C_H^2}{T^2} \int_0^T \int_0^T e^{-2as-2at} I_{st}^2 ds\, dt$$

$$+ \frac{2C_H}{T^2} \int_0^T \int_0^T (\sigma_0 e^{-as} + b(1 - e^{-as}))(\sigma_0 e^{-at} + b(1 - e^{-at})) I_{st} ds\, dt.$$

$$(16.28)$$

Using (16.18) and (16.27), we can proceed directly to the price of volatility swap.

Proposition 16.6.

$$P_{vol} = E\{e^{-rT}(\sigma_R(S) - K_{vol})\} = e^{-rT} E\left\{\sqrt{\sigma_R^2(S)} - K_{vol}\right\}$$

$$= e^{-rT}\left(\left(\frac{(\sigma_0 - b)^2}{2aT}(1 - e^{-2aT}) + b^2 + \frac{2b(\sigma_0 - b)}{aT}(1 - e^{-aT})\right.\right.$$

$$+ \frac{\gamma^2 C_H}{2aT}\left(J_{TT}(a) - e^{-2aT} I_{TT}(a)\right)\bigg)^{1/2} - \left(\frac{2\gamma^4 C_H^2}{T^2} \int_0^T \int_0^T e^{-2as-2at} I_{st}^2 ds\, dt\right.$$

$$+ \frac{2C_H}{T^2} \int_0^T \int_0^T (\sigma_0 e^{-as} + b(1 - e^{-as}))(\sigma_0 e^{-at}$$

$$+ b(1 - e^{-at})) I_{st} ds\, dt\bigg) 2^{-3}\left(\frac{(\sigma_0 - b)^2}{2aT}(1 - e^{-2aT}) + b^2\right.$$

$$+ \frac{2b(\sigma_0 - b)}{aT}(1 - e^{-aT}) + \frac{\gamma^2 C_H}{2aT}\left(J_{TT}(a) - e^{-2aT} I_{TT}(a)\right)\bigg)^{-3/2} - K_{vol}\bigg).$$

16.6.3 Volatility Swaps for Markets with Stochastic Volatility Driven by Geometric fBm

Now we consider model (16.7), (16.12) and, as usual, take formulae (16.2)–(16.3) to price volatility swap for the market with stochastic volatility driven by geometric fBm. With the help of (16.13),

$$Var(V) = \frac{\sigma_0^4}{T^2} \int_0^T \int_0^T e^{at+as} e^{\gamma^2(2t^{2H}+2s^{2H}-|t-s|^{2H})/2} ds\, dt$$

$$- \frac{\sigma_0^4}{T^2}\left(\int_0^T e^{at+\gamma^2 t^{2H}/2} dt\right)^2.$$

$$(16.29)$$

Now, substituting (16.20) and (16.29) into formula (16.3), we get the price of volatility swap for markets with stochastic volatility driven by geometric fBm.

Proposition 16.7

$$P_{vol} = E\{e^{-rT}(\sigma_R(S) - K_{vol})\} = e^{-rT}E\left\{\sqrt{\sigma_R^2(S)} - K_{vol}\right\}$$

$$= e^{-rT}\left(\left(\frac{\sigma_0^2}{T}\int_0^T e^{at+\gamma^2t^{2H}/2}dt\right)^{1/2}\right.$$

$$-\left(\frac{\sigma_0^4}{T^2}\int_0^T\int_0^T e^{at+as}e^{\gamma^2(2t^{2H}+2s^{2H}-|t-s|^{2H})/2}dsdt\right.$$

$$-\frac{\sigma_0^4}{T^2}\left(\int_0^T e^{at+\gamma^2t^{2H}/2}dt\right)^2\right)2^{-3}\left(\frac{\sigma_0^2}{T}\int_0^T e^{at+\gamma^2t^{2H}/2}dt\right)^{-3/2} - K_{vol}\right).$$

Remark 1. As to positivity of the Brockhaus-Long approximation, it will be discussed in the next subsection.

16.6.4 *Volatility Swaps for Markets with Stochastic Volatility Driven by Fractional Continuous-Time GARCH Process*

At last, we return to the model (16.7), (16.14) and use (16.15) to calculate $Var\{V\}$:

$$Var\{V\}$$

$$= \frac{\sigma_0^4}{T^2}\int_0^T\int_0^T e^{-at-as}e^{\gamma^2(2t^{2H}+2s^{2H}-|t-s|^{2H})/2}dsdt$$

$$+ \frac{ab\sigma_0^2}{T^2}\int_0^T\int_0^T e^{-as-at}\int_0^t e^{au}e^{\gamma^2(s^{2H}+t^{2H}-u^{2H}-|t-s|^{2H}+|u-s|^{2H}+|t-u|^{2H})/2}dudsdt$$

$$+ \frac{ab\sigma_0^2}{T^2}\int_0^T\int_0^T e^{-as-at}\int_0^s e^{av}e^{\gamma^2(s^{2H}+t^{2H}-v^{2H}-|t-s|^{2H}+|v-s|^{2H}+|t-v|^{2H})/2}dvdsdt$$

$$+ \frac{a^2b^2}{T^2}\int_0^T\int_0^T e^{-as-at}$$

$$\times \int_0^t\int_0^s e^{au+av}e^{\gamma^2(|s-u|^{2H}+|s-v|^{2H}+|t-u|^{2H}-|s-t|^{2H}-|u-v|^{2H}+|t-v|^{2H})/2}dudvdsdt$$

$$- \frac{\sigma_0^4}{T^2}\int_0^T\int_0^T e^{-at-as+\gamma^2t^{2H}/2+\gamma^2s^{2H}/2}ds\,dt$$

$$- \frac{2\sigma_0^2ab}{T^2}\int_0^T\int_0^T e^{-at-as+\gamma^2s^{2H}/2}\int_0^t e^{au+\gamma^2(t-u)^{2H}/2}du\,ds\,dt$$

$$- \frac{a^2b^2}{T^2}\int_0^T\int_0^T e^{-at-as}\int_0^t\int_0^s e^{au+av+\gamma^2(t-u)^{2H}/2+\gamma^2(s-v)^{2H}/2}du\,dv\,ds\,dt.$$

$$(16.30)$$

Now, substituting (16.22) and (16.30) into formula (16.3) we get the price of volatility swaps for markets with stochastic volatility driven by continuous-time fractional GARCH process. We omit the corresponding formulae since they are comparatively complicated but can be easily obtained, similar to previous case.

Remark 2. As to positivity of the Brockhaus-Long approximation, we must check the inequality (16.25), or, the equivalent inequality

$$E\{V\}^2 \le 9(E\{V\})^2. \tag{16.31}$$

For the last one, we can compare in pairs positive and negative terms in (16.29), for example, establish the inequality

$$\frac{\sigma_0^4}{T^2} \int_0^T \int_0^T e^{-at-as} e^{\gamma^2(2t^{2H}+2s^{2H}-|t-s|^{2H})/2} ds dt$$

$$< \frac{9\sigma_0^4}{T^2} \int_0^T \int_0^T e^{-at-as} e^{\gamma^2 t^{2H}/2+\gamma^2 s^{2H}/2} ds\, dt \tag{16.32}$$

and so on. In turn, it is sufficient to compare the integrands in the left- and right-hand sides of (16.32), and we easily obtain that

$$e^{\gamma^2(2t^{2H}+2s^{2H}-|t-s|^{2H})/2} < 9e^{\gamma^2 t^{2H}/2+\gamma^2 s^{2H}/2}$$

for any $0 \le s, t \le T$ if $\gamma^2 T^{2H} < 2\ln 3$. The same is true for the two other pairs in (16.32). We can give the recommendation that the good approximation can be obtained when the parameter γ is comparatively small in comparison with the maturity date T. Evidently, the same is true for geometric fBm model also.

16.7 Discussion: Asymptotic Results for the Pricing of Variance Swaps with Zero Risk-Free Rate when the Expiration Date Increases

Let the expiration date increase. We are interested now how the prices of variance swaps behave in this case and if the pricing can be simplified. At first, consider fractional Orhstein-Uhlenbeck and Vasiček models. It is not hard to see that in the case when $r > 0$ all the prices tend to zero. However, if we suppose that $r = 0$ some less trivial results can be obtained. Indeed, since all the mathematical expectations $E\sigma_t^2$ are continuous, we have the equality

$$\lim_{T\to\infty} E\{V\} = \lim_{T\to\infty} E\sigma_T^2 \tag{16.33}$$

whenever the latter limit exists. (Here, as before, $V = V_T = \frac{1}{T}\int_0^T E\sigma_t^2 dt$.)

1) We have for fractional Ornstein-Uhlenbeck process (see (16.8)) that

$$\lim_{T \to \infty} E\sigma_T^2 = \gamma^2 C_H \lim_{T \to \infty} e^{-2aT} I_{TT}(a)$$

$$= \gamma^2 C_H \lim_{T \to \infty} \frac{1}{a} e^{-2aT} \int_0^T e^{aT+av} (T-v)^{2\alpha-1} dv$$

$$= \frac{\gamma^2 C_H}{a^{2H}} \Gamma(2\alpha).$$

Hence, the limit price P_{var}^∞ has the following form.
Proposition 16.8.

$$P_{var}^\infty = \frac{\gamma^2 C_H}{a^{2H}} \Gamma(2\alpha) - K_{var}.$$

2) Similarly, the limit price for Vasiček model (see (16.19)) equals.
Proposition 16.9.

$$P_{var}^\infty = \frac{\gamma^2 C_H}{a^{2H}} \Gamma(2\alpha) + b^2 - K_{var}.$$

Note that we have, to some extent, the similar situation in the case when stochastic volatility is standard Ornstein-Uhlenbeck or Vasiček process, because their second moment is bounded in time.

If we consider geometric fractional Brownian motion and fractional GARCH process, then it is easy to see that the limit price is infinite for any values of parameters r, a, γ due to the multiplier $e^{\gamma^2 t^{2H}/2}$ (recall that $2H > 1$). This situation differs from the case $H = 1/2$ where limit price can be 0, ∞ or some non-zero constant, depending on the values of the parameters.

16.8 Summary

– In this Chapter, we studied financial markets with stochastic volatilities driven by fractional Brownian motion with Hurst index $H > 1/2$.
– Our models for stochastic volatility include new fractional versions of Ornstein-Uhlenbeck, Vasiček, geometric Brownian motion and continuous-time GARCH models.
– The novelty of the Chapter is in pricing of variance and volatility swaps for the above-mentioned models.
– We also stated some results on asymptotic behavior of variance swaps with zero risk-free rate when the expiration date increases.

Bibliography

Breidt, J. F., Crato, N. and Lima, Pedro de (1998). The detection and estimation of long memory in stochastic volatility. *Journal of Econometrics*, 83, 325-348.

Brockhaus, O. and Long, D. (2000). Volatility swaps made simple. *Risk Magazine*, January, 92-96.

Deo, R. S. and Hurvich, C. M. (2003). Estimation of long-memory in volatility. In Doukhan, P., Oppenheim, G. and Taqqu, M. (eds.), *Theory and Applications of Long-Range Dependence*, 313-325. Boston: Birkhäuser.

Doukhan, P. (2003). Models, inequalities and limit theorems for stationary sequences. In Doukhan, P., Oppenheim, G. and Taqqu, M. (eds.), *Theory and Applications of Long-Range Dependence*, 43-100. Boston: Birkhäuser.

Javaheri, A., Wilmott, P. and Haug, E. (2002). GARCH and volatility swaps. *Wilmott Technical Article*, January, 17 pages.

Johnson, H. and Shanno, D. (1987). Option pricing when the variance is changing. *Journal of Finance and Quantitative Analysis*, 22, 143-151.

Kallianpur, G. and Karandikar, R. (2000). *Introduction to Option Pricing Theory*. Boston: Birkhauser.

Klüppelberg, C. and Kühn, C. (2002). Fractional Brownian motion as a weak limit of Poisson shot noise processes-with applications to finance. Preprint, Munich University of Technology.

Mandelbrot, B. and Van Ness, J. (1968). Fractional Brownian motions, fractional noises and applications. *Bell Journal of Economic Management Science*, 4, 141-183.

Mishura Y. and Swishchuk A. (2010). Modeling and pricing of variance and volatility swaps for stochastic volatilities driven by fractional Brownian motion. *Applied Statistics, Actuarial and Financial Mathematics*, 1-2, 52-67.

Mishura, Yu. (2008). *Stochastic Calculus for Fractional Brownian Motion and Related Processes*. Berlin-Heidelberg: Springer-Verlag.

Molchan, G. (2002). Linear problems for a fractional Brownian motion: Group approach. *Theory of Probability and Its Applications*, 47, 69-78.

Norros, I., Valkeila, E. and Virtamo, J. (1999). An elementary approach to a Girsanov formula and other analytical results on fractional Brownian motions. *Bernoulli*, 5, 4, 571-587.

Nualart, D. and Răşcanu, A. (2000). Differential equations driven by fractional Brownian motion. *Collectanea Mathematica*, 53, 55-81.

Taqqu, M. (2003). Fractional Brownian motion and long-range dependence. In Doukhan, P., Oppenheim, G. and Taqqu, M. (eds.), *Theory and Applications of Long-Range Dependence*. 5-38. Boston: Birkhäuser.

Zähle, M. (1998). Integration with respect to fractal functions and stochastic calculus. *Probability Theory and Related Fields*, 111, 333-374.

Zähle, M. (2001). Integraton with respect to fractal functions and stochastic calculus. *Mathematische Nachrichten*, 225, 145-183.

Chapter 17

Variance and Volatility Swaps in Energy Markets

17.1 Introduction

This Chapter is devoted to the pricing of variance and volatility swaps in energy markets. We found explicit variance swap formula and closed form volatility swap formula (using change of time) for energy asset with stochastic volatility that follows continuous-time mean-reverting GARCH (1,1) model. A numerical example is presented for AECO Natural Gas Index (1 May 1998–30 April 1999).

Variance swaps are quite common in commodity, e.g., in energy markets, and they are commonly traded. We consider Ornstein-Uhlenbeck process for commodity asset with stochastic volatility following continuous-time GARCH model or Pilipovic (1998) one-factor model. The classical stochastic process for the spot dynamics of commodity prices is given by the Schwartz' model (1997). It is defined as the exponential of an Ornstein-Uhlenbeck (OU) process, and has become the standard model for energy prices possessing mean-reverting features.

In this chapter, we consider a risky asset in energy markets with stochastic volatility following a mean-reverting stochastic process satisfying the following SDE (continuous-time GARCH(1,1) model):

$$d\sigma^2(t) = a(L - \sigma^2(t))dt + \gamma\sigma^2(t)dW_t,$$

where a is a speed of mean reversion, L is the mean-reverting level (or equilibrium level), γ is the volatility of volatility $\sigma(t)$, W_t is a standard Wiener process. Using a change of time method we find an explicit solution of this equation and using this solution we are able to find the variance and volatility swaps pricing formula under the physical measure. Then, using the same argument, we find the option pricing formula under risk-neutral measure. We applied Brockhaus-Long (2000) approximation to find the value of volatility swap. A numerical example for the AECO Natural Gas Index for the period 1 May 1998 to 30 April 1999 is presented.

Commodities are emerging as an asset class in their own. The range of products offered to investors range from exchange traded funds (ETFs) to sophisticated products including principal protected structured notes on individual commodities or baskets of commodities and commodity range-accrual or varinace swap.

241

More and more institutional investors are including commodities in their asset allocation mix and hedge funds are also increasingly active players in commodities. Example: Amaranth Advisors lost USD 6 billion during September 2006 from trading natural gas futures contracts, leading to the fund's demise.

Concurrent with these developments, a number of recent papers have examined the risk and return characteristics of investments in individual commodity futures or commodity indices composed of baskets of commodity futures. See, e.g., Erb and Harvey (2006), Gorton and Rouwenhorst (2006), Ibbotson (2006), Kan and Oomen (2007).

However, since all but the most plain-vanilla investments contain an exposure to volatility, it is equally important for investors to understand the risk and return characteristics of commodity volatilities.

Our focus on energy commodities derives from two resons:

(1) energy is the most important commodity sector, and crude oil and natural gas constitute the largest components of the two most widely tracked commodity indices: the Standard & Poors Goldman Sachs Commodity Index (S&P GSCI) and the Dow Jones-AIG Commodity Index (DJ-AIGCI).

(2) existence of a liquid options market: crude oil and natural gas indeed have the deepest and most liquid options marketss among all commodities.

The idea is to use variance (or volatility) swaps on futures contracts.

At maturity, a variance swap pays off the difference between the realized variance of the futures contract over the life of the swap and the fixed variance swap rate.

And since a variance swap has zero net market value at initiation, absence of arbitrage implies that the fixed variance swap rate equals to conditional risk-neutral expectation of the realized variance over the life of swap.

Therefore, e.g., the time-series average of the payoff and/or excess return on a variance swap is a measure of the variance risk premium.

Variance risk premia in energy commodities, crude oil and natural gas, has been considered by Trolle and Schwartz (2010).

The same methodology as in Trolle and Schwartz (2010) was used by Carr and Wu (2009) in their study of equity variance risk premia. The idea was to use variance swaps on futures contracts.

The study in Trolle and Schwartz (2010) is based on daily data from January 2, 1996 until November 30, 2006-a total of 2750 business days. The source of the data is NYMEX.

Trolle and Schwartz (2010) found that:

(1) the average variance risk premia are negative for both energy commodities but more strongly statistically significant for crude oil than for natural gas;

(2) the natural gas variance risk premium (defined in dollars terms or in return terms) is higher during the cold months of the year (seasonality and peaks for natural gas variance during the cold months of the year);

(3) energy risk premia in dollar terms are time-varying and correlated with the

level of the variance swap rate. In contrast, energy variance risk premia in return terms, particularly in the case of natural gas, are much less correlated with the varinace swap rate.

The *S&P* GSCI is comprised of 24 commodities with the weight of each commodity determined by their relative levels of world production over the past five years.

The DJ-AIGCI is comprised of 19 commodities with the weight of each component detrmined by liquidity and world production values, with liquidity being the dominant factor.

Crude oil and natural gas are the largest components in both indices. In 2007, their weight were 51.30% and 6.71%, respectively, in the S&P GSCI and 13.88% and 11.03%, respectively, in the DJ-AIGCI.

The Chicago Board Optiopns Exchange (CBOE) recently introduced a Crude Oil Volatility Index (ticker symbol OVX). This index also measures the conditional risk-neutral expectation of crude oil variance, but is computed from a cross-section of listed options on the United States Oil Fund (USO), which tracks the price of WTI as closely as possible. The CBOE Crude Oil ETF Volatility Index ('Oil VIX', Ticker — OVX) measures the market's expectation of 30-day volatility of crude oil prices by applying the VIX methodology to United States Oil Fund, LP (Ticker — USO) options spanning a wide range of strike prices. We have to notice that crude oil and natural gas trade in units of 1,000 barrels and 10,000 British thermal units (mmBtu), respectively. Price are quoted as US dollars and cents per barrel or mmBtu. The continuous-time GARCH model has also been exploited by Javaheri, Wilmott and Haug (2002) to calculate volatility swap for *S&P*500 index. They used PDE approach and mentioned (page 8, Sec. 3.3) that 'it would be interesting to use an alternative method to calculate $F(v, t)$ and the other above quantities'. This paper exactly contains the alternative method, namely, 'change of time method', to get varinace and volatility swaps. The change of time method was also applied by Swishchuk (2004) for pricing variance, volatility, covariance and correlation swaps for Heston model. The first paper on pricing of commodity contracts was published by Black (1976).

17.2   Mean-Reverting Stochastic Volatility Model (MRSVM)

In this Section we introduce MRSVM and study some properties of this model that we can use later for calculating variance and volatility swaps.

Let $(\Omega, \mathcal{F}, \mathcal{F}_t, P)$ be a probability space with a sample space Ω, σ-algebra of Borel sets $\mathcal{F}$ and probability P. The filtration $\mathcal{F}_t$, $t \in [0, T]$, is the natural filtration of a standard Brownian motion W_t, $t \in [0, T]$, such that $\mathcal{F}_T = \mathcal{F}$.

We consider a risky asset in energy markets with stochastic volatility following a mean-reverting stochastic process the following stochastic differential equation:

$$d\sigma^2(t) = a(L - \sigma^2(t))dt + \gamma\sigma^2(t)dW_t, \qquad (17.1)$$

where $a > 0$ is a speed (or 'strength') of mean reversion, $L > 0$ is the mean reverting level (or equilibrium level, or long-term mean), $\gamma > 0$ is the volatility of volatility $\sigma(t)$, W_t is a standard Wiener process.

17.2.1 *Explicit Solution of MRSVM*

Let

$$V_t := e^{at}(\sigma^2(t) - L). \tag{17.2}$$

Then, from (17.1) and (17.2) we obtain

$$dV_t = ae^{at}(\sigma^2(t) - L)dt + e^{at}d\sigma^2(t) = \sigma(V_t + e^{at}L)dW_t. \tag{17.3}$$

Using change of time approach to the equation (17.3) (see Ikeda and Watanabe (1981) or Elliott (1982)) we obtain the following solution of the equation (17.3)

$$V_t = \sigma^2(0) - L + \tilde{W}(\phi_t^{-1}),$$

or (see (17.2)),

$$\sigma^2(t) = e^{-at}[\sigma^2(0) - L + \tilde{W}(\phi_t^{-1})] + L, \tag{17.4}$$

where $\tilde{W}(t)$ is an $\mathcal{F}_t$-measurable standard one-dimensional Wiener process, ϕ_t^{-1} is an inverse function to ϕ_t :

$$\phi_t = \gamma^{-2} \int_0^t (\sigma^2(0) - L + \tilde{W}(s) + e^{a\phi_s}L)^{-2}ds. \tag{17.5}$$

We note that

$$\phi_t^{-1} = \gamma^2 \int_0^t (\sigma^2(0) - L + \tilde{W}(\phi_t^{-1}) + e^{as}L)^2ds, \tag{17.6}$$

which follows from (17.5).

17.2.2 *Some Properties of the Process $\tilde{W}(\phi_t^{-1})$*

We note that process $\tilde{W}(\phi_t^{-1})$ is $\tilde{\mathcal{F}}_t := \mathcal{F}_{\phi_t^{-1}}$-measurable and $\tilde{\mathcal{F}}_t$-martingale.
Then

$$E\tilde{W}(\phi_t^{-1}) = 0. \tag{17.7}$$

Let's calculate the second moment of $\tilde{W}(\phi_t^{-1})$ (see (17.6)):

$$E\tilde{W}^2(\phi_t^{-1}) = E\langle \tilde{W}(\phi_t^{-1})\rangle = E\phi_t^{-1}$$

$$= \gamma^2 \int_0^t E(\sigma^2(0) - L + \tilde{W}(\phi_s^{-1}) + e^{as}L)^2ds$$

$$= \gamma^2[(\sigma^2(0) - L)^2t + \frac{2L(\sigma^2(0) - L)(e^{at} - 1)}{a} + \frac{L^2(e^{2at} - 1)}{2a}$$

$$+ \int_0^t E\tilde{W}^2(\phi_s^{-1})ds]. \tag{17.8}$$

From (17.8), solving this linear ordinary nonhomogeneous differential equation with respect to $E\tilde{W}^2(\phi_t^{-1})$,

$$\frac{dE\tilde{W}^2(\phi_t^{-1})}{dt} = \gamma^2[(\sigma^2(0) - L)^2 + 2L(\sigma^2(0) - L)e^{at} + L^2 e^{2at} + E\tilde{W}^2(\phi_t^{-1})],$$

we obtain

$$E\tilde{W}^2(\phi_t^{-1}) = \gamma^2\left[(\sigma^2(0) - L)^2\frac{e^{\gamma^2 t} - 1}{\gamma^2} + \frac{2L(\sigma^2(0) - L)(e^{at} - e^{\gamma^2 t})}{a - \gamma^2} + \frac{L^2(e^{2at} - e^{\gamma^2 t})}{2a - \gamma^2}\right].$$

We note, that

$$E\tilde{W}(\phi_s^{-1})\tilde{W}(\phi_t^{-1}) = \gamma^2\left(\sigma^2(0) - L)^2\frac{e^{\gamma^2(t \wedge s)} - 1}{\gamma^2}\right.$$

$$\left. + \frac{2L(\sigma^2(0) - L)(e^{a(t \wedge s)} - e^{\gamma^2(t \wedge s)})}{a - \gamma^2} + \frac{L^2(e^{2a(t \wedge)} - e^{\gamma^2(t \wedge s)})}{2a - \gamma^2}\right],$$

$$(17.9)$$

and the second moment for $\tilde{W}^2(\phi_t^{-1})$ above follows from (17.9).

17.2.3   *Explicit Expression for the Process* $\tilde{W}(\phi_t^{-1})$

It turns out that we can find the explicit expression for the process $\tilde{W}(\phi_t^{-1})$.

From the expression (see Section 17.1)

$$V_t = \sigma^2(0) - L + \tilde{W}(\phi_t^{-1}),$$

we have the following relationship between $W(t)$ and $\tilde{W}(\phi_t^{-1})$:

$$d\tilde{W}(\phi_t^{-1}) = \gamma\int_0^t [S(0) - L + Le^{at} + \tilde{W}(\phi_s^{-1})]dW(t).$$

It is a linear SDE with respect to $\tilde{W}(\phi_t^{-1})$ and we can solve it explicitly. The solution has the following look:

$$\tilde{W}(\phi_t^{-1}) = \sigma^2(0)(e^{\gamma W(t) - \frac{\gamma^2 t}{2}} - 1) + L(1 - e^{at}) + aLe^{\gamma W(t) - \frac{\gamma^2 t}{2}}\int_0^t e^{as}e^{-\gamma W(s) + \frac{\gamma^2 s}{2}}ds.$$

$$(17.10)$$

It is easy to see from (17.10) that $\tilde{W}(\phi_t^{-1})$ can be presented in the form of a linear combination of two zero-mean martingales $m_1(t)$ and $m_2(t)$:

$$\tilde{W}(\phi_t^{-1}) = m_1(t) + Lm_2(t),$$

where

$$m_1(t) := \sigma^2(0)(e^{\gamma W(t) - \frac{\gamma^2 t}{2}} - 1)$$

and

$$m_2(t) = (1 - e^{at}) + ae^{\gamma W(t) - \frac{\gamma^2 t}{2}}\int_0^t e^{as}e^{-\gamma W(s) + \frac{\gamma^2 s}{2}}ds.$$

Indeed, process $\tilde{W}(\phi_t^{-1})$ is a martingale (see Section 17.2.2), also it is well-known that process $e^{\gamma W(t) - \frac{\gamma^2 t}{2}}$ and, hence, process $m_1(t)$ is a martingale. Then the process $m_2(t)$, as the difference between two martingales, is also martingale.

17.2.4   *Some Properties of the Mean-Reverting Stochastic Volatility $\sigma^2(t)$: First Two Moments, Variance and Covariation*

From (17.4) we obtain the mean value of the first moment for mean-reverting stochastic volatility $\sigma^2(t)$:

$$E\sigma^2(t) = e^{-at}[\sigma^2(0) - L] + L. \qquad (17.11)$$

It means that $E\sigma^2(t) \rightarrow L$ when $t \rightarrow +\infty$. We need this moment to value the variance swap.

Using formula (17.4) and (17.9) we can calculate the second moment of $\sigma^2(t)$:

$$E\sigma^2(t) = (e^{-at}(\sigma^2(0) - L) + L)^2$$

$$+ \gamma^2 e^{-2at}\left[(\sigma^2(0) - L)^2\frac{e^{\gamma^2 t} - 1}{\gamma^2} + \frac{2L(\sigma^2(0) - L)(e^{at} - e^{\gamma^2 t})}{a - \gamma^2} + \frac{L^2(e^{2at} - e^{\gamma^2 t})}{2a - \gamma^2}\right].$$

Combining the first and the second moments we have the variance of $\sigma^2(t)$:

$$Var(\sigma^2(t)) = E\sigma^2(t)^2 - (E\sigma^2(t))^2$$

$$= \gamma^2 e^{-2at}\left[(\sigma^2(0) - L)^2\frac{e^{\gamma^2 t} - 1}{\gamma^2} + \frac{2L(\sigma^2(0) - L)(e^{at} - e^{\gamma^2 t})}{a - \gamma^2} + \frac{L^2(e^{2at} - e^{\gamma^2 t})}{2a - \gamma^2}\right].$$

From the expression for $\tilde{W}(\phi_t^{-1})$ (see (17.10)) and for $\sigma^2(t)$ in (17.4) we can find the explicit expression for $\sigma^2(t)$ through $W(t)$:

$$\sigma^2(t) = e^{-at}[\sigma^2(0) - L + \tilde{W}(\phi_t^{-1})] + L$$

$$= e^{-at}[\sigma^2(0) - L + m_1(t) + Lm_2(t)] + L$$

$$= \sigma^2(0)e^{-at}e^{\gamma W(t) - \frac{\gamma^2 t}{2}} + aLe^{-at}e^{\gamma W(t) - \frac{\gamma^2 t}{2}}\int_0^t e^{as}e^{-\gamma W(s) + \frac{\gamma^2 s}{2}}ds,$$

$$\qquad (17.12)$$

where $m_1(t)$ and $m_2(t)$ are defined as in Section 17.2.3.

From (17.12) it follows that $\sigma^2(t) > 0$ as long as $\sigma^2(0) > 0$.

The covariation for $\sigma^2(t)$ may be obtained from (17.4), (17.7) and (17.9):

$$E\sigma^2(t)\sigma^2(s) = e^{-a(t+s)}(\sigma^2(0) - L)^2 + e^{-a(t+s)}\left\{\gamma^2\left[(\sigma^2(0) - L)^2\frac{e^{\gamma^2(t \wedge s)} - 1}{\gamma^2}\right.\right.$$

$$\left.\left. + \frac{2L(\sigma^2(0) - L)(e^{a(t \wedge s)} - e^{\gamma^2(t \wedge s)})}{a - \gamma^2} + \frac{L^2(e^{2a(t \wedge s)} - e^{\gamma^2(t \wedge s)})}{2a - \gamma^2}\right]\right\}$$

$$+ e^{-at}(\sigma^2(0) - L)L + e^{-as}(\sigma^2(0) - L)L + L^2. \qquad (17.13)$$

We need this covariance to value the volatility swap.

17.3 Variance Swap for MRSVM

To calculate the variance swap for $\sigma^2(t)$ we need $E\sigma^2(t)$. From (17.11) it follows that

$$E\sigma^2(t) = e^{-at}[\sigma^2(0) - L] + L.$$

Then $E\sigma_R^2 := EV$ takes the following form:

$$E\sigma_R^2 := EV := \frac{1}{T}\int_0^T E\sigma^2(t)dt = \frac{(\sigma^2(0) - L)}{aT}(1 - e^{-aT}) + L. \qquad (17.14)$$

Recall, that $V := \frac{1}{T}\int_0^T \sigma^2(t)dt$.

17.4 Volatility Swap for MRSVM

To calculate the volatility swap for $\sigma^2(t)$ we need $E\sqrt{V} = E\sqrt{\sigma_R}$ and it means that we need more than just $E\sigma^2(t)$, because the realized volatility $\sigma_R := \sqrt{V} = \sqrt{\sigma_R^2}$. Using Brockhaus-Long approximation we then get

$$E\sqrt{V} \approx \sqrt{EV} - \frac{Var(V)}{8(EV)^{3/2}}. \qquad (17.15)$$

We have EV calculated in (17.14). We need

$$Var(V) = EV^2 - (EV)^2. \qquad (17.16)$$

From (17.14) it follows that $(EV)^2$ has the form:

$$(EV)^2 = \frac{(\sigma^2(0) - L)^2}{a^2 T^2}(1 - e^{-aT})^2 + 2\frac{(\sigma^2(0) - L)}{aT}(1 - e^{-aT})L + L^2. \qquad (17.17)$$

Let us calculate EV^2 using (17.9) and (17.13):

$$EV^2 = \frac{1}{T^2}\int_0^T\int_0^T E\sigma^2(t)\sigma^2(s)dtds$$

$$= \frac{1}{T^2}\int_0^T\int_0^T \left[e^{-a(t+s)}(\sigma^2(0) - L)^2 + e^{-a(t+s)}\left\{\gamma^2\left[(\sigma^2(0) - L)^2\frac{e^{\gamma^2(t\wedge s)} - 1}{\gamma^2}\right.\right.\right.$$

$$\left.\left.+ \frac{2L(\sigma^2(0) - L)(e^{a(t\wedge s)} - e^{\gamma^2(t\wedge s)})}{a - \gamma^2} + \frac{L^2(e^{2a(t\wedge s)} - e^{\gamma^2(t\wedge s)})}{2a - \gamma^2}\right]\right\}$$

$$\left. + e^{-at}(\sigma^2(0) - L)L + e^{-as}(\sigma^2(0) - L)L + L^2\right]dtds. \qquad (17.18)$$

After calculating the interals in the second, forth and fifth lines in (17.18) we have:

$$EV^2 = \frac{1}{T^2} \int_0^T \int_0^T E\sigma^2(t)\sigma^2(s)dtds$$

$$= \frac{(\sigma^2(0) - L)^2}{a^2 T^2}(1 - e^{-aT})^2$$

$$+ \frac{1}{T^2} \int_0^T \int_0^T e^{-a(t+s)} \left\{ \gamma^2 \left[(\sigma^2(0) - L)^2 \frac{e^{\gamma^2(t\wedge s)} - 1}{\gamma^2} \right. \right.$$

$$+ \frac{2L(\sigma^2(0) - L)(e^{a(t\wedge s)} - e^{\gamma^2(t\wedge s)})}{a - \gamma^2} + \frac{L^2(e^{2a(t\wedge s)} - e^{\gamma^2(t\wedge s)})}{2a - \gamma^2} \right] \bigg\} dtds$$

$$+ \frac{(\sigma^2(0) - L)L}{aT}(1 - e^{-aT}) + \frac{(\sigma^2(0) - L)L}{aT}(1 - e^{-aT}) + L^2. \qquad (17.19)$$

Taking into account (17.16), (17.17) and (17.19) we arrive at the following expression for $Var(V)$:

$$Var(V) = EV^2 - (EV)^2$$

$$= \frac{1}{T^2} \int_0^T \int_0^T e^{-a(t+s)} \left\{ \gamma^2 \left[(\sigma^2(0) - L)^2 \frac{e^{\gamma^2(t\wedge s)} - 1}{\gamma^2} \right. \right.$$

$$+ \frac{2L(\sigma^2(0) - L)(e^{a(t\wedge s)} - e^{\gamma^2(t\wedge s)})}{a - \gamma^2} + \frac{L^2(e^{2a(t\wedge s)} - e^{\gamma^2(t\wedge s)})}{2a - \gamma^2} \right] \bigg\} dtds$$

$$= \frac{\sigma^2(0) - L}{T^2} \int_0^T \int_0^T e^{-a(t+s)}(e^{\gamma^2(t\wedge s)} - 1)dtds$$

$$+ \frac{2L\gamma^2(\sigma^2(0) - L)}{(a^2 - \gamma^2)T^2} \int_0^T \int_0^T e^{-a(t+s)}(e^{a(t\wedge s)} - e^{\gamma^2(t\wedge s)})dtds$$

$$+ \frac{\gamma^2 L^2}{(2a - \gamma^2)T^2} \int_0^T \int_0^T e^{-a(t+s)}(e^{2a(t\wedge s)} - e^{\gamma^2(t\wedge s)})dtds. \qquad (17.20)$$

After calculating the three integrals in (17.20) we obtain:

$$Var(V) = EV^2 - (EV)^2$$

$$= \frac{1}{T^2} \int_0^T \int_0^T e^{-a(t+s)} \left\{ \gamma^2 \left[(\sigma^2(0) - L)^2 \frac{e^{\gamma^2(t\wedge s)} - 1}{\gamma^2} \right. \right.$$

$$+ \frac{2L(\sigma^2(0) - L)(e^{a(t\wedge s)} - e^{\gamma^2(t\wedge s)})}{a - \gamma^2} + \frac{L^2(e^{2a(t\wedge s)} - e^{\gamma^2(t\wedge s)})}{2a - \gamma^2} \right] \bigg\} dtds.$$

$$\qquad (17.21)$$

From (17.15) and (17.21) we get the volatility swap:

$$E\sqrt{V} \approx \sqrt{EV} - \frac{Var(V)}{8(EV)^{3/2}}. \qquad (17.22)$$

17.5 Mean-Reverting Risk-Neutral Stochastic Volatility Model

In this Section, we are going to obtain the values of variance and volatility swaps under risk-neutral measure P^*, using the same arguments as in Sections 17.3 and 17.4, where in place of a and L we are going to take a^* and L^*

$$a \to a^* := a + \lambda\sigma, \quad L \to L^* := \frac{aL}{a + \lambda\sigma},$$

where λ is a *market price of risk*.

17.5.1 *Risk-Neutral Stochastic Volatility Model (SVM)*

Consider our model (17.1)

$$d\sigma^2(t) = a(L - \sigma^2(t))dt + \gamma\sigma^2(t)dW_t. \tag{17.23}$$

Let λ be 'market price of risk' and defind the following constants:

$$a^* := a + \lambda\sigma, \quad L^* := aL/a^*.$$

Then, in the risk-neutral world, the drift paramater in (17.23) has the following form:

$$a^*(L^* - \sigma^2(t)) = a(L - \sigma^2(t)) - \lambda\gamma\sigma^2(t). \tag{17.24}$$

If we define the following process (W_t^*) by

$$W_t^* := W_t + \lambda t, \tag{17.25}$$

where W_t is a standard Brownian motion, then the risk-neutral stochastic volatility model has the following form

$$d\sigma^2(t) = (aL - (a + \lambda\gamma)\sigma^2(t))dt + \gamma\sigma^2(t)dW_t^*,$$

or, equivalently,

$$d\sigma^2(t) = a^*(L^* - \sigma^2(t))dt + \gamma\sigma^2(t)dW_t^*, \tag{17.26}$$

where

$$a^* := a + \lambda\gamma, \quad L^* := \frac{aL}{a + \lambda\gamma}, \tag{17.27}$$

and W_t^* is defined in (17.25).

Now, we have the same model in (17.26) as in (17.1), and we are going to apply our change of time method to this model (17.26) to obtain the values of variance and volatility swaps.

17.5.2   *Variance and Volatility Swaps for Risk-Neutral SVM*

Using the same arguments as in the previous section (where in place of (17.4) we have to take (17.26)) we get the following expressions for variance and volatility swaps taking into account (17.27).

For the variance swaps we have (see (17.14) and (17.27)):

$$E^* \sigma_R^2 := EV := \frac{1}{T} \int_0^T E\sigma^2(t)dt = \frac{(\sigma^2(0) - L^*)}{a^*T}(1 - e^{-a^*T}) + L^*. \qquad (17.28)$$

For the volatility swap we obtain (see (17.22) and (17.27))

$$E^* \sqrt{V} \approx \sqrt{E^*V} - \frac{Var^*(V)}{8(E^*V)^{3/2}}. \qquad (17.29)$$

17.5.3   *Numerical Example: AECO Natural GAS Index* *(1 May 1998–30 April 1999)*

We shall calculate the value of variance and volatility swaps prices of a daily natural gas contract. To apply our formula for calculating these values we need to calibrate the parameters a, L, σ_0^2 and γ (T is monthly). These parameters may be obtained from futures prices for the AECO Natural Gas Index for the period 1 May 1998 to 30 April 1999 (see Bos, Ware and Pavlov (2002)). The parameters are the following:

Parameters			
a	γ	L	λ
4.6488	1.5116	2.7264	0.18

For variance swap we use formula (17.14) and for volatility swap we use formula (17.22).

From this table we can calculate the values for risk adjusted parameters a^* and L^*:

$$a^* = a + \lambda\gamma = 4.9337,$$

and

$$L^* = \frac{aL}{a + \lambda\gamma} = 2.5690.$$

For the value of $\sigma^2(0)$ we can take $\sigma^2(0) = 2.25$.

For variance swap and for volatility swap with risk adjusted papameters we use formula (17.28) and (17.29), respectively.

Figure 17.1 depicts variance swap (price vs. maturity) for AECO Natural Gas Index (1 May 1998 to 30 April 1999), using formula (17.14).

Figure 17.2 depicts volatility swap (price vs. maturity) for AECO Natural Gas Index (1 May 1998 to 30 April 1999), using formula (17.22).

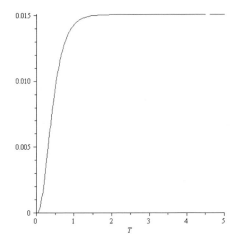

Fig. 17.1 Variance Swap.

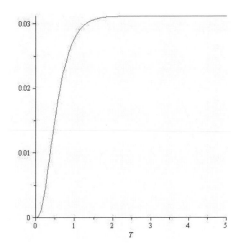

Fig. 17.2 Volatility Swap.

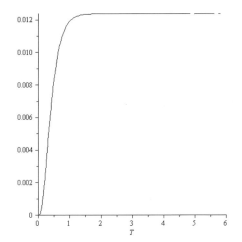

Fig. 17.3 Variance Swap (Risk Adjusted Parameters).

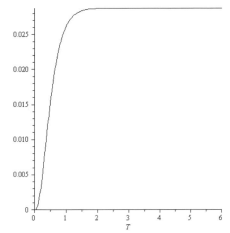

Fig. 17.4 Volatility Swap (Risk Adjusted Parameters).

Figure 17.3 depicts varinace swap with risk adjusted parameters (price vs. maturity) for AECO Natural Gas Index (1 May 1998 to 30 April 1999), using formula (17.28).

Figure 17.4 depicts volatility swap with risk adjusted parameters (price vs. maturity) for AECO Natural Gas Index (1 May 1998 to 30 April 1999), using formula (17.29).

Figure 17.5 depicts comparison of adjusted (green line) and non-adjusted price (red line) (naive strike vs. adjusted strike).

Figure 17.6 depicts convexity adjustment. It's decreasing with swap maturity (the volatility of volatility over a long period of time is low).

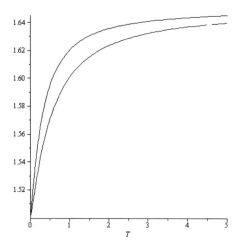

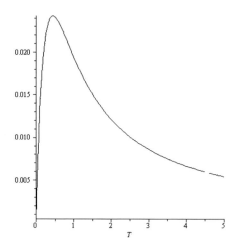

Fig. 17.5 Comparison: Adjusted and Non-Adjusted Price.

Fig. 17.6 Convexity Adjustment.

17.6 Summary

- This Chapter was devoted to the pricing of variance and volatility swaps in energy markets.
- We found explicit variance swap formula and closed form volatility swap formula (using change of time) for energy asset with stochastic volatility that follows continuous-time mean-reverting GARCH (1,1) model.
- Numerical example was presented for AECO Natural Gas Index (1 May 1998–30 April 1999)

Bibliography

Bos, L. P., Ware, A. F. and Pavlov, B. S. (2002). On a semi-spectral method for pricing an option on a mean-reverting asset. *Quantitative Finance*, 2, 337-345.

Black, F. (1976). The pricing of commodity contracts. *Journal of Financial Economics*, 3, 167-179.

Carr, P. and Wu, L. (2009). Variance risk premia. *Review of Financial Studies*, 22, 1311-1341

Brockhaus, O. and Long, D. (2000). Volatility swaps made simple. *Risk Magazine*, January, 2000.

Elliott, R. (1982). *Stochastic Calculus and Applications*. New York: Springer-Verlag.

Erb, C. and Harvey, C. (2006). The strategic and tactical value of commodity futures. *Financial Analysts Journal*, 62, 69-97.

Gorton, G. and Rouwenhorst, G. (2006). Facts and fantasies about commodity futures. *Financial Analysts Journal*, 62, 47-68.

Ibbotson (2006). Strategic asset allocation and commodities, March 27.

Ikeda, N. and Watanabe, S. (1981). *Stochastic Differential Equations and Diffusion Processes*. Tokyo: Kodansha Ltd.

Javaheri, A., Wilmott, P. and Haug, E. (2002). GARCH and volatility swaps. *Wilmott Magazine*, January.

Kan, H. and Oomen, R. (2007). What every investor should know about commodities, Part I. *Journal of Investment Management*, 5, 4-28.

Kan, H. and Oomen, R. (2007). What every investor should know about commodities, Part II. *Journal of Investment Management*, 5, 29-64.

Lamperton, D. and Lapeyre, B. (1996). *Introduction to Stochastic Calculus Applied to Finance*. New York: Chapmann & Hall.

Pilipović, D. (1998). *Energy Risk. Valuing and Managing Energy Derivatives*. New York: McGraw-Hill.

Schwartz, E. (1997). The stochastic behaviour of commodity prices: Implications for pricing and hedging. *Journal of Finance*, 52, 923-973.

Swishchuk, A. (2004). Modeling and valuing of variance and volatility swaps for financial markets with stochastic volatilities. *Wilmott Magazine*, Technical Article No. 2, September Issue, 64-72.

Swishchuk, A. (2012). Varinace and volatility swaps in energy markets. *Journal of Energy Markets*. (In-press).

Trolle, A. and Schwartz, E. (2010). Varinace risk premia in energy commodities. *Journal of Derivatives*, 17, 3, 15-32.

Wilmott, P., Howison, A. and Dewynne, J. (1995). *The Mathematics of Financial Derivatives*. Cambridge: Cambridge University Press.

Wilmott, P. (2000). *Paul Wilmott on Quantitative Finance*. New York: Wiley.

Chapter 18

Explicit Option Pricing Formula for a Mean-Reverting Asset in Energy Markets

18.1 Introduction

In this Chapter we consider a risky asset S_t following the mean-reverting stochastic process. We obtain an explicit expression for a European option price based on S_t, using a change of time method (see Chapter 4). A numerical example for the AECO Natural Gas Index (1 May 1998–30 April 1999) is presented.

Some commodity prices, like oil and gas, exhibit the mean reversion, unlike stock price. It means that they tend over time to return to some long-term mean. We consider a risky asset S_t following the mean-reverting stochastic process given by the following stochastic differential equation

$$dS_t = a(L - S_t)dt + \sigma S_t dW_t,$$

where W is a standard Wiener process, $\sigma > 0$ is the volatility, the constant L is called the 'long-term mean' of the process, to which it reverts over time, and $a > 0$ measures the 'strength' of mean reversion.

This mean-reverting model is a one-factor version of the two-factor model made popular in the context of energy modelling by Pilipovic (1997). Black's model (1976) and chwartz's model (1997) have become a standard approach to the problem of pricing options on commodities. These models have the advantage of mathematical convenience, in that they give rise to closed-form solutions for some types of options (See Wilmott (2000)).

Bos, Ware and Pavlov (2002) presented a method for evaluation of the price of a European option based on S_t, using a semi-spectral method. They did not have the convenience of a closed-form solution, however, they showed that values for certain types of options may nevertheless be found extremely efficiently. They used the following partial differential equation (e.g., Wilmott, Howison and Dewynne (1995))

$$C'_t + R(S,t)C'_S + \sigma^2 S^2 C''_{SS}/2 = rC$$

for option prices $C(S,t)$, where $R(S,t)$ depends only on S and t, and corresponds to the drift induced by the risk-neutral measure, and r is the risk-free interest rate. Simplifying this equation to the singular diffusion equation they were able to calculate numerically the solution.

The aim of this paper is to obtain an explicit expression for a European option price, $C(S, t)$, based on S_t, using a change of time method (see Swishchuk (2007)). This method was once applied by the author to price variance, volatility, covariance and correlation swaps for the Heston model (see Swishchuk (2004)).

18.2   Mean-Reverting Asset Model (MRAM)

Let $(\Omega, \mathcal{F}, \mathcal{F}_t, P)$ be a probability space with a sample space Ω, σ-algebra of Borel sets $\mathcal{F}$ and probability P. The filtration $\mathcal{F}_t$, $t \in [0, T]$, is the natural filtration of a standard Brownian motion W_t, $t \in [0, T]$, such that $\mathcal{F}_T = \mathcal{F}$.

Some commodity prices, like oil and gas, exhibit the mean reversion, unlike stock price. It means that they tend over time to return to some long-term mean. In this paper we consider a risky asset S_t following the mean-reverting stochastic process given by the following stochastic differential equation

$$dS_t = a(L - S_t)dt + \sigma S_t dW_t, \tag{18.1}$$

where W_t is an $\mathcal{F}_t$-measurable one-dimensional standard Wiener process, $\sigma > 0$ is the volatility, constant L is called the 'long-term mean' of the process, to which it reverts over time, and $a > 0$ measures the 'strength' of mean reversion.

18.3   Explicit Option Pricing Formula for European Call Option for MRAM under Physical Measure

In this Section, we are going to obtain an explicit expression for a European option price, $C(S, t)$, based on S_t, using a change of time method and physical measure.

18.3.1   *Explicit Solution of MRAM*

Let

$$V_t := e^{at}(S_t - L). \tag{18.2}$$

Then, from (18.2) and (18.1) we obtain

$$dV_t = ae^{at}(S_t - L)dt + e^{at}dS_t = \sigma(V_t + e^{at}L)dW_t. \tag{18.3}$$

Using change of time approach to the equation (18.3) (see Ikeda and Watanabe (1981) or Elliott (1982)) we obtain the following solution of the equation (18.3)

$$V_t = S_0 - L + \tilde{W}(\phi_t^{-1}),$$

or (see (18.2)),

$$S_t = e^{-at}[S_0 - L + \tilde{W}(\phi_t^{-1})] + L, \tag{18.4}$$

where $\tilde{W}(t)$ is an $\mathcal{F}_t$-measurable standard one-dimensional Wiener process, ϕ_t^{-1} is an inverse function to ϕ_t :

$$\phi_t = \sigma^{-2} \int_0^t (S_0 - L + \tilde{W}(s) + e^{a\phi_s} L)^{-2} ds. \tag{18.5}$$

We note that

$$\phi_t^{-1} = \sigma^2 \int_0^t (S_0 - L + \tilde{W}(\phi_t^{-1}) + e^{as} L)^2 ds, \tag{18.6}$$

which follows from (18.5) and the following transformations:

$$d\phi_t = \sigma^{-2}(S_0 - L + \tilde{W}(t) + e^{a\phi_t} L)^{-2} dt \Rightarrow \sigma^2 (S_0 - L + \tilde{W}(t) + e^{a\phi_t} L)^2 d\phi_t = dt \Rightarrow$$

$$t = \sigma^2 \int_0^t (S_0 - L + \tilde{W}(s) + e^{a\phi_s} L)^2 d\phi_s \Rightarrow$$

$$\phi_t^{-1} = \sigma^2 \int_0^{\phi_t^{-1}} (S_0 - L + \tilde{W}(s) + e^{a\phi_s} L)^2 d\phi_s$$

$$= \sigma^2 \int_0^t (S_0 - L + \tilde{W}(\phi_s^{-1}) + e^{as} L)^2 ds.$$

18.3.2 *Properties of the Process* $\tilde{W}(\phi_t^{-1})$

We note that process $\tilde{W}(\phi_t^{-1})$ is $\tilde{\mathcal{F}}_t := \mathcal{F}_{\phi_t^{-1}}$-measurable and $\tilde{\mathcal{F}}_t$-martingale.
    Then

$$E\tilde{W}(\phi_t^{-1}) = 0. \tag{18.7}$$

Let's calculate the second moment of $\tilde{W}(\phi_t^{-1})$ (see (18.6)):

$$E\tilde{W}^2(\phi_t^{-1}) = E\langle \tilde{W}(\phi_t^{-1}) \rangle = E\phi_t^{-1}$$

$$= \sigma^2 \int_0^t E(S_0 - L + \tilde{W}(\phi_s^{-1}) + e^{as} L)^2 ds$$

$$= \sigma^2 \left[(S_0 - L)^2 t + \frac{2L(S_0 - L)(e^{at} - 1)}{a} \right.$$

$$\left. + \frac{L^2(e^{2at} - 1)}{2a} + \int_0^t E\tilde{W}^2(\phi_s^{-1}) ds \right]. \tag{18.8}$$

From (18.8), solving this linear ordinary nonhomogeneous differential equation with respect to $E\tilde{W}^2(\phi_t^{-1})$,

$$\frac{dE\tilde{W}^2(\phi_t^{-1})}{dt} = \sigma^2 [(S_0 - L)^2 + 2L(S_0 - L)e^{at} + L^2 e^{2at} + E\tilde{W}^2(\phi_t^{-1})],$$

we obtain

$$E\tilde{W}^2(\phi_t^{-1}) = \sigma^2 \left[(S_0 - L)^2 \frac{e^{\sigma^2 t} - 1}{\sigma^2} + \frac{2L(S_0 - L)(e^{at} - e^{\sigma^2 t})}{a - \sigma^2} + \frac{L^2(e^{2at} - e^{\sigma^2 t})}{2a - \sigma^2} \right]. \tag{18.9}$$

18.3.3   *Explicit Expression for the Process* $\tilde{W}(\phi_t^{-1})$.

It is turns out that we can find the explicit expression for the process $\tilde{W}(\phi_t^{-1})$.
From the expression (see Section 18.3.1)

$$V_t = S_0 - L + \tilde{W}(\phi_t^{-1}),$$

we have the following relationship between $W(t)$ and $\tilde{W}(\phi_t^{-1})$:

$$d\tilde{W}(\phi_t^{-1}) = \sigma \int_0^t [S(0) - L + Le^{at} + \tilde{W}(\phi_s^{-1})]dW(t).$$

It is a linear SDE with respect to $\tilde{W}(\phi_t^{-1})$ and we can solve it explicitly. The
solution has the following look:

$$\tilde{W}(\phi_t^{-1}) = S(0)(e^{\sigma W(t) - \frac{\sigma^2 t}{2}} - 1) + L(1 - e^{at}) + aLe^{\sigma W(t) - \frac{\sigma^2 t}{2}} \int_0^t e^{as} e^{-\sigma W(s) + \frac{\sigma^2 s}{2}} ds.$$

$$(18.10)$$

It is easy to see from (18.10) that $\tilde{W}(\phi_t^{-1})$ can be presented in the form of a
linear combination of two zero-mean martingales $m_1(t)$ and $m_2(t)$:

$$\tilde{W}(\phi_t^{-1}) = m_1(t) + Lm_2(t),$$

where

$$m_1(t) := S(0)(e^{\sigma W(t) - \frac{\sigma^2 t}{2}} - 1)$$

and

$$m_2(t) = (1 - e^{at}) + ae^{\sigma W(t) - \frac{\sigma^2 t}{2}} \int_0^t e^{as} e^{-\sigma W(s) + \frac{\sigma^2 s}{2}} ds.$$

Indeed, process $\tilde{W}(\phi_t^{-1})$ is a martingale (see Section 18.3.2), also it is well-known
that process $e^{\sigma W(t) - \frac{\sigma^2 t}{2}}$ and, hence, process $m_1(t)$ is a martingale. Then the process
$m_2(t)$, as the difference between two martingales, is also martingale. In this way,
we have

$$Em_1(t) = 0,$$

since

$$Ee^{\sigma W(t) - \frac{\sigma^2 t}{2}} = 1.$$

As for $m_2(t)$ we have

$$Em_2(t) = 0,$$

since from Itô's formula we have

$$d(ae^{\sigma W(t) - \frac{\sigma^2 t}{2}} \int_0^t e^{as} e^{-\sigma W(s) + \frac{\sigma^2 s}{2}} ds) = a\sigma e^{\sigma W(t) - \frac{\sigma^2 t}{2}} \int_0^t e^{as} e^{-\sigma W(s) + \frac{\sigma^2 s}{2}} dsdW(t)$$

$$+ ae^{\sigma W(t) - \frac{\sigma^2 t}{2}} e^{at} e^{-\sigma W(t) + \frac{\sigma^2 t}{2}} dt$$

$$= a\sigma e^{\sigma W(t) - \frac{\sigma^2 t}{2}} \int_0^t e^{as} e^{-\sigma W(s) + \frac{\sigma^2 s}{2}} dsdW(t)$$

$$+ ae^{at} dt,$$

and, hence,

$$Eae^{\sigma W(t)-\frac{\sigma^2 t}{2}}\int_0^t e^{as}e^{-\sigma W(s)+\frac{\sigma^2 s}{2}}ds = e^{at}-1.$$

It is interesting to see that the last expression, the first moment for

$$\eta(t) := ae^{\sigma W(t)-\frac{\sigma^2 t}{2}}\int_0^t e^{as}e^{-\sigma W(s)+\frac{\sigma^2 s}{2}}ds,$$

does not depend on σ.

It is true not only for the first moment but for all the moments of the process $\eta(t) = ae^{\sigma W(t)-\frac{\sigma^2 t}{2}}\int_0^t e^{as}e^{-\sigma W(s)+\frac{\sigma^2 s}{2}}ds$.

Indeed, using Itô's formula for $\eta^n(t)$ we obtain

$$d\eta^n(t) = na^n\sigma e^{n\sigma W(t)-\frac{n\sigma^2 t}{2}}\left(\int_0^t e^{as}e^{-\sigma W(s)+\frac{\sigma^2 s}{2}}ds\right)^n dW(t) + an(\eta_2(t))^{n-1}e^{at}dt,$$

and

$$dE\eta^n(t) = nae^{at}E\eta^{n-1}(t)dt, \quad n \geq 1.$$

This is a recursive equation with initial function $(n = 1)$ $E\eta(t) = e^{at}-1$. After calculations we obtain the following formula for $E\eta^n(t)$:

$$E\eta^n(t) = (e^{at}-1)^n.$$

18.3.4   *Some Properties of the Mean-Reverting Asset S_t*

From (18.4) we obtain the mean value of the first moment for mean-reverting asset S_t:

$$ES_t = e^{-at}[S_0 - L] + L.$$

It means that $ES_t \to L$ when $t \to +\infty$.

Using formulae (18.4) and (18.9) we can calculate the second moment of S_t:

$$ES_t^2 = (e^{-at}(S_0 - L) + L)^2$$

$$+ \sigma^2 e^{-2at}\left[(S_0 - L)^2\frac{e^{\sigma^2 t}-1}{\sigma^2} + \frac{2L(S_0 - L)(e^{at}-e^{\sigma^2 t})}{a - \sigma^2} + \frac{L^2(e^{2at}-e^{\sigma^2 t})}{2a - \sigma^2}\right].$$

Combining the first and the second moments we have the variance of S_t:

$$Var(S_t) = ES_t^2 - (ES_t)^2$$

$$= \sigma^2 e^{-2at}\left[(S_0 - L)^2\frac{e^{\sigma^2 t}-1}{\sigma^2} + \frac{2L(S_0 - L)(e^{at}-e^{\sigma^2 t})}{a - \sigma^2} + \frac{L^2(e^{2at}-e^{\sigma^2 t})}{2a - \sigma^2}\right].$$

From the expression for $\tilde{W}(\phi_t^{-1})$ (see (18.10)) and for $S(t)$ in (18.4) we can find the explicit expression for $S(t)$ through $W(t)$:

$$S(t) = e^{-at}[S_0 - L + \tilde{W}(\phi_t^{-1})] + L$$

$$= e^{-at}[S_0 - L + m_1(t) + Lm_2(t)] + L$$

$$= S(0)e^{-at}e^{\sigma W(t) - \frac{\sigma^2 t}{2}} + aLe^{-at}e^{\sigma W(t) - \frac{\sigma^2 t}{2}} \int_0^t e^{as}e^{-\sigma W(s) + \frac{\sigma^2 s}{2}} ds, \quad (18.11)$$

where $m_1(t)$ and $m_2(t)$ are defined as in Section 18.3.3.

18.3.5 *Explicit Option Pricing Formula for European Call Option for MRAM under Physical Measure*

The payoff function f_T for European call option equals

$$f_T = (S_T - K)^+ := \max(S_T - K, 0),$$

where S_T is an asset price defined in (18.4), T is an expiration time (maturity) and K is a strike price.

In this way (see (18.11)),

$$f_T = [e^{-aT}(S_0 - L + \tilde{W}(\phi_T^{-1})) + L - K]^+$$

$$= [S(0)e^{-aT}e^{\sigma W(T) - \frac{\sigma^2 T}{2}} + aLe^{-aT}e^{\sigma W(T) - \frac{\sigma^2 T}{2}} \int_0^T e^{as}e^{-\sigma W(s) + \frac{\sigma^2 s}{2}} ds - K]^+.$$

To find the option pricing formula we need to calculate

$$C_T = e^{-rT} E f_T$$

$$= e^{-rT} E[e^{-aT}(S_0 - L + \tilde{W}(\phi_T^{-1})) + L - K]^+$$

$$= \frac{1}{\sqrt{2\pi}} e^{-rT} \int_{-\infty}^{+\infty} \max[S(0)e^{-aT}e^{\sigma y\sqrt{T} - \frac{\sigma^2 T}{2}}$$

$$+ aLe^{-aT}e^{\sigma y\sqrt{T} - \frac{\sigma^2 T}{2}} \int_0^T e^{as}e^{-\sigma y\sqrt{s} + \frac{\sigma^2 s}{2}} ds - K, 0]e^{-\frac{y^2}{2}} dy. \quad (18.12)$$

Let y_0 be a solution of the following equation:

$$S(0) \times e^{-aT}e^{\sigma y_0\sqrt{T} - \frac{\sigma^2 T}{2}}$$

$$+ aLe^{-aT}e^{\sigma y_0\sqrt{T} - \frac{\sigma^2 T}{2}} \int_0^T e^{as}e^{-\sigma y_0\sqrt{s} + \frac{\sigma^2 s}{2}} ds = K \quad (18.13)$$

or

$$y_0 = \frac{\ln(\frac{K}{S(0)}) + (\frac{\sigma^2}{2} + a)T}{\sigma\sqrt{T}} - \frac{\ln(1 + \frac{aL}{S(0)} \int_0^T e^{as}e^{-\sigma y_0\sqrt{s} + \frac{\sigma^2 s}{2}} ds)}{\sigma\sqrt{T}} \quad (18.14)$$

From (18.12)–(18.13) we have:

$$
C_T = \frac{1}{\sqrt{2\pi}} e^{-rT} \int_{-\infty}^{+\infty} \max[S(0) e^{-aT} e^{\sigma y \sqrt{T} - \frac{\sigma^2 T}{2}}
$$

$$
+ aL e^{-aT} e^{\sigma y \sqrt{T} - \frac{\sigma^2 T}{2}} \int_0^T e^{as} e^{-\sigma y \sqrt{s} + \frac{\sigma^2 s}{2}} ds - K, 0] e^{-\frac{y^2}{2}} dy
$$

$$
= \frac{1}{\sqrt{2\pi}} e^{-rT} \int_{y_0}^{+\infty} \left[S(0) e^{-aT} e^{\sigma y \sqrt{T} - \frac{\sigma^2 T}{2}} \right.
$$

$$
\left. + aL e^{-aT} e^{\sigma y \sqrt{T} - \frac{\sigma^2 T}{2}} \int_0^T e^{as} e^{-\sigma y \sqrt{s} + \frac{\sigma^2 s}{2}} ds - K \right] e^{-\frac{y^2}{2}} dy
$$

$$
= \frac{1}{\sqrt{2\pi}} e^{-rT} \int_{y_0}^{+\infty} [S(0) e^{-aT} e^{\sigma y \sqrt{T} - \frac{\sigma^2 T}{2}} e^{-\frac{y^2}{2}} dy - e^{-rT} K[1 - \Phi(y_0)]
$$

$$
+ L e^{-(r+a)T} \frac{1}{\sqrt{2\pi}} \int_{y_0}^{+\infty} \left(a e^{\sigma y \sqrt{T} - \frac{\sigma^2 T}{2}} \int_0^T e^{as} e^{-\sigma y \sqrt{s} + \frac{\sigma^2 s}{2}} ds \right) e^{-\frac{y^2}{2}} dy
$$

$$
= BS(T) + A(T), \tag{18.15}
$$

where

$$
BS(T) := \frac{1}{\sqrt{2\pi}} e^{-rT} \int_{y_0}^{+\infty} [S(0) e^{-aT} e^{\sigma y \sqrt{T} - \frac{\sigma^2 T}{2}} e^{-\frac{y^2}{2}} dy - e^{-rT} K[1 - \Phi(y_0)],
$$

$$
\tag{18.16}
$$

$$
A(T) := L e^{-(r+a)T}
$$

$$
\times \frac{1}{\sqrt{2\pi}} \int_{y_0}^{+\infty} \left(a e^{\sigma y \sqrt{T} - \frac{\sigma^2 T}{2}} \int_0^T e^{as} e^{-\sigma y \sqrt{s} + \frac{\sigma^2 s}{2}} ds \right) e^{-\frac{y^2}{2}} dy, \tag{18.17}
$$

and

$$
\Phi(x) = \frac{1}{\sqrt{2\pi}} \int_{-\infty}^{x} e^{-\frac{y^2}{2}} dy. \tag{18.18}
$$

After calculation of $BS(T)$ we obtain

$$
BS(T) = e^{-(r+a)T} S(0) \Phi(y_+) - e^{-rT} K \Phi(y_-), \tag{18.19}
$$

where

$$
y_+ := \sigma \sqrt{T} - y_0 \quad and \quad y_- := -y_0 \tag{18.20}
$$

and y_0 is defined in (18.14).

Consider $A(T)$ in (18.17).

Let $F_T(dz)$ be a distribution function for the process

$$\eta(T) = ae^{\sigma W(T) - \frac{\sigma^2 T}{2}} \int_0^T e^{as} e^{-\sigma W(s) + \frac{\sigma^2 s}{2}} ds,$$

which is a part of the integrand in (18.17).

As Yor (1992, 2005) mentioned there is still no closed form probability density function for time integral of an exponential Brownian motion, while the best result is a function with a double integral.

We can use Yor's result (1992) to get $F_T(dz)$ above. Using the scaling property of Wiener process and change of variables, we can rewrite our expression for $S(t)$ in (18.11) in the following way

$$S(T) = S(0)e^{-2B_{T_0}^v} + \frac{4}{\sigma^2} aLe^{-2B_{T_0}^v} A_{T_0}^v,$$

where $T_0 = \frac{\sigma^2}{4}T, v = \frac{2}{\sigma^2}a+1, B_t = -\frac{\sigma}{2}W(\frac{4}{\sigma^2}t), B_{T_0}^v = vT_0+B_{T_0}, A_{T_0}^v = \int_0^{T_0} e^{2B_s^v} ds$.

Also, the process $\eta(T)$ may be presented in the following way using these transformations

$$\eta(T) = \frac{4ae^{-aT}}{\sigma^2} e^{-2B_{\frac{\sigma^2 T}{4}}} A_{\frac{\sigma^2 T}{4}}.$$

We state here the result obtained by Yor [15] for the joint probability density function of $A_{T_0}^v$ and $B_{T_0}^v$.

Theorem 4.3.-1. (Yor (1992)). *The joint probability density function of $A_{T_0}^v$ and $B_{T_0}^v$ satisfies*

$$P(A_{T_0}^v \in du, B_{T_0}^v \in dx) = e^{vx - v^2 t/2} \exp{-\frac{1+e^{2x}}{2u}} \theta\left(\frac{e^x}{u}, t\right) \frac{dxdu}{u},$$

where $t > 0, u > 0, x \in R$ and

$$\theta(r, t) = \frac{r}{(2\pi^3 t)^{1/2}} e^{\frac{\pi^2}{2t}} \int_0^{+\infty} e^{-\frac{s^2}{2t} - r\cosh s} \sinh(s) \sin\left(\frac{\pi s}{t}\right) ds.$$

Using this result we can write the distribution function for $\eta(T)$ in the following way

$$P(\eta(T) \leq u) = P\left(\frac{4ae^{-aT}}{\sigma^2} e^{-2B_{\frac{\sigma^2 T}{4}}} A_{\frac{\sigma^2 T}{4}} \leq u\right)$$

$$= P\left(e^{-2B_{\frac{\sigma^2 T}{4}}} A_{\frac{\sigma^2 T}{4}} \leq \frac{\sigma^2 e^{aT}}{4a} u\right)$$

$$= F_T(u). \tag{18.21}$$

In this way, $A(T)$ in (18.17) may be presented in the following way:

$$A(T) = Le^{-(r+a)T} \int_{y_0}^{+\infty} zF_T(dz).$$

After calculation of $A(T)$ we obtain the following expression for $A(T)$:

$$A(T) = Le^{-(r+a)T}\left[(e^{aT} - 1) - \int_0^{y_0} zF_T(dz)\right],$$

since $E\eta(T) = e^{aT} - 1$.

Finally, summarizing (18.12)–(18.21), we have obtained the following Theorem.

Theorem 18.1. Option pricing formula for European call option for mean-reverting asset under physical measure has the following look:

$$C_T = e^{-(r+a)T}S(0)\Phi(y_+) - e^{-rT}K\Phi(y_-)$$

$$+ Le^{-(r+a)T}\left[(e^{aT} - 1) - \int_0^{y_0} zF_T(dz)\right], \qquad (18.22)$$

where y_0 is defined in (18.14), y_+ and y_- in (18.20), $\Phi(y)$ in (18.18), and $F_T(dz)$ is a distribution function in (18.21).

Remark. From (18.21)–(18.22) we find that European Call Option Price C_T for mean-reverting asset lies between the following boundaries:

$$BS(T) \le C_T \le BS(T) + Le^{-(r+a)T}[e^{aT} - 1],$$

or (see (18.19)),

$$e^{-(r+a)T}S(0)\Phi(y_+) - e^{-rT}K\Phi(y_-) \le C_T$$

$$\le e^{-(r+a)T}S(0)\Phi(y_+)$$

$$- e^{-rT}K\Phi(y_-) + Le^{-(r+a)T}[e^{aT} - 1].$$

18.4 Mean-Reverting Risk-Neutral Asset Model (MRRNAM)

Consider our model (18.1)

$$dS_t = a(L - S_t)dt + \sigma S_t dW_t. \qquad (18.23)$$

We want to find a probability P^* equivalent to P, under which the process $e^{-rt}S_t$ is a martingale, where $r > 0$ is a constant interest rate. The hypothesis we made on the filtration $(\mathcal{F}_t)_{t\in[0,T]}$ allows us to express the density of the probability P^* with respect to P. We denote this density by L_T.

It is well-known (see Lamperton and Lapeyre (1996), Proposition 6.1.1, p. 123), that there is an adopted process $(q(t))_{t\in[0,T]}$ such that, for all $t \in [0, T]$,

$$L_t = \exp\left[\int_0^t q(s)dW_s - \frac{1}{2}\int_0^t q^2(s)ds\right] \quad a.s.$$

In this case,

$$\frac{dP^*}{dP} = \exp\left[\int_0^T q(s)dW_s - \frac{1}{2}\int_0^T q^2(s)ds\right] = L_T.$$

In our case, with model (18.17), the process $q(t)$ is equal to

$$q(t) = -\lambda S_t, \tag{18.24}$$

where λ is the *market price of risk* and $\lambda \in R$. Hence, for our model

$$L_T = \exp\left[-\lambda \int_0^T S(u)dW_u - \frac{1}{2}\lambda \int_0^T S^2(u)du\right].$$

Under probability P^*, the process (W_t^*) defined by

$$W_t^* := W_t + \lambda \int_0^t S(u)du \tag{18.25}$$

is a standard Brownian motion (Girsanov theorem) (see Elliott and Kopp (1999)).
    Therefore, in a risk-neutral world our model (18.23) takes the following look:

$$dS_t = (aL - (a + \lambda\sigma)S_t)dt + \sigma S_t dW_t^*,$$

or, equivalently,

$$dS_t = a^*(L^* - S_t)dt + \sigma S_t dW_t^*, \tag{18.26}$$

where

$$a^* := a + \lambda\sigma, \quad L^* := \frac{aL}{a + \lambda\sigma}, \tag{18.27}$$

and W_t^* is defined in (18.25).
    Now, we have the same model in (18.26) as in (18.1), and we are going to apply our method of changing of time to this model (18.26) to obtain the explicit option pricing formula.

18.5   Explicit Option Pricing Formula for European Call Option for MRRNAM

In this Section, we are going to obtain explicit option pricing formula for European call option under risk-neutral measure P^*, using the same arguments as in sections 18.3–18.7, where in place of a and L we are going to take a^* and L^*

$$a \to a^* := a + \lambda\sigma, \quad L \to L^* := \frac{aL}{a + \lambda\sigma},$$

where λ is a *market price of risk* (see Section 18.3).

18.5.1   *Explicit Solution for the Mean-Reverting Risk-Neutral Asset Model*

Applying (18.2)–(18.6) to our model (18.26) we obtain the following explicit solution for our risk-neutral model (18.26):

$$S_t = e^{-a^*t}[S_0 - L^* + \tilde{W}^*((\phi_t^*)^{-1})] + L, \tag{18.28}$$

where $\tilde{W}^*(t)$ is an $\mathcal{F}_t$-measurable standard one-dimensional Wiener process under measure P^* and $(\phi_t^*)^{-1}$ is an inverse function to ϕ_t^* :

$$\phi_t^* = \sigma^{-2} \int_0^t (S_0 - L^* + \tilde{W}^*(s) + e^{a^* \phi_s^*} L^*)^{-2} ds. \qquad (18.29)$$

We note that

$$(\phi^*)_t^{-1} = \sigma^2 \int_0^t (S_0 - L^* + \tilde{W}^*((\phi_t^*)^{-1}) + e^{a^* s} L^*)^2 ds, \qquad (18.30)$$

where a^* and L^* are defined in (18.27).

18.5.2 Some Properties of the Process $\tilde{W}^*((\phi_t^*)^{-1})$

Using the same argument as in Section 18.4, we obtain the following properties of the process $\tilde{W}^*((\phi_t^*)^{-1})$ in (18.25). This is a zero-mean P^*-martingale and

$$E^* \tilde{W}^*((\phi_t^*)^{-1}) = 0,$$

$$E^* [\tilde{W}^*((\phi_t^*)^{-1})]^2 = \sigma^2 \left[(S_0 - L^*)^2 \frac{e^{\sigma^2 t} - 1}{\sigma^2} + \frac{2L^*(S_0 - L^*)(e^{a^* t} - e^{\sigma^2 t})}{a^* - \sigma^2} \right.$$

$$\left. + \frac{(L^*)^2 (e^{2a^* t} - e^{\sigma^2 t})}{2a^* - \sigma^2} \right], \qquad (18.31)$$

where E^* is the expectation with respect to the probability P^* and a^*, L^* and $(\phi_t^*)^{-1}$ are defined in (18.27) and (18.30), respectively.

18.5.3 Explicit Expression for the Process $\tilde{W}^*(\phi_t^{-1})$

It is turns out that we can find the explicit expression for the process $\tilde{W}^*(\phi_t^{-1})$.
    From the expression

$$V_t = S_0 - L + \tilde{W}^*(\phi_t^{-1}),$$

we have the following relationship between $W(t)$ and $\tilde{W}(\phi_t^{-1})$:

$$d\tilde{W}^*(\phi_t^{-1}) = \sigma \int_0^t [S(0) - L + Le^{at} + \tilde{W}^*(\phi_s^{-1})] dW^*(t).$$

It is linear SDE with respect to $\tilde{W}^*(\phi_t^{-1})$ and we can solve it explicitly. The solution has the following look:

$$\tilde{W}^*(\phi_t^{-1}) = S(0)(e^{\sigma W^*(t) - \frac{\sigma^2 t}{2}} - 1) + L(1 - e^{at})$$

$$+ aL e^{\sigma W^*(t) - \frac{\sigma^2 t}{2}} \int_0^t e^{as} e^{-\sigma W^*(s) + \frac{\sigma^2 s}{2}} ds. \qquad (18.32)$$

It is easy to see from (18.32) that $\tilde{W}^*(\phi_t^{-1})$ can be presented in the form of a linear combination of two zero-mean P^*-martingales $m_1^*(t)$ and $m_2^*(t)$:

$$\tilde{W}^*(\phi_t^{-1}) = m_1^*(t) + L^* m_2^*(t),$$

where

$$m_1^*(t) := S(0)(e^{\sigma W^*(t) - \frac{\sigma^2 t}{2}} - 1)$$

and

$$m_2^*(t) = (1 - e^{a^* t}) + a^* e^{\sigma W^*(t) - \frac{\sigma^2 t}{2}} \int_0^t e^{a^* s} e^{-\sigma W^*(s) + \frac{\sigma^2 s}{2}} ds.$$

Indeed, process $\tilde{W}^*(\phi_t^{-1})$ is a martingale (see equation (18.32)), also it is well-known that process $e^{\sigma W^*(t) - \frac{\sigma^2 t}{2}}$ and, hence, process $m_1^*(t)$ is a martingale. Then the process $m_2^*(t)$, as the difference between two martingales, is also martingale. In this way, we have

$$E_{P^*} m_1^*(t) = 0,$$

since

$$E_{P^*} e^{\sigma W^*(t) - \frac{\sigma^2 t}{2}} = 1.$$

As for $m_2(t)$ we have

$$E_{P^*} m_2(t) = 0,$$

since from Itô's formula we have

$$d(a^* e^{\sigma W^*(t) - \frac{\sigma^2 t}{2}} \int_0^t e^{a^* s} e^{-\sigma W^*(s) + \frac{\sigma^2 s}{2}} ds)$$

$$= a^* \sigma e^{\sigma W^*(t) - \frac{\sigma^2 t}{2}} \int_0^t e^{a^* s} e^{-\sigma W^*(s) + \frac{\sigma^2 s}{2}} ds dW^*(t)$$

$$+ a^* e^{\sigma W^*(t) - \frac{\sigma^2 t}{2}} e^{a^* t} e^{-\sigma W^*(t) + \frac{\sigma^2 t}{2}} dt$$

$$= a^* \sigma e^{\sigma W^*(t) - \frac{\sigma^2 t}{2}} \int_0^t e^{a^* s} e^{-\sigma W^*(s) + \frac{\sigma^2 s}{2}} ds dW^*(t)$$

$$+ a^* e^{a^* t} dt,$$

and, hence,

$$E_{P^*} a^* e^{\sigma W^*(t) - \frac{\sigma^2 t}{2}} \int_0^t e^{a^* s} e^{-\sigma W^*(s) + \frac{\sigma^2 s}{2}} ds = e^{a^* t} - 1.$$

It is interesting to see that in the last expression, the first moment for

$$\eta^*(t) := a^* e^{\sigma W^*(t) - \frac{\sigma^2 t}{2}} \int_0^t e^{a^* s} e^{-\sigma W^*(s) + \frac{\sigma^2 s}{2}} ds,$$

does not depend on σ.

This is true not only for the first moment but for all the moments of the process $\eta^*(t) = a^* e^{\sigma W^*(t) - \frac{\sigma^2 t}{2}} \int_0^t e^{a^* s} e^{-\sigma W^*(s) + \frac{\sigma^2 s}{2}} ds.$

Indeed, using the Itô's formula for $(\eta^*(t))^n$ we obtain

$$d(\eta^*(t))^n = n(a^*)^n \sigma e^{n\sigma W^*(t) - \frac{n\sigma^2 t}{2}} \left(\int_0^t e^{a^* s} e^{-\sigma W^*(s) + \frac{\sigma^2 s}{2}} ds \right)^n dW^*(t)$$

$$+ a^* n(\eta_2(t))^{n-1} e^{a^* t} dt,$$

and

$$dE(\eta^*(t))^n = na^* e^{a^* t} E(\eta^*(t))^{n-1} dt, \quad n \geq 1.$$

This is a recursive equation with initial function $(n = 1)$ $E\eta^*(t) = e^{a^* t} - 1$. After calculations we obtain the following formula for $E(\eta^*(t))^n$:

$$E(\eta^*(t))^n = (e^{a^* t} - 1)^n.$$

18.5.4 Some Properties of the Mean-Reverting Risk-Neutral Asset S_t

Using the same argument as in Section 18.4, we obtain the following properties of the mean-reverting risk-neutral asset S_t in (18.18):

$$E^* S_t = e^{-a^* t}[S_0 - L^*] + L^*$$

$$Var^*(S_t) := E^* S_t^2 - (E^* S_t)^2$$

$$= \sigma^2 e^{-2a^* t} \left[(S_0 - L^*)^2 \frac{e^{\sigma^2 t} - 1}{\sigma^2} + \frac{2L^*(S_0 - L^*)(e^{a^* t} - e^{\sigma^2 t})}{a^* - \sigma^2} \right.$$

$$\left. + \frac{(L^*)^2(e^{2a^* t} - e^{\sigma^2 t})}{2a^* - \sigma^2} \right], \tag{18.33}$$

where E^* is the expectation with respect to the probability P^* and a^*, L^* and $(\phi_t^*)^{-1}$ are defined in (18.27) and (18.30), respectively.

From the expression for $\tilde{W}^*(\phi_t^{-1})$ (see (18.32)) and for $S(t)$ in (18.28) (see also (18.29)–(18.30)) we can find the explicit expression for $S(t)$ through $W^*(t)$:

$$S(t) = e^{-a^* t}[S_0 - L^* + \tilde{W}^*(\phi_t^{-1})] + L^*$$

$$= e^{-a^* t}[S_0 - L^* + m_1^*(t) + L^* m_2^*(t)] + L^*$$

$$= S(0)e^{-at}e^{\sigma W^*(t) - \frac{\sigma^2 t}{2}} + aLe^{-at}e^{\sigma W^*(t) - \frac{\sigma^2 t}{2}} \int_0^t e^{as} e^{-\sigma W^*(s) + \frac{\sigma^2 s}{2}} ds,$$

$$\tag{18.34}$$

where $m_1^*(t)$ and $m_2^*(t)$ are defined as in Section 18.5.3.

18.5.5   Explicit Option Pricing Formula for European Call Option for MRAM under Risk-Neutral Measure

Proceeding with the same calculations (18.15)–(18.22) as in Section 18.3, where in place of a and L we take a^* and L^* in (18.27), we obtain the following Theorem.

Theorem 18.2. Explicit option pricing formula for European call option under risk-neutral measure has the following look:

$$C_T^* = e^{-(r+a^*)T}S(0)\Phi(y_+) - e^{-rT}K\Phi(y_-)$$

$$+ L^* e^{-(r+a^*)T}\left[(e^{a^*T} - 1) - \int_0^{y_0} zF_T^*(dz)\right], \qquad (18.35)$$

where y_0 is the solution of the following equation

$$y_0 = \frac{\ln(\frac{K}{S(0)}) + (\frac{\sigma^2}{2} + a^*)T}{\sigma\sqrt{T}} - \frac{\ln(1 + \frac{a^*L^*}{S(0)}\int_0^T e^{a^*s}e^{-\sigma y_0\sqrt{s}+\frac{\sigma^2 s}{2}}ds)}{\sigma\sqrt{T}}, \qquad (18.36)$$

$$y_+ := \sigma\sqrt{T} - y_0 \quad and \quad y_- := -y_0, \qquad (18.37)$$

$$a^* := a + \lambda\sigma, \quad L^* := \frac{aL}{a + \lambda\sigma},$$

and $F_T^*(dz)$ is the probability distribution as in (18.21), where instead of a we have to take $a^* = a + \lambda\sigma$.

Remark. From (18.35) we can find that European Call Option Price C_T^* for mean-reverting asset under risk-neutral measure lies between the following boundaries:

$$e^{-(r+a^*)T}S(0)\Phi(y_+) - e^{-rT}K\Phi(y_-) \le C_T$$

$$\le e^{-(r+a^*)T}S(0)\Phi(y_+) - e^{-rT}K\Phi(y_-)$$

$$+ L^* e^{-(r+a^*)T}[e^{a^*T} - 1], \qquad (18.38)$$

where y_0, y_-, y_+ are defined in (18.36)–(18.37).

18.5.6   Black-Scholes Formula Follows: $L^* = 0$ and $a^* = -r$

If $L^* = 0$ and $a^* = -r$ we obtain from (18.35)

$$C_T = S(0)\Phi(y_+) - e^{-rT}K\Phi(y_-), \qquad (18.39)$$

where

$$y_+ := \sigma\sqrt{T} - y_0 \quad and \quad y_- := -y_0, \qquad (18.40)$$

and y_0 is the solution of the following equation (see (18.36))

$$S(0)e^{-rT}e^{\sigma y_0\sqrt{T}-\frac{\sigma^2 T}{2}} = K$$

or

$$y_0 = \frac{\ln(\frac{K}{S(0)}) + (\frac{\sigma^2}{2} - r)T}{\sigma\sqrt{T}}. \qquad (18.41)$$

But (18.39)–(18.41) is exactly the well-known Black-Scholes result!

18.6 Numerical Example: AECO Natural GAS Index (1 May 1998–30 April 1999)

We shall calculate the value of a European call option on the price of a daily natural gas contract. To apply our formula for calculating this value we need to calibrate the parameters a, L, σ and λ. These parameters may be obtained from futures prices for the AECO Natural Gas Index for the period 1 May 1998 to 30 April 1999 (see Bos, Ware and Pavlov (2002), p.340). The parameters pertaining to the option are the following:

Price and Option Process Parameters							
T	a	σ	L	λ	r	K	
6 months	4.6488	1.5116	2.7264	0.1885	0.05	3	

From this table we can calculate the values for a^* and L^* :
$$a^* = a + \lambda\sigma = 4.9337,$$
and
$$L^* = \frac{aL}{a + \lambda\sigma} = 2.5690.$$
For the value of S_0 we can take $S_0 \in [1, 6]$.

Figure 18.1 (see below) depicts the dependence of mean value ES_t on the maturity T for AECO Natural Gas Index (1 May 1998 to 30 April 1999).

Figure 18.2 (see below) depicts the dependence of mean value ES_t on the initial value of stock S_0 and maturity T for AECO Natural Gas Index (1 May 1998 to 30 April 1999).

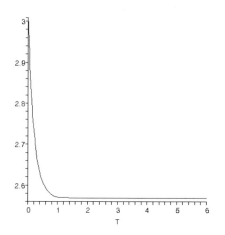

Fig. 18.1 Dependence of ES_t on T (AECO Natural Gas Index (1 May 1998–30 April 1999))

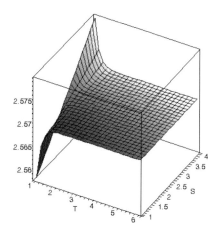

Fig. 18.2 Dependence of ES_t on S_0 and T (AECO Natural Gas Index (1 May 1998–30 April 1999))

Figure 18.3 (see Figure) depicts the dependence of variance of S_t on the initial value of stock S_0 and maturity T for AECO Natural Gas Index (1 May 1998 to 30 April 1999).

Figure 18.4 (see Figure) depicts the dependence of volatility of S_t on the initial value of stock S_0 and maturity T for AECO Natural Gas Index (1 May 1998 to 30 April 1999).

Figure 18.5 (see Figure) depicts the dependence of European Call Option Price for MRRNAM on the maturity (months) for AECO Natural Gas Index (1 May 1998 to 30 April 1999) with $S(0) = 1$ and $K = 3$.

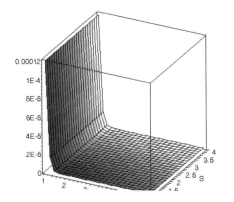 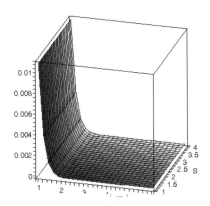

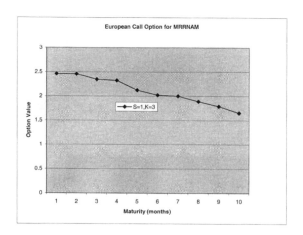

Fig. 18.5 Dependence of European Call Option Price on Maturity (months) ($S(0) = 1$ and $K = 3$) (AECO Natural Gas Index (1 May 1998–30 April 1999))

18.7 Summary

- In this Chapter we considered a risky asset S_t following the mean-reverting stochastic process.
- We obtained an explicit expression for a European option price based on S_t, using a change of time method (see Chapter 4).
- A numerical example for the AECO Natural Gas Index (1 May 1998–30 April 1999) was also presented.

Bibliography

Bos, L. P., Ware, A. F. and Pavlov, B. S. (2002). On a semi-spectral method for pricing an option on a mean-reverting asset. *Quantitative Finance*, 2, 337-345.

Black, F. (1076). The pricing of commodity contracts. *Journal of Financial Economics*, 3, 167-179.

Chen, Z. and Forsyth, P. (2006). *Stochastic Models of Natural Gas Prices and Applications to Natural Gas Storage Valuation*, Technical Report, University of Waterloo, Waterloo, Canada, November 24, 32 pages. Available at: http://www.cs.uwaterloo.ca/ paforsyt/regimestorage.pdf.

Elliott, R. (1982). *Stochastic Calculus and Applications*. New York: Springer-Verlag.

Elliott, R. and Kopp, E. (1999). *Mathematics of Financial Markets*. New York: Springer-Verlag.

Ikeda, N. and Watanabe, S. (1981). *Stochastic Differential Equations and Diffusion Processes*. Tokyo: Kodansha Ltd.

Lamperton, D. and Lapeyre, B. (1996). *Introduction to Stochastic Calculus Applied to Finance*. New York: Chapmann & Hall.

Pilipovic, D. (1997). *Valuing and Managing Energy Derivatives*. New York: McGraw-Hill.

Schoutens, W. (2003). *Lévy Processes in Finance: Pricing Financial Derivatives*. England: Wiley.

Schwartz, E. (1997). The stochastic behaviour of commodity prices: Implications for pricing and hedging. *Journal of Finance*, 52, 923-973.

Swishchuk, A. (2004). *Modeling and Valuing of Variance and Volatility Swaps for Financial Markets with Stochastic Volatilities*. Wilmott Magazine, Technical Article No. 2, September Issue, 64-72.

Swishchuk, A. (2007). Change of time method in mathematical finance. *Canadian Applied Mathematics Quarterly*, 15, 3, 299-336.

Swishchuk, A. (2009). Multi-factor Levy models for pricing financial and energy derivatives. *Canadian Applied Mathematics Quarterly*, 17, 4, Winter 2009.

Wilmott, P., Howison, S. and Dewynne, J. (1995). *The Mathematics of Financial Derivatives*. New York: Cambridge University Press.

Wilmott, P. (2000). *Paul Wilmott on Quantitative Finance*. New York: Wiley.

Yor, M. (1992). On some exponential functions of Brownian motion. *Advances in Applied Probability*, 24, 3, 509-531.

Yor, M. and Matsumoto, H. (2005). Exponential functionals of Brownian motion, I: Probability laws at fixed time. *Probability Surveys*, 2, 312-347.

Chapter 19

Forward and Futures in Energy Markets: Multi-Factor Lévy Models

19.1 Introduction

In this Chapter, we introduce new one-factor and multi-factor α-stable Lévy-based models to price energy derivatives, such as forwards and futures. For example, we introduce new multi-factor models such as Lévy-based Schwartz-Smith and Schwartz models. Using change of time method for SDEs driven by α-stable Lévy processes we present the solutions of these equations in simple and compact forms.

Black's model (1976) and Schwartz's model (1997) have become a standard approach to the problem of pricing options on commodities. These models have the advantage of mathematical convenience, in that they give rise to closed-form solutions for some types of options (See Wilmott (2000)).

A drawback of single-factor mean-reverting models lies in the case of options pricing: the fact the long-term rate is fixed results in a model-implied volatility term structure that has the volatilities going to zero as expiration time increases.

Using single-factor non-mean-reverting models has also a drawback: it will impact valuation and hedging. The differences between the distributions are particularly obvious when pricing out-of-the-money options, where the tails of the distribution play a very important role. Thus, if a lognormal model, for example, is used to price a far out-of-the-money option, the price can be very different from a mean-reverting model's price (see Pilipović (1997)). A popular model used for modeling energy and agricultural commodities and introduced by Schwartz (1990) aims at resembling the geometric Brownian motion while introducing mean-reversion to a long-term value in the drift term (see Schwartz (1997). This mean-reverting model is a one-factor version of the two-factor model made popular in the context of energy modelling by Pilipović (1997).

Villaplana (2004) proposed the introduction of two sources of risk X and Y representing, respectively, short-term and long-term shocks, and describes the spot price S_t. Geman and Roncoroni (2002) introduced a jump-reversion model for electricity prices. The two-factor model for oil-contingent claim pricing was proposed by Gibson and Schwartz (1990). Eydeland and Geman (1998) proposed extending the Heston (1993) stochastic volatility model to gas or electricity prices by introducing mean-reversion in the spot price and proposing two-factor model. Geman (2005) introduced three-factor model for commodity prices taking into account stochastic equilibrium level and stochastic volatility. Björk and Landen (2002) investigated the term structure of forward and futures prices for models where the price processes are allowed to be driven by a general market point process as well as by a multi-dimensional Wiener process. Benth *et al.* (2008) applied independent increments processes (see Skorokhod (1964), Lévy (1965)) to model and price electricity, gas and temperature derivatives (forwards, futures, swaps, options). Swishchuk (2008) considers a risky asset in energy markets following mean-reverting stochastic process. An explicit expression for a European option price based on this asset, using a change of time method, is derived. A numerical example for the AECO Natural Gas Index (1 May 1998–30 April 1999) is presented.

19.2 α-Stable Lévy Processes and Their Properties

19.2.1 *Lévy Processes*

Definition 1. By Lévy process we mean a stochastically continuous process with stationary and independent increments (see Sato (2005), Applebaum (2003), Schoutens (2003)).

Examples of Lévy Processes $L(t)$ include: a linear deterministic function $L(t) = \gamma t$; Brownian motion with drift; Poisson process, compound Poisson process; jump-diffusion process; variance-gamma (VG), inverse Gaussian (IG), normal inverse Gaussian (NIG), generalized hyperbolic and α-stable processes (see Sato (2005)).

19.2.2 *Lévy-Khintchine Formula and Lévy-Itô Decomposition for Lévy Processes L(t)*

The characteristic function of the Lévy process follows the following formula (so-called Lévy-Khintchine formula)

$$E(e^{i(u,L(t))}) = \exp\left\{t[i(u,\gamma) - \frac{1}{2}(u,Au)\right.$$

$$\left. + \int_{R^d - \{0\}} [e^{i(u,y)} - 1 - i(u,y)\mathbf{1}_{B_1(0)}]\nu(dy)]\right\}$$

where (γ, A, ν) is the Lévy-Khintchine triplet.

If L is a Lévy process, then there exists $\gamma \in R^d$, a Brownian motion B_A with covariance matrix A and an independent Poisson random measure N on $R^+ \times (R^d - \{0\})$ such that, for each $t \geq 0$, $L(t)$ has the following decomposition (Lévy-Itô decomposition)

$$L(t) = \gamma t + B_A(t) + \int_{|x|<1} x\tilde{N}(t,dx) + \int_{|x|\geq 1} xN(t,dx),$$

where N is a Poisson counting measure and $\tilde{N}$ is a compensated Poisson measure (see Applebaum (2003)).

Remark (Lévy Processes in Finance). The most commonly used Lévy processes in finance include Brownian motion with drift (the only continuous Lévy process), the Merton process=Brownian motion+drift+Gaussian jumps, the Kou process=Brownian motion+drift+exponential jumps and variance gamma (VG), inverse Gaussian (IG), normal inverse Gaussian (NIG), generalized hyperbolic (GH) and α-stable Lévy processes.

19.2.3 α-Stable Distributions and Lévy Processes

In this Section, we introduce α-stable distributions and Lévy processes, and describe their properties.

19.2.3.1   *Symmetric α-Stable (SαS) Distribution*

The characteristic function of the $S\alpha S$ distribution is defined as follows:

$$\phi(u) = e^{(i\delta u - \sigma |u|^\alpha)},$$

where α is the *characteristic exponent* $(0 < \alpha \leq 2)$, $\delta \in (-\infty, +\infty)$ is the *location* parameter, and $\sigma > 0$ is the *dispersion*.

For values of $\alpha \in (1, 2]$ the location parameter δ corresponds to the mean of the α-stable distribution, while for $0 < \alpha \leq 1$, δ corresponds to its median. The dispersion parameter σ corresponds to the spread of the distribution around its location parameter δ. The characteristic exponent α determines the shape of the distribution.

A stable distribution is called *standard* if $\delta = 0$ and $\sigma = 1$. If a random variable L is stable with parameters α, δ, σ, then $(L-\delta)/\sigma^{1/\alpha}$ is standard with characteristic exponent α. By letting α take the values $1/2$, 1 and 2, we get three important special cases: the Lévy $(\alpha = 1/2)$, Cauchy $(\alpha = 1)$ and the Gaussian $(\alpha = 2)$ distributions:

$$f_{1/2}(\gamma, \delta; x) = \left(\frac{t}{2\sqrt{\pi}}\right) x^{-3/2} e^{-t^2/(4x)}$$

$$f_1(\gamma, \delta; x) = \frac{1}{\pi}\frac{\gamma}{\gamma^2 + (x-\delta)^2},$$

$$f_2(\gamma, \delta; x) = \frac{1}{\sqrt{4\pi\gamma}}\exp[-\frac{(x-\delta)^2}{4\gamma}].$$

Unfortunately, no closed form expression exists for general α-stable distributions other than the Lévy, the Cauchy and the Gaussian. However, power series expansions can be derived for the density $f_\alpha(\delta, \sigma; x)$. Its tails (algebraic tails) decay at a lower rate than the Gaussian density tails (exponential tails).

The smaller the characteristic exponent α is, the heavier the tails of the α-stable density.

This implies that random variables following α-stable distribution with small characteristic exponent are *highly impulsive*, and it is this heavy-tail characteristic that makes this density appropriate for modeling noise which is impulsive in nature, for example, electricity prices or volatility.

Only moments of order less than α exist for the non-Gaussian family of α-stable distribution. The fractional lower order moments with zero location parameter and dispersion σ are given by

$$E|X|^p = D(p,\alpha)\sigma^{p/\alpha}, \quad for \quad 0 < p < \alpha,$$

$$D(p,\alpha) = \frac{2^p \Gamma(\frac{p+1}{2})\Gamma(1-\frac{p}{\alpha})}{\alpha\sqrt{\pi}\Gamma(1-\frac{p}{2})},$$

where $\Gamma(\cdot)$ is the Gamma function (Sato (2005)).

Since the $S\alpha S$ r.v. has 'infinite variance', the covariation of two jointly $S\alpha S$ real r.v. with dispersions γ_x and γ_y defined by

$$[X,Y]_\alpha = \frac{E[X|Y|^{p-2}Y]}{E[|Y|^p]}\gamma_y$$

has often been used instead of the covariance (and correlation), where $\gamma_y = [Y,Y]_\alpha$ is the dispersion of r.v. Y.

19.2.3.2 α-Stable Lévy Processes

Definiton 2. Let $\alpha \in (0,2]$. An α-stable Lévy process L such that L_1 (or equivalently any L_t) has a strictly α-stable distribution (i.e., $L_1 \equiv S_\alpha(\sigma,\beta,\delta)$) for some $\alpha \in (0,2] \setminus \{1\}$, $\sigma \in R_+$, $\beta \in [-1,1]$, $\delta = 0$ or $\alpha = 1$, $\sigma \in R_+$, $\beta = 0$, $\delta \in R$). We call L a symmetric α-stable Lévy process if the distribution of L_1 is even symmetric α-stable (i.e., $L_1 \equiv S_\alpha(\sigma,0,0)$ for some $\alpha \in (0,2]$, $\sigma \in R_+$.) A process L is called $(T_t)_{t\in R_+}$-adapted if L is constant on $[T_{t-}, T_t]$ for any $t \in R_+$ (see Sato (2005)).

19.2.3.3   *Properties of α-Stable Lévy Processes*

The α-stable Lévy processes are the only self-similar Lévy processes such that $L(at) \overset{Law}{=} a^{1/\alpha}L(t), a \geq 0$. They are either Brownian motion or pure jump. They have characteristic exponent and Lévy-Khintchine triplet known in closed form. They also have only 4 parameters, but infinite variance (except for Brownian motion). The α-stable Lévy Processes are semimartingales (in this way, $\int_0^t f_s dL_s$ can be defined) and α-stable Lévy Processes are pure discontinuous Markov processes with generator

$$Af(x) = \int_{R-\{0\}} [f(x+y) - f(x) - yf'(y)\mathbf{1}_{|y|<1}(y)]\frac{K_\alpha}{|y|^{1+\alpha}}dy.$$

$E|L(t)|^p$ is finite or infinite according as $0 < p < \alpha$ or $p > \alpha$, respectively. In particular, for an α-stable process, $EL(t) = \delta t$ $(1 < \alpha < 2)$ (Sato (2005)).

19.3 Stochastic Differential Equations Driven by α-Stable Lévy Processes

Consider the following SDE driven by an α-stable Lévy process $L(t)$:

$$dZ_t = b(t, Z_{t-})dt + \sigma(t, Z_{t-})dL(t). \tag{19.1}$$

Janicki *et al.* (1996) proved that this equation has a weak solution for continuous coefficients a and b.

We consider below one-factor and multi-factor models described by SDEs driven by α-stable Lévy process $L(t)$.

19.3.1 *One-Factor α-Stable Lévy Models*

$L(t)$ below is a symmetric α-stable Lévy process. We define below various processes via SDE driven by α-stable Lévy process.

1. The Geometric α-stable Lévy motion: $dS(t) = \mu S(t-)dt + \sigma S(t-)dL(t)$.

2. The Ornstein-Uhlenbeck Process Driven by α-stable Lévy motion: $dS(t) = -\mu S(t-)dt + \sigma dL(t)$.

3. The Vasićek Process Driven by α-stable Lévy motion: $dS(t) = \mu(b - S(t-))dt + \sigma dL(t)$.

4. The Continuous-Time GARCH Process Driven by α-stable Lévy motion: $dS(t) = \mu(b - S(t-))dt + \sigma S(t-)dL(t)$.

5. The Cox-Ingersoll-Ross Process Driven by α-stable Lévy motion: $dS(t) = k(\theta - S(t-))dt + \gamma\sqrt{S(t-)}dL(t)$.

6. The Ho and Lee Process Driven by α-stable Lévy motion: $dS(t) = \theta(t-)dt + \sigma dL(t)$.

7. The Hull and White Process Driven by α-stable Lévy motion: $dS(t) = (a(t-) - b(t-)S(t-))dt + \sigma(t)dL(t)$

8. The Heath, Jarrow and Morton Process Driven by α-stable Lévy motion: Define the forward interest rate $f(t, s)$, for $t \leq s$, that represents the instantaneous interest rate at time s as 'anticipated' by the market at time t. The process $f(t, u)_{0 \leq t \leq u}$ satisfies an equation

$$f(t, u) = f(0, u) + \int_0^t a(v, u)dv + \int_0^t b(f(v, u))dL(v),$$

where the processes a and b are continuous.

We note that Eberlein and Raible (1999) considered Lévy-based term structure models.

19.3.2 *Multi-Factor α-Stable Lévy Models*

Multi-factor models driven by α-stable Lévy motions can be obtained using various combinations of the above-mentioned processes. We give one example of a two-factor continuous-time GARCH model driven by α-stable Lévy motions.

$$dS(t) = r(t-)S(t-))dt + \sigma S(t-)dL^1(t)$$

$$dr(t) = a(m - r(t-))dt + \sigma_2 r(t-)dL^2(t),$$

where L^1, L^2 may be correlated, $m \in R, \sigma_i, a > 0, i = 1, 2$.

Also, we can consider various combinations of models, presented above, i.e., mixed models containing Brownian and Lévy motions. For example,

$$dS(t) = \mu(b(t-) - S(t-))dt + \sigma S(t-)dL(t)$$

$$db(t) = \xi b(t)dt + \eta b(t)dW(t),$$

where the Brownian motion $W(t)$ and Lévy process $L(t)$ may be correlated.

19.4   Change of Time Method (CTM) for SDEs Driven by Lévy Processes

We denote by $L^{\alpha}_{a.s.}$ the family of all real measurable $\mathcal{F}_t$-adapted processes a on $\Omega \times [0, +\infty)$ such that for every $T > 0, \int_0^T |a(t, \omega)|^{\alpha} dt < +\infty$ a.s. We consider the following SDE driven by a Lévy motion:

$$dX(t) = a(t, X(t-))dL(t),$$

where $L(t)$ is an α-stable Lévy process.

Theorem. (Rosinski and Woyczynski (1986), Theorem 3.1., p. 277). Let $a \in L^{\alpha}_{a.s.}$ be such that $T(u) := \int_0^u |a|^{\alpha} dt \to +\infty$ a.s. as $u \to +\infty$. If $\hat{T}(t) := \inf\{u : T(u) > t\}$ and $\hat{\mathcal{F}}_t = \mathcal{F}_{\hat{T}(t)}$, then the time-changed ctochastic integral $\hat{L}(t) = \int_0^{\hat{T}(t)} adL(t)$ is an $\hat{\mathcal{F}}_t - \alpha$-stable Lévy process, where $L(t)$ is $\mathcal{F}_t$-adapted and $\mathcal{F}_t$-α-stable Lévy process. Consequently, a.s. for each $t > 0$ $\int_0^t adL = \hat{L}(T(t))$, i.e., the stochastic integral with respect to a α-stable Lévy process is nothing but another α-stable Lévy process with randomly changed time scale.

19.4.1   *Solutions of One-Factor Lévy Models using the CTM*

Below we give the solutions to the one-factor Lévy models decribed by SDEs driven by α-stable Lévy process introduced in Section 19.2.3.

Proposition 1. Let $L(t)$ be a symmetric α-stable Lévy process, and $\hat{L}$ is a $(\hat{T}_t)_{t \in R_+}$-adapted symmetric α-stable Levy process on $(\Omega, \mathcal{F}, (\hat{\mathcal{F}}_t)_{t \in R_+}, P))$. Then, we have the following solutions for the above-mentioned one-factor Lévy models 1-8 (Section 19.3.1):

1.   The Geometric α-stable Lévy Motion: $dS(t) = \mu S(t-)dt + \sigma S(t-)dL(t)$. Solution $S(t) = e^{\mu t}[S(0) + \hat{L}(\hat{T}_t)]$,   where $\hat{T}_t = \sigma^{\alpha} \int_0^t [S(0) + \hat{L}(\hat{T}_s)]^{\alpha} ds$.

2.   The Ornstein-Uhlenbeck Process Driven by α-stable Lévy Motion: $dS(t) = -\mu S(t-)dt + \sigma dL(t)$. Solution $S(t) = e^{-\mu t}[S(0) + \hat{L}(\hat{T}_t)]$, where $\hat{T}_t = \sigma^{\alpha} \int_0^t (e^{\mu s}[S(0) + \hat{L}(\hat{T}_s)])^{\alpha} ds$.

3. The Vasiček Process Driven by α-stable Lévy Motion: $dS(t) = \mu(b - S(t-))dt + \sigma dL(t)$. Solution $S(t) = e^{-\mu t}[S(0) - b + \hat{L}(\hat{T}_t)]$, where $\hat{T}_t = \sigma^\alpha \int_0^t (e^{\mu s}[S(0) - b + \hat{L}(\hat{T}_s)] + b)^\alpha ds$.

4. The Continuous-Time GARCH Process Driven by α-stable Lévy process: $dS(t) = \mu(b - S(t-))dt + \sigma S(t-)dL(t)$. Solution $S(t) = e^{-\mu t}(S(0) - b + \hat{L}(\hat{T}_t)) + b$, where $\hat{T}_t = \sigma^\alpha \int_0^t [S(0) - b + \hat{L}(\hat{T}_s) + e^{\mu s}b]^\alpha ds$.

5. The Cox-Ingersoll-Ross Process Driven by α-stable Lévy Motion: $dS(t) = k(\theta^2 - S(t-))dt + \gamma\sqrt{S(t-)}dL(t)$. Solution $S^2(t) = e^{-kt}[S_0^2 - \theta^2 + \hat{L}(\hat{T}_t)] + \theta^2$, where $\hat{T}_t = \gamma^\alpha \int_0^t [e^{k\hat{T}_s}(S_0^2 - \theta^2 + \hat{L}(\hat{T}_s)) + \theta^2 e^{2k\hat{T}_s}]^{\alpha/2} ds$.

6. The Ho and Lee Process Driven by α-stable Lévy Motion: $dS(t) = \theta(t-)dt + \sigma dL(t)$. Solution $S(t) = S(0) + \hat{L}(\sigma^\alpha t) + \int_0^t \theta(s)ds$.

7. The Hull and White Process Driven by α-stable Lévy Motion: $dS(t) = (a(t-) - b(t-)S(t-))dt + \sigma(t-)dL(t)$. Solution $S(t) = \exp[-\int_0^t b(s)ds][S(0) - \frac{a(s)}{b(s)} + \hat{L}(\hat{T}_t)]$, where $\hat{T}_t = \int_0^t \sigma^\alpha(s)[S(0) - \frac{a(s)}{b(s)} + \hat{L}(\hat{T}_s) + \exp[\int_0^s b(u)du]\frac{a(s)}{b(s)}]^\alpha ds$.

8. The Heath, Jarrow and Morton Process Driven by α-stable Lévy Motion: $f(t,u) = f(0,u) + \int_0^t a(v,u)dv + \int_0^t b(f(v,u))dL(v)$. Solution $f(t,u) = f(0,u) + \hat{L}(\hat{T}_t) + \int_0^t a(v,u)dv$, where $\hat{T}_t = \int_0^t b^\alpha(f(0,u) + \hat{L}(\hat{T}_s) + \int_0^s a(v,u)dv)ds$.

Proof. The approach is to eliminate drift, reduce obtained SDE to the above-mentioned form $dX(t) = a(t,X(t-))dL(t)$ and then to use the above-mentioned Rosinski-Woyczynski (1983) result.

19.4.2 Solution of Multi-Factor Lévy Models using CTM

Solutions of multi-factor models driven by α-stable Lévy motions (see Section 19.3.1) can be obtained using various combinations of solutions of the above-mentioned processes and the CTM. We give one example of two-factor continuous-time GARCH model driven by α-stable Lévy motions.

Proposition 2. Let we have the following two-factor Lévy-based model:

$$dS(t) = r(t-)S(t-))dt + \sigma_1 S(t-)dL^1(t)$$

$$dr(t) = a(m - r(t-))dt + \sigma_2 r(t-)dL^2(t),$$

where L^1, L^2 may be correlated, $m \in R, \sigma_i, a > 0, i = 1, 2$.

Then the solution of the two-factor Lévy model using the CTM is (applying CTM for the first and the second equations, respectively):

$$S(t) = e^{\int_0^t r_s ds}[S_0 + \hat{L}^1(\hat{T}_t^1)]$$

$$= e^{\int_0^t e^{-as}[r_0 - m + \hat{L}^2(\hat{T}_s^2)]ds}[S_0 + \hat{L}^1(\hat{T}_t^1)],$$

where $\hat{T}^i$ are defined in 1. and 4., respectively, Section 19.3.1.

Proof. The approach is to eliminate drifts in both equations, reduce the obtained SDEs to the above-mentioned form $dX(t) = a(t, X(t-))dL(t)$ and then to use the above-mentioned Rosinski-Woyczynski (1986) result.

Important Remark. Kallsen and Shiryaev (2002) showed that the Rosiński and Woyczyński (1986) statement cannot be extended to any other Lévy process but α-stable processes. If one considers only nonnegative integrands a in $dX(t) = a(t, X(t-))dL(t)$, then we can extend their statement to asymmetric α-stable Lévy processes.

19.5   Applications in Energy Markets

In this Section, we consider applications of the change of time method for Lévy-based SDEs arising in energy markets, namely pricing forward and futures. We also give an idea how to use this technique for Lévy-based SABR/LIBOR market model.

19.5.1   *Energy Forwards and Futures*

Random variables following α-stable distribution with small characteristic exponent are *highly impulsive*, and it is this heavy-tail characteristic that makes this density appropriate for modeling noise which is impulsive in nature, for example, energy prices such as electricity. Here, we introduce two Lévy-based models in energy market: two-factor Lévy-based Schwartz-Smith and three-factor Schwartz models. We show how solve them using the change of time method.

19.5.1.1   *Lévy-Based Schwartz-Smith Model*

We introduce the Lévy-based Schwartz-Smith model:

$$\begin{cases} \ln(S_t) = \kappa_t + \xi_t \\ d\kappa_t = (-k\kappa_t - \lambda_\kappa)dt + \sigma_\kappa dL_\kappa \\ d\xi_t = (\mu_\xi - \lambda_\xi)dt + \sigma_\xi dW_\xi, \end{cases}$$

where S_t is the current spot price, κ_t is the short-term deviation in prices, and ξ_t is the equilibrium price level.

Let $F_{t,T}$ denote the market price for a futures contract with maturity T, then:

$$\ln(F_{t,T}) = e^{-k(T-t)}\kappa_t + \xi_t + A(T - t),$$

where $A(T - t)$ is a deterministic function with explicit expression. We note that κ_t, using change of time for α-stable processes, can be presented in the following form:

$$\kappa_t = e^{-kt}\left[\kappa_0 + \frac{\lambda_\kappa}{k} + \hat{L}_\kappa(\hat{T}_t)\right],$$

$$\hat{T}_t = \sigma_\kappa^\alpha \int_0^t \left(e^{-ks}\left[\kappa_0 + \frac{\lambda_\kappa}{k} + \hat{L}_\kappa(\hat{T}_s)\right] - \frac{\lambda_\kappa}{k}\right)^\alpha ds.$$

In this way, the market price for a futures contract with maturity T has the following form:

$$\ln(F_{t,T}) = e^{-kT}\left[\kappa_0 + \frac{\lambda_\kappa}{k} + \hat{L}_\kappa(\hat{T}_t)\right]$$
$$+ \xi_0 + (\mu_\xi - \lambda_\xi)t + \sigma_\xi W_\xi + A(T-t),$$

where the Lévy process $\hat{L}_\kappa$ and Wiener process W_ξ may be correlated.

If $\alpha \in (1,2]$, then we can calculate the value of Lévy-based futures contracts.

19.5.1.2 *Lévy-Based Schwartz Model*

We also introduce a Lévy-based Schwartz model:

$$\begin{cases} d\ln(S_t) = (r_t - \delta_t)S_t dt + S_t \sigma_1 dW_1 \\ d\delta_t = k(a - \delta_t)dt + \sigma_2 dL \\ dr_t = a(m - r_t)dt + \sigma_3 dW_2, \end{cases}$$

where the Wiener processes W_1, W_2 and α-stable Lévy process L may be correlated. δ_t and r_t are the instantaneous convenience yield and interest rate, respectively.

We note that:

$$\delta_t = e^{kt}(\delta_0 - a + \hat{L}(\hat{T}_t)),$$

$$\hat{T}_t = \sigma_2^\alpha \int_0^t (e^{ks}[\delta_0 - a + \hat{L}(\hat{T}_s)] + a)^\alpha ds$$

and

$$r_t = e^{at}(r_0 - m + \hat{W}_2(\hat{T}_t)),$$

$$\hat{T}_t = \sigma_3^2 \int_0^t (e^{as}[r_0 - m + \hat{W}_2(\hat{T}_s)] + m)^2 ds.$$

The solution for $\ln[S_t]$:

$$\ln[S_t] = e^{\int_0^t [e^{as}(r_0-m+\hat{W}_2(\hat{T}_s^2))-e^{ks}(\delta_0-a+\hat{L}(\hat{T}_s))]ds}[\ln S_0 + \hat{W}_1(\hat{T}_t^1)].$$

In this way, the futures contract has the following form:

$$\ln(F_{t,T}) = \frac{1-e^{-k(T-t)}}{k}\delta_t + \frac{1-e^{-a(T-t)}}{a}r_t + \ln(S_t) + C(T-t)$$

$$= \frac{1-e^{-k(T-t)}}{k}[e^{kt}(\delta_0 - a + \hat{L}(\hat{T}_t))]$$

$$+ \frac{1-e^{-a(T-t)}}{a}e^{at}(r_0 - m + \hat{W}_2(\hat{T}_t^2))$$

$$+ \exp\left\{\int_0^t (e^{as}(r_0 - m + \hat{W}_2(\hat{T}_s^2))\right.$$

$$\left. - e^{ks}(\delta_0 - a + \hat{L}(\hat{T}_s)))ds\right\}[\ln(S_0) + \hat{W}_1(\hat{T}_t^1)]$$

$$+ C(T-t),$$

where $C(T - t)$ is a deterministic explicit function. If $\alpha > 1$, then we can calculate the value of a futures contract.

19.5.2   *Gaussian- and Lévy-Based SABR/LIBOR Market Models*

SABR model (see Hagan, Kumar, Lesniewski and Woodward (2002)) and the Libor Market Model (LMM) (Brace, Gatarek and Musiela (BGM, 1996), Piterbarg (2003)) have become industry standards for pricing plain-vanilla and complex interest rate products, respectively.

The Gaussian-based SABR model (Hagan, Kumar, Lesniewski and Woodward (2002)) is a stochastic volatility model in which the forward value satisfies the following SDE:

$$\begin{cases} dF_t = \sigma_t F_t^\beta dW_t^1 \\ d\sigma_t = \nu \sigma_t dW_t^2. \end{cases}$$

In a similar way, we introduce the Lévy-based SABR model, a stochastic volatility model in which the forward value satisfies the following SDE:

$$\begin{cases} dF_t = \sigma_t F_t^\beta dW_t \\ d\sigma_t = \nu \sigma_t dL_t, \end{cases}$$

where $L(t)$ is an α-stable Lévy process.

The solution of Lévy-based SABR model using a change of time method has the following expression:

$$F_t = F_0 + \hat{W}(\hat{T}_t^1),$$

$$T_t^1 = \int_0^t \sigma_{T_s^1}^{-2}(F_0 + \hat{W}(s))^{-2\beta} ds,$$

$$\sigma_t = \sigma_0 + \hat{L}(\hat{T}_t^2),$$

$$T_t^2 = \nu^{-\alpha} \int_0^t (\sigma_0 + \hat{L}(s))^{-\alpha} ds.$$

The expressions for F_t and σ_t give the possibility to calculate many financial derivatives.

19.6   Summary

- In this Chapter, we introduced one-factor and multi-factor α-stable Lévy-based models to price energy derivatives.
- These models include, in particular, Lévy-based SABR/LIBOR models, Schwartz-Smith and Schwartz models.
- Using the change of time method for Lévy-based SDEs we present the solutions of these equations in very simple and compact form.
- We apply this method to price forward and futures contracts.

Bibliography

Applebaum, D. (2003). *Lévy Processes and Stochastic Calculus*. Cambridge: Cambridge University Press.

Benth, F., Benth, J. and Koekebakker, S. (2008). *Stochastic Modelling of Electricity and Related Markets*. Singapore: World Scientific.

Björk, T. and Landen, C. (2002). On the term structure of futures and forward prices. In Geman, H., Madan, D., Pliska, S. and Vorst, T. (eds.), *Mathematical Finance-Bachelier Congress 2000*, 111-149. Berlin: Springer.

Black, F. (1976). The pricing of commodity contracts. *Journal of Financial Economics*, 3, 167-179.

Brace, A., Gatarek, D. and Musiela, M. (1991). The market model of interest rate dynamics. *Mathematical Finance*, 7, 2, 127-155.

Eberlin, E. and Raible, S. (1999). Term structure models driven by general Lévy processes. *Mathematical Finance*, 9, 1, 31-53.

Eydeland, E. and Geman, H. (1998). Pricing power derivatives. *Risk Magazine*, September.

Geman, H. (2005). *Commodities and Commodity Derivatives: Modelling and Pricing for Agricaltural, Metals and Energy*. USA: Wiley.

Geman, H. and Roncoroni, R. (2005). Understanding the fine structure of electricity prices. *Journal of Business*, 79, 3, 1225-1261.

Gibson and Schwartz, E. (1990). Stochastic convenience yield and the pricing of oil contigent claims. *Journal of Finance*, 45, 959-976.

Hagan, P., Kumar, D. and Lesniewski, A. (2002). Managing smile risk. *Wilmott Magazine*, 84-108.

Heston, S. (1993). A closed-form solution for options with stochastic volatility with applications to bond and currency options. *Review of Financial Studies*, 6, 327-343.

Lévy, P. (1965). *Processus Stochastiques et Mouvement Brownian* (2nd edition). Paris: Gauthier-Villars.

Pilipović, D. (1997). *Valuing and Managing Energy Derivatives*. New York: McGraw-Hill.

Rosiński, J. and Woyczyński, W. A. (1986). On Ito stochastic integration with respect to *p*-stable motion: Inner clock, integrability of sample paths, double and multiple integrals. *Annals of Probability*, 14, 1, 271-286.

Sato, K. (1999). *Lévy Processes and Infinitely Divisible Distributions*. Cambridge, UK: Cambridge University Press.

Schoutens, W. (2003). *Lévy Processes in Finance. Pricing Financial Derivatives*. England: Wiley & Sons.

Schwartz, E. (1997). The stochastic behaviour of commodity prices: Implications for pricing and hedging. *Journal of Finance*, 52, 923-97.

Schwartz, E. (1990). Short-term variations and long-term dynamics in commodity prices. *Management Science*, 46, 7.

Skorokhod, A. (1964). *Random Processes with Independent Increments*. Nauka, Moscow (English translation: Kluwer AP, 1991).

Swishchuk, A. (2008). Explicit option pricing formula for a mean-reverting asset in energy market. *Journal of Numerical and Applied Mathematics*, 96, 1, 216-233.

Swishchuk, A. (2009). Multi-factor Lévy models for pricing financial and energy derivatives. *Canadian Applied Mathematics Quarterly*, 17, 4, 777-806.

Villaplana, (2004). *A Two-State Variables Model for Electricity Prices*. Third World Congress of the Bachelier Finance Society, Chicago.

Wilmott, P. (2000). *Paul Wilmott on Quantitative Finance*. New York: Wiley.

Chapter 20

Generalization of Black-76 Formula: Markov-Modulated Volatility

20.1 Introduction

In this Chapter, we invoke the Markov-modulated volatility and apply it to generalize Black-76 formula (see Black (1976)). Black formulas for Markov-modulated markets with and without jumps are derived. Application is given using Nordpool weekly electricity forward prices.

Markov-modulated models have been extensively studied in the literature. The term regime-switching is used to describe such models as in Benth *et al.* (2008), Guo (2001) and Yao *et al.* (2003). In this article, we consider renewable resources which are non-storable. Such resources can be mainly represented by energy products. When produced, these resources are traded on markets for a certain spot price, however, because of non-storability, these markets can be very short lived.[1] In order to allow companies who use the resource in their production to reduce the price risk, the electricity market has been deregulated since early 1990, the forwards and futures markets for these resources are then have been created. One of the most famous markets to trade electricity is the Nordpool located in Norway.

We will argue that when modeling forward prices in the context of such renewable resources, it does not appear to be appropriate to specify forward prices as a geometric Brownian motion, and apply classical Black-Scholes analysis as in Black (1976). We will also assume that we have a Markov-modulated volatility (MMV), instead of the original deterministic volatility.

For constant volatility, Black and Scholes (1973) obtains the option pricing formula for the Brownian market. Cox and Ross (1976) valued options for alternative stochastic processes. Hamilton (1989) introduced Markov switching into the econometric mainstream. Di Masi *et al.* (1994) obtained option pricing formula for stochastic volatility driven by a Markov chain in continuous time. Hofmann (1996) studied option pricing under incompleteness and with stochastic volatility. Griego and Swishchuk (2000) contained the Black-Scholes formula for a market in a Markov random environment. Swishchuk and Elliot (2007) studied option

---

[1]As short as one hour or even less than one hour, in the case of electricity.

pricing formula and swaps for Markov-Modulated Brownian and fractional Brownian Markets with jumps. To our best knowledge, the MMV model was first applied to electricity model in Ethier and Mount (1998). A two-state specification was proposed, in which in both regimes the log-prices were governed by autoregressive processes of order one, i.e. AR(1). In de Jong (2006), spikes in spot electricity prices can be better captured by regime-switching models than by a Poisson jump model. Huisman *et al.* (2003) proposed a regime-switching model with three possible regimes in which the initial jump regime was immediately followed by the reversing regime and then moved back to the base regime. Consequently, their model did not allow for consecutive high prices (in fact of log-prices) and hence did not offer any obvious advantage over jumpdiffusion models. In de Jong and Huisman (2002) and Deng (2000), different regime-switching models for electricity prices were developed. Evidence shows that the third regime was not needed to pull prices back to stable levels, because the prices were independent from each other in the two regimes.

This paper is organized as follows. In Section 20.2, we re-state the famous Black-76 formula for pricing European call option first. Then use the martingale characterization of Markov processes, we derive the generalization of the Black-76 formula for Markov-modulated markets with and without jumps. The explicit formula from Di Masi *et al.* (1994) are then used for a two-state MMV model. In Section 20.3, we apply our option pricing formulas to the weekly electricity forward prices from Nordpool. Finally, in Section 20.4, we conclude.

20.2   Generalization of Black-76 Formula with Markov-Modulated Volatility

20.2.1   *Black-76 Formula*

In his 1976 article, Black postulates that forward prices should evolve according to a geometric Brownian motion with zero drift term. Based on this assumption he derives a version of the Black-Scholes formula for commodity prices. It has long been discussed, whether given the fact that geometric Brownian motion is certainly not an adequate model for commodity prices, it can be used as a realistic model for forward prices of commodities.

The following Theorem states the price of a call option, known as the Black-76 formula (see Black (1976)).

Theorem 20.1. (Black-76 Formula) The price of a call option at time $t \leq T$, written on a forward with delivery at time τ, where the option has exercise time $T \leq \tau$ and strike price K, is

$$C(t; T, K, \tau) = e^{-r(T-t)}\{f(t, \tau)\Phi(d_1) - K\Phi(d_2)\}. \tag{20.1}$$

Here,

$$d_1 = \frac{ln(f(t,\tau)/K) + \frac{1}{2}\sigma^2(T-t)}{\sigma\sqrt{T-t}},$$

$$d_2 = \frac{ln(f(t,\tau)/K) - \frac{1}{2}\sigma^2(T-t)}{\sigma\sqrt{T-t}},$$

and Φ is the cumulative standard normal probability distribution function.

20.2.2 *Pricing Options for Markov-Modulated Markets*

20.2.2.1 *Pricing Options for Markov-Modulated Markets without Jumps*

In here, we are going to use the risk-neutral dynamics for forward contract with Markov-Modulated Volatility

$$\frac{df(t,\tau)}{f(t,\tau)} = \sigma(Z(t))dW(t), \tag{20.2}$$

where $W(t)$ is the Brownian motion under the risk-neutral probability Q, to get the generalization of the famous Black-76 Formula (see Black (1976)).

Theorem 20.2. The price of a call option at time $t \leq T$, written on a forward with delivery at time τ, where the option has exercise time $T \leq \tau$ and strike price K is

$$C_T^Z(\tau) = \int_0^\infty C_T^B\left(\left(\frac{z}{T}\right)^{1/2}, T\right) F_T^Z(dz) \tag{20.3}$$

where $F_T^Z(dz)$ is the distribution of the random variable $Z_T \equiv \int_0^T \sigma^2(Z_s)ds$ and $C_T^B(\sigma, T)$ is the Black-76 formula for call option written on forward at delivery τ, as defined in (20.1).[2]

Corollary 20.1. For the case where $E = \{1,2\}$ the distribution F_T of the random variable Z_T can be expressed in an explicit form. Indeed, let $\nu(t)$ be a counting process of jumps of Y. Then

$$Z_T = \int_0^T [\sigma^2(1)I(Z_t = 1) + \sigma^2(2)I(Z_t = 2)]dt = a \cdot T + b \cdot J_T$$

where

$$J_T = \int_0^T (-1)^{\nu(t)} dt$$

$$a = \frac{1}{2}(\sigma^2(1) + \sigma^2(2)),$$

$$b = \frac{1}{2}(\sigma^2(1) - \sigma^2(2)).$$

---

[2]Proof of Theorem 20.2 follows from Theorem 20.3 by setting the jump term equal to 0.

We consider the symmetric case where transition intensities are equal: $q_{12} = q_{21} = \lambda$. The distribution G_T of the random variable J_T is given by

$$G_T(dz) = e^{-\lambda T} \epsilon_T(dz) + h_{T,\lambda}(z)dz,$$

$$h_{T,\lambda}(z) = \lambda/2e^{-\lambda T}\left[I_0(\lambda(T^2 - z^2)^{\frac{1}{2}})\right.$$

$$\left. + \left(\frac{T+z}{T-z}\right)^{\frac{1}{2}} I_1(\lambda(T^2 - z^2)^{\frac{1}{2}})\right]\mathbb{I}_{(-T,T)}(z),$$

where ϵ_T is the unit mass at T and the modified Bessel functions of the first kind I_0 and I_1 are defined by

$$I_0(s) = \sum_{k=0}^{\infty} \frac{(s^2/4)^k}{k!^2}, \qquad I_1(s) = \frac{s}{2}\sum_{k=0}^{\infty} \frac{(s^2/4)^k}{k!(k+1)!}$$

Corollary 20.2. For the case where $E = \{1, 2\}$ the distribution F_T of the random variable Z_T can be expressed in an explicit form. Indeed, let $\nu(t)$ be a counting process of jumps of Y. Then

$$Z_T = \int_0^T [\sigma^2(1)I(Z_t = 1) + \sigma^2(2)I(Z_t = 2)]dt = a \cdot T + b \cdot J_T$$

where

$$J_T = \int_0^T (-1)^{\nu(t)} dt$$

$$a = \frac{1}{2}(\sigma^2(1) + \sigma^2(2)),$$

$$b = \frac{1}{2}(\sigma^2(1) - \sigma^2(2)).$$

We consider the non-symmetric case where transition intensities are: $q_{12} = \lambda$ and $q_{21} = \kappa$. The distribution G_T of the random variable J_T is given by

$$G_T(dz) = e^{-\lambda T} \epsilon_T(dz) + h_{T,\lambda,\kappa}(z)dz,$$

$$h_{T,\lambda,\kappa}(z) = \lambda/2e^{-(\lambda+\kappa)T/2 - (\lambda-\kappa)z/2}\left[I_0(\lambda\kappa(T^2 - z^2)^{\frac{1}{2}})\right.$$

$$\left. + \left(\frac{\kappa}{\lambda}\right)^{\frac{1}{2}}\left(\frac{T+z}{T-z}\right)^{\frac{1}{2}} I_1(\lambda\kappa(T^2 - z^2)^{\frac{1}{2}})\right]\mathbb{I}_{(-T,T)}(z),$$

where ϵ_T is the unit mass at T and the modified Bessel functions of the first kind I_0 and I_1 are defined by

$$I_0(s) = \sum_{k=0}^{\infty} \frac{(s^2/4)^k}{k!^2}, \qquad I_1(s) = \frac{s}{2}\sum_{k=0}^{\infty} \frac{(s^2/4)^k}{k!(k+1)!}.$$

20.2.2.2 *Pricing Options for Markov-Modulated Markets with Jumps*

Let $Y_1, Y_2, \ldots, Y_{\nu(t)}$ be independent identically distributed random variables with values in $(-1, +\infty)$, $\nu(t)$ is a Poisson process with intensity $\lambda > 0$, and $\tau_1, \tau_2, \ldots,$ $\tau_{\nu(t)}$ are random moments of time. $\nu(t)$, $(Y_i; i \geq 1)$ and $(\tau_i; i \geq 1)$ are independent on $Z(t)$ and $W(t)$. Let $H(dy)$ is a probability distribution on $(-1, +\infty)$, that respects to $(Y_i; i \geq 0)$, and $\nu(dt, dy)$ is a random measure, that equals the number of jumps of the process $\nu(t)$ with values in dy up to the moment dt. Hence, $(\lambda, H(dy))$ is a local characteristic of measure $\nu(dt, dy)$ and $\tilde{\nu}(dt, dy) := \nu(dt, dy) - \lambda H(dy)$ is a local martingale. A compound geometric Poisson process is called the following process:

$$\sum_{k=1}^{\nu(t)} Y_k. \tag{20.4}$$

We note that

$$\sum_{k=1}^{\nu(t)} Y_k = \sum_{k \geq 1} Y_k \mathbb{I}_{\tau_k \leq t} = \int_0^t \int_{-1}^{+\infty} y \nu(dt, dy). \tag{20.5}$$

Let the process $f^d(t, \tau)$ be a solution of the equation:

$$\frac{df^d(t, \tau)}{f^d(t, \tau)} = \int_0^t \int_{-1}^{+\infty} y \nu(dt, dy). \tag{20.6}$$

Then the solution of this equation has the representation:

$$f^d(t, \tau) = f(0, \tau) \prod_{k=1}^{\nu(t)} (1 + Y_k). \tag{20.7}$$

Let

$$L_t := L_0 \prod_{k=1}^{\nu(t)} h(Y_k), \tag{20.8}$$

where L_0 is an $\mathcal{F}_0$-measurable random variable, $\mathbb{E}[L_0] = 1$, and nonnegative function $h(y)$ satisfies the equalities:

$$\int_0^t h(y) H(dy) = 1 \quad \text{and} \quad \int_0^t y h(y) H(dy) = 0. \tag{20.9}$$

Let us take the measure $\mathbb{Q}$ such that $\frac{d\mathbb{Q}}{dP} = L_T$, where $\mathbb{E}[L_T] = 1$, and

$$Q^*(A) = \int_A L_T(\omega) dP(\omega). \tag{20.10}$$

Then jump process $f^d(t, \tau)$ is $(\mathbb{Q}, \mathcal{F}_t)$-martingale, that follows from the properties (20.9), and L_t in (20.8) is $(P, \mathcal{F}_t)$-martingale. We note that $\nu(dt, dy)$ admits on $[0, T]$ and $(\mathbb{Q}, \mathcal{F}_t)$ local characteristics $(\lambda, h(y) H(dy))$, where function $h(y)$ is defined in (20.9). Let us denote $h(y) H(dy)$ by

$$H^*(dy) := h(y) H(dy). \tag{20.11}$$

Here, we are going to use the risk-neutral dynamics for forward contract with Markov-modulated volatility with compound geometric Poisson process.

Such dynamics is described by the following stochastic equation:

$$\frac{df(t,\tau)}{f(t,\tau)} = \sigma(Z(t))dW(t) + \int_{-1}^{+\infty} y\nu(dt,dy), \tag{20.12}$$

where $W(t)$ is the Brownian motion under the risk-neutral probability Q.

Solution of this equation may be given in the form

$$f(t,\tau) = X \exp\left\{ \left(\lambda \int_{-1}^{+\infty} \ln(1+y)H(dy)ds + \sigma(Z(s)) \right) \right.$$

$$\left. - \frac{1}{2} \int_0^t \sigma^2(Z(s)))dW(s) + \int_{-1}^{+\infty} \ln(1+y)\tilde{\nu}(ds,dy) \right\}, \tag{20.13}$$

where $X = f(0,\tau)$, $\tilde{\nu}(ds,dy) := \nu(dt,dy) - \lambda H(dy)$. Let's denote by $f^*(t,\tau)$ a discounted process of stock price $f(t,\tau)$:

$$f^*(t,\tau) = f(t,\tau) \exp\left\{ \left(\lambda \int_{-1}^{+\infty} \ln(1+y)H(dy)ds + \sigma(Z(s)) \right) \right.$$

$$\left. - \frac{1}{2} \int_0^t \sigma^2(Z(s)))dW(s) + \int_{-1}^{+\infty} \ln(1+y)\tilde{\nu}(ds,dy) \right\} \tag{20.14}$$

where $\tilde{\nu}(ds,dy) := \nu(dt,dy) - \lambda H(dy)$. Let's denote by $f^*(t,\tau)$ a discounted process of stock price $f(t,\tau)$:

$$f^*(t,\tau) = f(t,\tau) \exp\left(-\int_0^t r(Z(s)) \right) ds, \quad t \in [0,t]. \tag{20.15}$$

Then $f^*(t,\tau)$ may be represented in the form:

$$f^*(t,\tau) = X \exp\left\{ \left(-\int_0^t r(Z(s)) - \frac{1}{2}\sigma^2(Z(s)) \right) ds \right.$$

$$\left. + \int_0^t \sigma(Z(s))dW(s) \right\} \prod_{k=1}^{N(t)} (1+Y_k). \tag{20.16}$$

We note that process

$$W^*(t) := W(t) - \int_0^t \frac{r(Z(s))}{\sigma(Z(s))} ds \tag{20.17}$$

is $(\mathbb{Q},\mathcal{F}_t)$ — standard Wiener process, where $\mathcal{P}$ is a such measure that $\frac{d\mathbb{Q}}{dP} = \rho_T$, where

$$\rho_t = \rho_0 \exp\left\{ \int_0^t \frac{r(Z(s))}{\sigma(Z(s))} dW(s) + \frac{1}{2} \int_0^t \left(\frac{r(Z(s))}{\sigma(Z(s))}\right)^2 ds \right\} \times \prod_{k=1}^{N(t)} h(Y_k), \tag{20.18}$$

$$\mathbb{E}[\rho_0] = 1, \tag{20.19}$$

where function $h(y)$ is defined in (20.9). Taking into account (20.17) for $W^*(t)$ and (20.19), process $f^*(t, \tau)$ in (20.16) may also be represented in the form:

$$f^*(t, \tau) = X \exp \left\{ \int_0^t \sigma(Z(s)) dW^*(s) - \frac{1}{2} \int_0^t \sigma^2(Z(s)) ds \right\} \prod_{k=1}^{N(t)} (1 + Y_k^*) \quad (20.20)$$

where $(Y_k^*; k \geq 1)$ has the distribution $H^*(dy) := h(y) H(dy)$.

20.2.2.3 Formula for the Price of Contingent Claim $g_T(f(t, \tau))$

In here, we focus on the pricing of a contingent claim written on a forward contract expires at time τ.

Theorem 20.3. The price $C_{T,Z,s}^g$ of contingent claim $g_T(f(t, \tau))$ in zero moment of time with expiry date T has the form:

$$C_{T,Z,s}^g = \mathbb{E}_{T,Z,s}^* \left[g_T(f(t, \tau)) \exp \left\{ - \int_0^T r(Z(s)) ds \right\} \right]$$

$$= \mathbb{E}_{T,Z,s}^* \left[g_T \left(f^*(t, \tau) \exp \left\{ \int_0^T r(Z(s)) ds \right\} \right) \exp \left\{ - \int_0^T r(Z(s)) ds \right\} \right],$$

$$(20.21)$$

In the case $g_T(f(t, \tau)) = (f(t, \tau) - K)^+$, where K is a strike price. Inserting the function $g_T(f(t, \tau))$ in the expression (20.21) we obtain the result.

Theorem 20.4. The price $C_{T,Z,s}$ of contingent claim $g_T(f(t, \tau)) = (f(t, \tau) - K)^+$ of the European call option has the form:

$$C_{T,Z,s} = \mathbb{E}_{T,Z,s}^* \left[(f(t, \tau) - K)^+ \exp \left\{ - \int_0^T r(Z(s)) ds \right\} \right]$$

$$= \mathbb{E}_{T,Z,s}^* \left[(f^*(t, \tau) \exp \left\{ - \int_0^T r(Z(s)) ds \right\} - K)^+ \times \exp \left\{ - \int_0^T r(Z(s)) ds \right\} \right],$$

$$(20.22)$$

where process $f^*(t, \tau)$ is defined in (20.20).[3]

The value $C_{T,Z,s}^g (C_{T,Z,s})$ in some cases may be calculated more simply. For example, let us take $r \equiv 0$. Then

$$C_{T,Z,s}^g = \mathbb{E}_{T,Z,s}^* [g_T(f(t, \tau))] = \mathbb{E}_{T,Z,s}^* [g(f^*(t, \tau))] \quad (20.23)$$

where $f^*(t, \tau)$ is defined in (20.20) with $r \equiv 0$. We note that function $C_{T,Z,s}^g = \mathbb{E}_{Z,s} [g_T(f(T - t, \tau))]$ is the solution of Cauchy problem

$$\begin{cases} \frac{\partial C}{\partial t} + \frac{1}{2} \sigma^2(Z) f^2 \times \frac{\partial^2 C}{\partial^2 f} + \lambda \int_{-1}^{+\infty} (C(t, f(1 + y)) - C(t, x)) h(y) H(dy) + QC = 0 \\ C(T, f) = g_T(f). \end{cases}$$

$$(20.24)$$

---

[3] Proof of Theorem 20.4 is a direct substitution.

Let F_t^Z be a distribution of random variable $Z_T^z := \int_0^T \sigma^2(Z(s))ds$.

Theorem 20.5. If $r \equiv 0$, then the price $C_{T,Z,s}^g$ of contingent claim $g_T(f(t,\tau))$ is calculated by the formula

$$C_{T,Z,s}^g = \sum_{k=0}^{+\infty} \frac{\exp(-\lambda T)(\lambda T)^k}{k!}$$

$$\times \int_{-1}^{+\infty} \cdots \int_{-1}^{+\infty} \left(\int \left(\int g(y)y^{-1} \times \psi(z, \ln \frac{y}{f \prod_{i=1}^k (1+y_i)} + \frac{1}{2}z)dy \right) F_T(dz) \right)$$

$$\times H^*(dy_1) \times \ldots H^*(dy_k) \tag{20.25}$$

where $H^*(dy) = h(y)H(dy)$, and $\psi(z, \nu) := (2\pi z)^{-\frac{1}{2}} \exp(-\frac{\nu^2}{2z})$.

Corollary 20.3. Let $g_T(f(t,\tau)) = (f(t,\tau) - K)^+$, and let $r \equiv 0$. Then from Theorem 20.4, formula (20.25), and Black-76 value C_T^{BS} for European call option it follows that the price $C_{T,Z,s}$ of contingent claim has the form

$$C_{T,Z,s} = \sum_{k=0}^{+\infty} \frac{\exp -\lambda T(\lambda T)^k}{k!}$$

$$\times \int_{-1}^{+\infty} \cdots \int_{-1}^{+\infty} C_T^{BS}\left(\left(\frac{z}{T}\right)^{\frac{1}{2}}, T, f \prod_{i=1}^k (1+y_i) \right) F_T(dz) \right)$$

$$\times H^*(dy_1) \times \ldots H^*(dy_k) \tag{20.26}$$

where function C_T^{BS} is a Black-76 value for European call option.

20.2.3   *Proof of Theorem 20.3*

From Ito formula it follows that $f^*(t,\tau)$ is the solution of the following equation:

$$df^*(t,\tau) = r(Z(t))f^*(t,\tau)dt + \sigma(Z(t))f^*(t,\tau)d^*W(t) + f^*(t,\tau)\int_{-1}^{+\infty} y\nu(dt, dy).$$
$$\tag{20.27}$$

We note that the following Cauchy problem

$$\begin{cases} \frac{\partial C}{\partial t} + r(z)f\frac{\partial C}{\partial f} + \frac{1}{2}\sigma^2(z)f^2\frac{\partial^2 C}{\partial^2 f} \\ \quad + r(z)C + \lambda \int_{-1}^{+\infty}(C(t, f(1+y)) - C(t,z))h(y)H(dy) + QC = 0 \quad (20.28) \\ C(T,f) = g_T(f). \end{cases}$$

has the solution

$$C_{T,Z,s}^g = \mathbb{E}_{T,Z,s}\left[g_T(f(t,\tau))\exp\left\{\int_0^t r(z(s))ds\right\} \right], \tag{20.29}$$

that follows from (20.22). Taking into account (20.23) and (20.24), we obtain the proof of the Theorem 20.3.

20.2.4 *Proof of Theorem 20.5*

It follows from representation (20.20) for $f^*(t, \tau)$, formula (20.21) and iterations on function $g_T(f(t, \tau)) = g_T^*(f(t, \tau))$, taking into account a distribution of Z_T.

20.3 Numerical Results for Synthetic Data

20.3.1 *Case Without Jumps*

In the case without jumps, we extract jumps terms from each state and then use Corollary 20.1 to get the numerical results for symmetric case. The strike price K is taken to be 60. The forward price $f(0, \tau)$ at time 0 is 50.

In Figure 20.1,[4] the upper line represents the classical Black-76 model with volatility $\sigma_1 = 1.2$, the lower line represents the classical Black-76 model with volatility $\sigma_2 = 0.8$, and the middle two lines represent the European call option price given by equation (20.3) for the symmetric case, starting from state 1 and state 2 respectively. We noticed that the prices given by (20.3) are within the two classical Black-76 models.

We then use Corollary 20.3 to get the numerical results for the non-symmetric case. The strike price K is again taken to be 60. The forward price $f(0, \tau)$ at time 0 is 50.

In Figure 20.2, the upper line represents the classical Black-76 model with volatility $\sigma_1 = 1.2$, the lower line represents the classical Black-76 model with volatility $\sigma_2 = 0.8$, and the middle two lines represent the European call option price given by equation (20.3) for the non-symmetric case, starting from state 1 and state 2 respectively. We noticed that the prices given by (20.3) are within the two classical Black-76 models. Compare with Figure 20.1, we see that the two lines in the middle are getting closer to the two Black-76 lines.

20.3.2 *Case with Jumps*

In the case with jumps, we keep the jumps terms and then use Corollary 20.1 to get the numerical results for symmetric case. The strike price K, this time, is taken to be 100. The forward price $f(0, \tau)$ at time 0 is 90.

In Figure 20.3, the upper two lines represent the European call option price given by (20.26) for the symmetric case, starting from state 1 and state 2 respectively. The lower two lines represent the classic Black-76 model with volatility $\sigma_1 = 12.5$ and $\sigma_2 = 8$ respectively. We notice that the option prices given by (20.26) converge

---

[4]All figures are prepared using Matlab R2010(a).

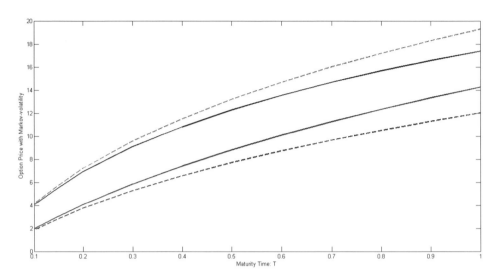

Fig. 20.1   Option Price with Markov Volatility. The Parameter Values: $f_0 = 50$, $K = 60$, $\lambda = 1$, $r = 0.04$, $\sigma_1 = 1.2$, $\sigma_2 = 0.8$.

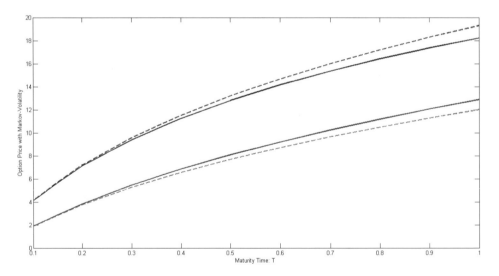

Fig. 20.2   Option Price with Markov Volatility. The Parameter Values are: $f_0 = 50$, $K = 60$, $\lambda = 0.3$, $\kappa = 0.1$, $r = 0.04$, $\sigma_1 = 1.2$, $\sigma_2 = 0.8$.

to a fixed option price when T is large enough. This empirical phenomenon also provide the rational of the generalized formula for large exercise time.

We then use Corollary 20.3 to get the numerical results for the non symmetric case. The The strike price K is again taken to be 100. The forward price $f(0, \tau)$ at time 0 is 90.

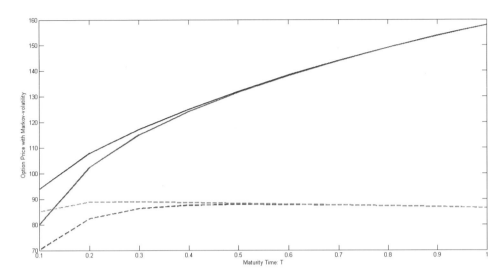

Fig. 20.3 Option Price with Markov Volatility. The Parameter Values: $f_0 = 90$, $K = 100$, $\lambda = 0.4$, $\kappa = 0.1$, $r = 0.04$, $\sigma_1 = 12.5$, $\sigma_2 = 8$.

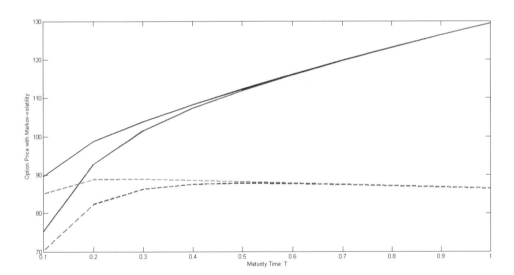

Fig. 20.4 Option Price with Markov Volatility. The Parameter Values are: $f_0 = 90$, $K = 100$, $\lambda = 0.4$, $\kappa = 0.1$, $r = 0.04$, $\sigma_1 = 12.5$, $\sigma_2 = 8$.

Figure 20.4 has a very similar shape as Figure 20.3. The only advantage is that we seem to make a more precise prediction when we consider the transition probabilities to be different going from state 1 to state 2 and vice versa.

20.4   Applications: Data from Nordpool

After testing our model using toy data, we are ready to move on to test our model using real time data. The data we use here are the weekly electricity forward prices from Nordpool, the world's only multinational exchange for trading electric power, for the period from January 01, 2009 to August 30, 2011.

In our study, we use the daily electricity spot prices from Nordpool to calculate the weekly electricity forward prices. We then assuming that our forward prices belongs to either one of the two states: state 1, the normal state and state 2, the jump state. To calculate the option price, we first need to explain how to compute the stationary distribution π. Let $\{Z(t), t \geq 0\}$ represents the continuous time Markov process. Denote the transition probability from state i to state j in the time period t as

$$p_{ij} = P(Z(t) = j | Z(0) = i).$$

The transition probability matrix Q contains the probabilities p_{ij} of switching from regime i at time t to regime j at time $t+1$, for $i, j = 1, 2$ is:

$$Q = (p_{ij}) = \begin{pmatrix} p_{11} & p_{12} \\ p_{21} & p_{22} \end{pmatrix} = \begin{pmatrix} p_{11} & 1 - p_{11} \\ 1 - p_{22} & p_{22} \end{pmatrix}$$

In the following two figures, we extract the jumps from the original series of electricity spot prices by writing a numerical algorithm that filters spot prices with absolute values greater than three times the standard derivation of the spot prices at that specific iteration. On the second iteration, the standard derivation of the remaining series is again calculated; those spot prices which are now greater than 3 times this last standard derivation are filtered again. The process is repeated until no further spot prices can be filtered. This algorithm allows us to estimate the cumulative frequency of jumps and other statistical information of relevance for calibrating the model. Once we have separated the jumps, we group our jump and non-jump spot prices into two states. We then calculate the volatilities for each of the two states.

In Figure 20.5, the upper two lines represent the European call option prices given by (20.26) for the symmetric case. The jump rate, λ, jumping from state 1 to state 2 and vice versa in this case is equal to 0.5. The lower two lines represent the classical Black-76 model with volatility $\sigma_1 = 7.3683$ and $\sigma_2 = 3.3916$ respectively. We notice that one of the price calculated by the classical Black-76 model has a decreasing trend. This further proved that the constant volatility assumption made by the classical Black-76 model is not appropriate. When the market is very volatile, the classical Black-76 model fails to capture the characteristic of the market.

In Figure 20.6, the upper two lines represent the European call option prices given by (20.26) for the nonsymmetric case. The jump rate, The jump rate, λ, jumping from state 1 to state 2 is equal to 0.823, and κ, jumping from state 2 to state 1 is equal to 0.177. The lower two lines represent the classical Black-76 model with volatility $\sigma_1 = 6.3747$ and $\sigma_2 = 3.8898$ respectively.

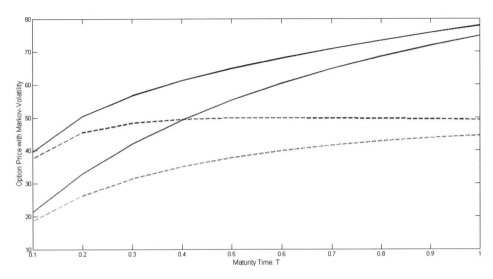

Fig. 20.5 Option Price with Markov Volatility. Symmetric Case with Real Time Data. The Parameter Values are: $f_0 = 51.3375$, $K = 60$, $\lambda = 0.5$, $r = 0.04$, $\sigma_1 = 7.3683$, $\sigma_2 = 3.3916$.

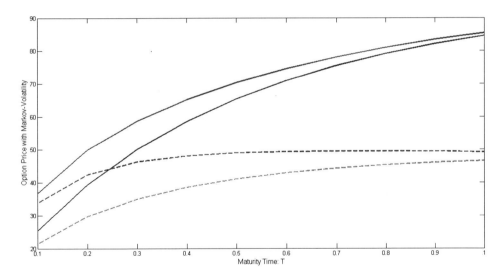

Fig. 20.6 Option Price with Markov Volatility. Nonsymmetric Case with Real Time Data. The Parameter Values are: $f_0 = 50$, $K = 60$, $\lambda = 0.823$, $\kappa = 0.177$, $r = 0.04$, $\sigma_1 = 6.3747$, $\sigma_2 = 3.8889$.

Once we have fit our model to the real data. Our next question is whether our Markov-volatility model can capture all the characteristics of electricity market, or does our model has any advantage compare to the combination of traditional Black-76 plus jumps? We then produce the following figure in Matlab to compare.

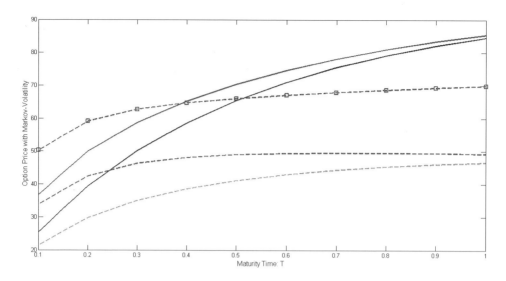

Fig. 20.7   Comparison between Markov Driven Volatility and Black-76 + Jumps.

In Figure 20.7, the line with square dots represents the option price calculated by the combination of traditional Black-76 plus jumps we observed from our weekly electricity forward prices. As we can see from the figure, when we have a relatively longer maturity time for our options, the Markov-driven-volatility model produces a better result than the combination of traditional Black-76 plus jumps.

20.5   Summary

- A closed form formula for generalize Black-76 with Markov-modulated volatility option pricing model has been developed in this chapter.
- In the case of two-state Markov chain, we have an exact analytic formula for both symmetric and nonsymmetric case.
- Numerical evaluation of the formula was studied using toy data.
- We then tested our model using real time electricity data from Nordpool.
- Our empirical study has the following important consequence: by adding an extra randomness into the volatility, we are able to capture the important characteristic of electricity prices, the spike.
- An exact closed formula provides useful insight for the European option pricing in the Black model with Markov-modulated volatility.
- It is not only explains the effect of markov model, but accelerates the computation of the European option pricing in the Black model with Markov-modulated volatility.

Bibliography

Benth, F., Benth, J. and Koekebakker, S. (2008). *Stochastic Modelling of Electricity and Related Markets*. Singapore: World Scientific.

Erlwein, C., Benth, F.E. and Mamon, R. (2010). HMM filtering and parameter estimation of an electricity spot price model. *Energy Economics*, 32, 5, 1034-1043.

Black, F. and Scholes, M. (1973). The pricing of options and corporate liabilities. *Journal of Political Economy*, 637-657.

Black, F. (1976). The pricing of commodity contracts. *Journal of Financial Economics*, 3, 167-179.

Buffington, J. and Elliott, R.J. (2002). American options with regime switching. *International Journal of Theoretical and Applied Finance*, 5, 497-514.

Cox, J. and Ross, S. (1976). Valuation of options for alternative stochastic processes. *Journal of Financial Economics*, 3, 145-166.

de Jong, C. (2006). The nature of power spikes: A regime-switch approach. *Studies in Nonlinear Dynamics and Econometric*, 10.

de Jong, C. and Huisman, R. (2002). Option formulas for mean-reverting power prices with spikes. Erasmus Research Institute of Management (ERIM). Erasmus University Rotterdam.

Deng, S. (2000). Stochastic Models of Energy Commodity Prices and their Appplications: Mean-reversion with Jumps and Spikes. University of California Energy Institute.

Di Masi, G.B., Kabanov, Yu. M. and Runggaldier, W. J. (1994) Mean-variance hedging of options on stocks with Markov volatilities. *Theory of Probability and Its Applications*, 39, 173-181.

Di Masi, G.B., Platen, E. and Runggaldier, W. (1994). Hedging of options under discrete observation on assets with stochastic volatility. In Bolthausen, E., Dozzi, M. and Russo, F. (eds.), *Stochastic Analysis, Random Fields and Applications*, 359-364. Boston: Birkhauser.

Elliott, R.J., Sick, G.A. and Stein, M. (2003). Modelling electricity price risk. Preprint, University of Calgary.

Elliott, R.J. and Valchev, S. (2004). Libor Market Model with Regime-Switching Volatility. Working Paper No. 228, Mational Centre of Competence in Research Financial Valuation and Risk Management.

Ethier, R. and Mount, T. (1998). Estimating the volatility of spot prices in restructed electricity markets and the implications for option values. PSERC Working Paper. 98-31.

Griego, R. and Swishchuk, A. (2000). Black-Scholes formula for a market in a Markov enviroment. *Theory of Probability and Mathematical Statistics*, 62, 9-18.

Guo, X. (2001). An explicit solution to an optimal stopping problem with regime switching. *Journal of Applied Probability*, 38, 1, 461-481.

Guo, X. (2002). Information and option pricing, *Journal of Quantitative Finance*, 1, 38-44.

Guo, X. and Zhang, Q. (2004). Closed-form solutions for perpetual American put options with regime switching. *SIAM Journal of Applied Mathematics*, 64, 1-12.

Hamilton, J. (1989). Rational-expectations econometric analysis of changes in regime. *Journal of Economic Dynamics and Control*, 12, 385-423.

Hofmann, N., Platen, E., Schweizer, M. (1996). Options pricing under incompleteness and stochastic volatility. *Mathematical Finance*, 2, 153-187.

Huisman, R. and Mahieu, R. (1996). Regime jumps in electricity prices. *Energy Economics*, 25, 425-434.

Janssen, A. (1990). The distance between the Kac Process and the Wiener process with applications to generalized telegraph equations. *Theory of Probability and Its Applications*, 3, 349-360.

Steutel, F.M. (1985). Poisson processes and Bessel function integral. *SIAM Review*, 27, 1, 73-75.

Swishchuk, A. and Elliott, R. (2007). Pricing options and variance swaps in Markov-modulated Brownian markets. *Hidden Markov Models in Finance: International Series in Operations Research and Management Science*, 104, 45-68.

Swishchuk, A. (1995). Hedging of options pricing under mean-square criterion and with semi-Markov volatility. *Ukrainian Mathematics Journal*, 47, 7, 1119-1127.

Weron, R. (2008). Heavy-tails and regime-switching in electricity prices. *Mathematical Methods in Operations Research*. Available at: 10.1007/s00186-008-0247-4.

Wiggins, J.B. (1987). Options value under stochastic volatility. Theory and empirical estimates. *Journal of Financial Economics*, 10, 351-372.

Yao, D., Zhang, Q. and Zhou, X. (2003). A Regime-Switching Model for European Option Pricing. Working Paper.

Index